Die Grundlehren der mathematischen Wissenschaften

in Einzeldarstellungen
mit besonderer Berücksichtigung
der Anwendungsgebiete

Band 170

J. L. Lions

Optimal Control of Systems Governed by Partial Differential Equations

Translated by Dr. S. K. Mitter

Springer-Verlag Berlin Heidelberg New York 1971

Prof. Dr. J. L. Lions

Faculté des Sciences
Département de Mathématiques
Paris

Translated by

Prof. Dr. S. K. Mitter

Electrical Engineering Department and Electronic Systems Laboratory
Massachusetts Institute of Technology
Cambridge, Massachusetts

This book is translated from the French edition "Contrôle optimal de systèmes gouvernés par des équations aux dérivées partielles", published by Dunod and Gauthier-Villars (Paris 1968) in the series "Etudes Mathématiques" edited by P. Lelong.

AMS Subject Classifications (1970)

Primary 49A20, 49B25, 49C10, 49C15, 49D30, 93C20
Secondary 35J40, 35K35, 35L35

ISBN 978-3-642-65026-0 ISBN 978-3-642-65024-6 (eBook)
DOI 10.1007/978-3-642-65024-6

Softcover reprint of the hardcover 1st edition 1971

Contents

Chapter II

Control of Systems Governed by Elliptic Partial Differential Equations

Chapter III

Control of Systems Governed by Parabolic Partial Differential Equations

Chapter IV

Control of Systems Governed by Hyperbolic Equations or by Equations which are well Posed in the Petrowsky Sense

Chapter V

Regularization, Approximation and Penalization

Introduction

1. The development of a theory of optimal control (deterministic) requires the following initial data:

(i) a control u belonging to some set $\mathcal{U}_{ad}$ (the set of 'admissible controls') which is at our disposition,

(ii) for a given control u, the state $y(u)$ of the system which is to be controlled is given by the solution of an equation

$$\Lambda y(u) = \text{given function of } u \tag{*}$$

where Λ is an operator (assumed known) which specifies the system to be controlled (Λ is the 'model' of the system),

(iii) the observation $z(u)$ which is a function of $y(u)$ (assumed to be known exactly; we consider only deterministic problems in this book),

(iv) the "cost function" $J(u)$ ("economic function") which is defined in terms of a numerical function $z \to \Phi(z) \geqslant 0$ on the "space of observations" by

$$J(u) = \Phi(z(u)). \tag{**}$$

It is required to find (problem of the Calculus of Variations)

$$\operatorname{Inf} J(u), \qquad u \in \mathcal{U}_{ad}.$$

The objectives of the theory are

(i) to obtain necessary (or possibly necessary and sufficient) conditions for u to be an extremum (or minimum),

(ii) to study the structure and properties of the equations expressing these conditions (where the 'model' Λ naturally intervenes),

(iii) to obtain constructive algorithms amenable to numerical computations for the approximation of a (the) control $u \in \mathcal{U}_{ad}$ which determines the Inf (such a control is termed an "optimal control").

2. Clearly the development of such a theory depends on the model Λ in a fundamental manner.

The theory described in the works of Pontryagin-Boltyanskii-Gamkrelidze-Mischenko [1] and Hestenes [1] is concerned with the study of points (i) and (ii) of 1 in the case where Λ is a family of ordinary differential (or with delay or integro-differential) operators.

In a variety of applications, due to the complexity of the system to be controlled, it is often advantageous to discard the above-mentioned mathematical model in favor of a model described by a family of partial differential operators (for examples, cf. Butkovskii [1], Wang [1] and the bibliography of these works). It is this situation that we propose to investigate in this book[1]. We thus consider systems whose state $y(u)$ is given by the solution of a *partial differential equation* to which we must add appropriate *boundary conditions*[2] and in the case of evolution equations *initial conditions.*

3. It is clear that unless we wish to restrict ourselves to results which are purely formal, the minimization of (**) presupposes that the boundary value problem (*) is formulated and solved in a precise mathematical setting. The results that are needed in this direction are proved in this book for the case where Λ is an operator which is elliptic or parabolic or hyperbolic or well-posed in the sense of Petrovsky. In order not to overburden this work we have restricted ourselves to comparatively simple examples. However the techniques we have used are quite general and hence more general problems can be solved using the same techniques. We refer the reader to Lions-Magenes [1], Chapter 6 where extensions to more general cases can be found.

Once we have in our possession a good theory for the solution of (*), it remains to obtain and analyse the necessary (or necessary and sufficient in favorable cases) conditions for (**) to have a minimum. In this manner we are led to a number of boundary value problems which appear to be novel in character. These problems however have a striking analogy with "multiphase" and "unilateral" problems of mechanics—in particular in plasticity.

4. We now give a brief analysis of the contents of the various chapters.

In Chapter 1 we study amongst other things, the minimization of positive definite or semi-definite quadratic forms defined on a closed, convex subset of a Hilbert space. Applications to unilateral problems are given. These unilateral problems are prototypes of problems which we encounter in the sequel.

In Chapter 2 we study the optimal control of systems governed by elliptic equations, while in Chapters 3 and 4 we examine the parabolic and hyperbolic or well-posed in the sense of Petrovsky cases. In each of these chapters we first consider the case of a linear system with a quadratic cost function and study the "unilateral problem" which this leads to. In Chapters 3 and 4 we study in detail the "feedback problem" and the related integro-differential equation of Riccati type. We then study

---

[1] We hasten to add that only very partial results have been obtained in a number of directions.

[2] Control may be exercised through the boundary condition, which in fact is the situation generally encountered in practice.

existence theorems for simple nonlinear systems (given the present state of the art of nonlinear partial differential equations, a general theory in this sense appears to be outside our reach for the moment) and first order necessary conditions.

Finally, in Chapter 5, we present various procedures of regularization, approximation and penalization. These procedures may be utilized in the numerical solution of optimal control problems which we have studied.

The chapter headings are the following:

Each chapter begins with a detailed plan indicating the scope of the chapter and closes with bibliographic notes and indications on problems which are unsolved (of which there are many) or on aspects which have not been considered in this book. The whole subject is clearly in a process of evolution.

The contents of the book have developed from a course given at the Faculty of Sciences, University of Paris since the academic year 1965–1966. An abbreviated version of the book was presented in a summer course at the University of California, Los Angeles in August 1967 (the course was organized by A. V. Balakrishnan).

I wish to express my sincere appreciation to Mr. P. Lelong for having graciously accepted this work as the first volume of a new series of mathematical treatises which is under his direction.

I also wish to thank Mr. A. Bensoussan whose remarks allowed me to clarify various points in the initial text and Mr. J. P. Ayrault who edited the first version of this course.

Finally it remains for me to thank Editions Dunod for their excellent typographical work.

Paris, January 1968.

In this English translation of the French edition, Remarks and bibliographical references have been added. I wish to thank very much Professor S. K. Mitter for his excellent translation and for his interesting remarks made during the process of translation.

Paris, November 1969. J. L. Lions

Principal Notations

$x=\{x_1,\ldots,x_n\}$ denotes the *space* variable; x ranges in an open set $\Omega\subset\mathbb{R}^n$ with boundary Γ.

t denotes time; in general $t\in]0,T[$, $T<\infty$.

We set

$$Q=\Omega\times]0,T[,\qquad \Sigma=\Gamma\times]0,T[\,.$$

The *controls* (or commands) are, in general, denoted by $u,v,w\ldots$; they are generally taken to be in a space $\mathscr{U}$ (quite generally a Hilbert space on $\mathbb{R}$); $\mathscr{U}_{\mathrm{ad}}$ (= set of *admissible* controls) is a *closed, convex subset of* $\mathscr{U}$.

The *state* of the system is denoted by $y(v)$; in the *elliptic* case (Chapter 2) $y(v)$ is a function of $x\in\Omega, y(x,v)$; in the *evolution* case (Chapters 3 and 4) $y(v)$ is a function of $x\in\Omega$ and $t\in]0,T[:y(x,t;v)$.

The *observation* is denoted by $z(v)=C\,y(v)$ (we do not study the case where there is noise present).

The *cost function* (or *criterion*, or *economic function*) is denoted by $J(v)$.

The $u\in\mathscr{U}_{\mathrm{ad}}$ such that $J(u)\leqslant J(v)\ \forall\, v\in\mathscr{U}_{\mathrm{ad}}$ are the optimal controls.

$p(v)$ denotes the *adjoint state*.

Main function spaces used.

$C^k(\Omega)$ = space of k-times continuously differentiable functions on $\bar{\Omega}$, k integer $\geqslant 0$[3].

$\mathscr{D}(\Omega)$ = space of infinitely differentiable functions in Ω, with *compact support in* Ω, endowed with the inductive limit topology of L. Schwartz [1].

$\mathscr{D}'(\Omega)$ = dual space of $\mathscr{D}(\Omega)$ = space of distributions on Ω[4].

If X is a Banach space, $\mathscr{D}'(]0,T[;X)$ denotes the space of distributions on $]0,T[$ with values in X (cf. Schwartz [3] and brief recapitulation in Chapter 3, section 1.1).

$L^2(\Omega)$ = space (equivalence class) of functions square integrable on Ω.

---

[3] Clearly we have analogous notation for Q, Γ, Σ. All functions considered are real-valued.

[4] In general, X' denotes the dual of X.

$H^m(\Omega)$ = *(Sobolev* [1] *space of order m)* = space of functions φ such that

$$\varphi \in L^2(\Omega), \frac{\partial \varphi}{\partial x_i} \in L^2(\Omega), \ldots, D^\alpha \varphi \in L^2(\Omega) \quad \forall \alpha, |\alpha| \leqslant m$$

$$\alpha = \{\alpha_1, \ldots, \alpha_n\}, \ |\alpha| = \alpha_1 + \cdots + \alpha_n .$$

$H_0^m(\Omega)$ = $\{\varphi | \varphi \in H^m(\Omega), D^\alpha \varphi = 0 \text{ on } \Gamma. \ |\alpha| \leqslant m-1\}$.

$H^s(\Omega)$ = *Fractional Sobolev space* of order s on Ω,
= space of restrictions on Ω of functions of $H^s(\mathbb{R}^n)$ defined (by Fourier Transforms) in Chapter 1, (3.12).

$L^2(S;E)$ = space (equivalence class) of functions defined on S (locally compact measure space with measure $\mu \geqslant 0$) with values in a Hilbert space E and such that

$$\int_S \|f(t)\|_E^2 \, d\mu(t) \leqslant \infty .$$

We use primarily $L^2(0, T; E)$, $d\mu(t) = dt$.

$L^\infty(S;E)$ = space (equivalence class) of functions f defined on S with values in E and essentially bounded:

$$\|f(t)\|_E \leqslant \|f\|_{L^\infty(S,E)} < \infty, \quad \text{a.e.}$$

CHAPTER I

Minimization of Functions and Unilateral Boundary Value Problems

1. Minimization of Coercive Forms

1.1. Notation

Let $\mathscr{U}$ be a real Hilbert space. The elements of $\mathscr{U}$ will be denoted by $u, v, w \ldots$ In the applications we have in mind in the following chapters, $\mathscr{U}$ will be the space of controls.

In this chapter $\|\cdot\|$ will denote the norm on $\mathscr{U}$; in general, if there is possible ambiguity, the norm in the space X will be denoted by $\|u\|_X$.

For the moment, we shall assume that the following data is given:

$$\text{(i)} \qquad \begin{cases} \text{a continuous bi-linear form on } \mathscr{U}, \text{ which is symmetric,} \\ u, v \to \pi(u,v), \qquad \pi(u,v) = \pi(v,u) \;\; \forall\, u, v \in \mathscr{U}, \end{cases}$$

$$\text{(ii)} \qquad \begin{cases} \text{a continuous linear form on } \mathscr{U}, \\ v \to L(v), \end{cases}$$

(iii) a closed, convex set $\mathscr{U}_{\mathrm{ad}}$ in $\mathscr{U}$.

In the applications considered in the sequel $\mathscr{U}_{\mathrm{ad}}$ will be the set of admissible controls.

The quadratic functional

$$J(v) = \pi(v,v) - 2L(v) \tag{1.1}$$

is required to be minimized over the set $\mathscr{U}_{\mathrm{ad}}$.

1.2. The Case when π is Coercive

π is said to be coercive on $\mathscr{U}$ if

$$\pi(v,v) \geqslant c\,\|v\|^2 \;\; \forall\, v \in \mathscr{U}, \qquad c > 0. \tag{1.2}$$

We then have

Theorem 1.1. *Let* $\pi(u,v)$ *be a continuous symmetric bilinear form on* $\mathscr{U}$ *which satisfies* (1.2). *Then there exists a unique element* u *in* $\mathscr{U}_{\mathrm{ad}}$ *such that* $J(u) = \operatorname*{Inf}_{v\in\mathscr{U}_{\mathrm{ad}}} J(v)$.

Proof of Existence. Let $\{v_n\}\in\mathscr{U}_{\mathrm{ad}}$ be a minimizing sequence such that

$$J(v_n)\to \operatorname*{Inf}_{v\in\mathscr{U}_{\mathrm{ad}}} J(v). \tag{1.3}$$

From (1.2) we have

$$J(v)\geqslant c\,\|v\|^2 - c_1\,\|v\|, \quad \text{where } c_1 \text{ is a constant.} \tag{1.4}$$

From (1.3) and (1.4), we obtain

$$\|v_n\|\leqslant \text{constant}. \tag{1.5}$$

From (1.5) we may extract a subsequence $\{v_\mu\}$ of $\{v_n\}$ such that

$$v_\mu\to w \quad \textit{weakly} \text{ in } \mathscr{U}. \tag{1.6}$$

Since $\mathscr{U}_{\mathrm{ad}}$ is a closed *convex* set, it is *weakly closed.* Hence (1.6) implies that

$$w\in\mathscr{U}_{\mathrm{ad}}. \tag{1.7}$$

But the function

$$v\to\pi(v,v)$$

is lower semi-continuous (l.s.c.) in the weak topology of $\mathscr{U}$ and the function $v\to L(v)$ is continuous in the weak topology. Thus the function $v\to J(v)$ is weakly lower semi-continuous and hence

$$\liminf J(v_\mu)\geqslant J(w). \tag{1.8}$$

Therefore from (1.3) and (1.7) we obtain,

$$w\in\mathscr{U}_{\mathrm{ad}} \quad \text{and } J(w)\leqslant \operatorname*{Inf}_{v\in\mathscr{U}_{\mathrm{ad}}} J(v). \tag{1.9}$$

Hence we must necessarily have $J(w) = \operatorname*{Inf}_{v\in\mathscr{U}_{\mathrm{ad}}} J(v)$ and we may take $u=w$ in the theorem. □

Proof of Uniqueness. The function $v\to\pi(v,v)$ is *strictly convex* (that is: $\pi((1-\theta)v_1+\theta v_2,\ (1-\theta)v_1+\theta v_2)<(1-\theta)\pi(v_1,v_1)+\theta\pi(v_2,v_2)$, if $\theta\in]0,1[$, unless $v_1=v_2$). Hence the function $v\to J(v)$ is also strictly convex.

Let u_1 and u_2 be two elements such that $J(u_1)=J(u_2)=\inf_{v\in\mathscr{U}_{ad}} J(v)$. Since $\mathscr{U}_{ad}$ is a convex set, $\frac{1}{2}(u_1+u_2)\in\mathscr{U}_{ad}$ and $J(\frac{1}{2}(u_1+u_2))<\inf_{v\in\mathscr{U}_{ad}} J(v)$, unless $u_1=u_2$ which must necessarily be the case. This demonstrates the uniqueness of the minimizing element[1]. □

Example 1.1. Let us consider

$$\pi(u,v)=(u,v)=\text{scalar product in } \mathscr{U}, \qquad L(v)=(g,v)_{\mathscr{U}},$$

where g is a given element in $\mathscr{U}$.

Then, $J(v)=\|g-v\|_{\mathscr{U}}^2-\|g\|_{\mathscr{U}}^2$ and the unique element $u\varepsilon\mathscr{U}_{ad}$ such that $J(v)=\inf_{v\in\mathscr{U}_{ad}} J(v)$ is characterized by

$$\|g-u\|_{\mathscr{U}}\leqslant\|g-v\|_{\mathscr{U}} \quad \forall v\in\mathscr{U}_{ad};$$

u is thus the projection of g on $\mathscr{U}_{ad}$.

Analysis of the Proof of Theorem 1.1. An analysis of the way the assumptions of Theorem 1.1 come into play in the proof of the theorem suggests the following remarks:

Remark 1.1. Theorem 1.1 is true if we assume that the bilinear form $\pi(u,v)$ is defined on $\mathscr{U}_{ad}\times\mathscr{U}_{ad}$ and satisfies (1.2), $\forall v\in\mathscr{U}_{ad}$.

The fact that the function $v\to J(v)$ is a quadratic form does not enter in any essential way in the proof of Theorem 1.1.

Remark 1.2. Let $v\to J(v)$ be a convex function from $\mathscr{U}_{ad}\to\mathbb{R}$, such that

$$J(v)\to+\infty \quad \text{as } \|v\|\to+\infty, \quad v\in\mathscr{U}_{ad}, \tag{1.10}$$

$$v\to J(v) \quad \text{is strongly l.s.c.} \tag{1.11}$$

Then there exists $u\in\mathscr{U}_{ad}$ such that

$$J(u)=\inf_v J(v). \tag{1.12}$$

This remarks also applies to functions $v\to J(v)$ defined on $\mathscr{U}_{ad}\subset\mathscr{U}$, where $\mathscr{U}$ is, for example, a reflexive Banach space.

With hypotheses (1.10), (1.11) only, we do not necessarily have uniqueness; clearly we have uniqueness if we assume that the function $v\to J(v)$ is strictly convex.

Remark 1.3. Assumption (1.10) is necessary to prove that every minimizing sequence is bounded (cf. 1.5). If $\mathscr{U}_{ad}$ is also bounded then we may dispense with Assumption (1.10).

---

[1] The sign □ indicates the end of a proof, remark, etc.

1.3. Characterization of the Minimizing Element. Variational Inequalities

Theorem 1.2. *Let the assumptions of Theorem 1.1 remain valid. The minimizing element u of* $\mathscr{U}_{ad}$ *is characterized by*

$$\pi(u, v-u) \geqslant L(v-u), \quad \forall v \in \mathscr{U}_{ad}. \tag{1.13}$$

Proof. 1. Let u be the minimizing element. Then for any $v \in \mathscr{U}_{ad}$ and $\theta \in]0,1[$, we have,

$$J(u) \leqslant J((1-\theta)u + \theta v),$$

whence,

$$\frac{1}{\theta}[J(u+\theta(v-u)) - J(u)] \geqslant 0. \tag{1.14}$$

Therefore, if we can pass to the limit, we obtain,

$$\lim_{\theta \to 0} \frac{1}{\theta}[J(u+\theta(v-u)) - J(u)] \geqslant 0. \tag{1.15}$$

If the limit in (1.15) exists, we may write (1.15) as,

$$J'(u) \cdot (v-u) \geqslant 0 \qquad (v \in \mathscr{U}_{ad}). \tag{1.16}$$

If $J(v)$ is given by (1.1), we may immediately verify that

$$J'(u) \cdot (v-u) = 2[\pi(u, v-u) - L(v-u)]$$

which proves (1.13)[2].

2) Alternatively suppose (1.13) (and hence 1.16) is true. Since the function $v \to J(v)$ is convex, we have for any $\theta \in]0,1[$:

$$J(v) - J(w) \geqslant \frac{1}{\theta}[J((1-\theta)w + v) - J(w)], \quad \forall v, w,$$

and hence if the limit in (1.15) exists, $\forall v, w$:

$$J(v) - J(w) \geqslant J'(w) \cdot (v-w). \tag{1.17}$$

Taking $w = u$, by virtue of (1.16) we deduce,

$$J(v) - J(u) \geqslant 0, \quad \forall v \in \mathscr{U}_{ad}$$

which proves the theorem. □

Inequalities of the type given by (1.13) are termed "*variational inequalities*".

---

[2] In this case $\frac{1}{2}J'(u)$ is the linear form $\varphi \to \pi(u, \varphi) - L(\varphi)$.

Remark 1.4. Let $\mathcal{U}$ be a Hilbert space over C (instead of $\mathbb{R}$) and let $\pi(u,v)$ be a sesquilinear hermitian form, that is,

$$\pi(u,v)=\overline{\pi(v,u)} \quad \forall u,v\in\mathcal{U}.$$

Assuming that (1.2) is satisfied, Theorem 1.1 remains valid without any change. Replacing (1.13) by

$$\operatorname{Re}\pi(u,v-u)\geqslant \operatorname{Re}L(v-u) \quad \forall v\in\mathcal{U}_{\mathrm{ad}}, \tag{1.18}$$

Theorem 1.2 remains valid. (In the above, $\operatorname{Re}\xi=$ real part of ξ.)

Remark 1.5. The Case $\mathcal{U}_{\mathrm{ad}}=\mathcal{U}$.[3]

If $U_{\mathrm{ad}}=\mathcal{U}$, in (1.13) we may take $v=u\pm\varphi$, where φ is any element of $\mathcal{U}$ and (1.13) becomes equivalent to

$$\pi(u,\varphi)=L(\varphi) \quad \forall\varphi\in\mathcal{U}. \tag{1.19}$$

This is the Euler equation of the problem.

Remark 1.6. The Case $\mathcal{U}_{\mathrm{ad}}=$ *Cone.*

Let us suppose

$$\mathcal{U}_{\mathrm{ad}}=\text{closed convex cone with vertex at the origin.} \tag{1.20}$$

Then (1.13) is equivalent to

$$\left.\begin{aligned}&\pi(u,v)\geqslant L(v) \quad \forall v\in\mathcal{U}_{\mathrm{ad}},\\ &\pi(u,u)=L(u).\end{aligned}\right\} \tag{1.21}$$

In fact, in (1.13) we may replace v by $v+u$ which gives the first inequality in (1.21); putting $v=0$ in (1.13), we obtain $\pi(u,v)\leq L(u)$ and hence the equality $\pi(u,v)\leq L(u)$. Conversely, it is obvious that (1.21) implies (1.13).

Remark 1.7. The case of a functional $v\to J(v)$ *which is not necessarily quadratic.*

Suppose that the function (or functional) $v\to J(v)$ is differentiable.[4]

The proof of Theorem 1.2 is also applicable to

Theorem 1.3. *Assume that the function* $v\to J(v)$ *is strictly convex, differentiable and satisfies* (1.10) *(the last hypothesis may be omitted if* $\mathcal{U}_{\mathrm{ad}}$ *is bounded). Then the unique element u in* $\mathcal{U}_{\mathrm{ad}}$ *satisfying* $J(u)=\operatorname*{Inf}_{v\in\mathcal{U}_{\mathrm{ad}}} J(v)$ *is characterized by*

$$J'(u)\cdot(v-u)\geqslant 0 \quad \forall v\in\mathcal{U}_{\mathrm{ad}}. \tag{1.22}$$

---

[3] In control problems this corresponds to the case where there are no constraints.

[4] Cf. J. Dieudonné [1], Chap. 8, section 1.

1.4. Alternative Form of Variational Inequalities

The following results are very useful in a technical sense:

Theorem 1.4. *Let all the hypotheses of Theorem 1.3 be satisfied. Then the characterization* (1.22) *is equivalent to:*

$$J'(v)\cdot(v-u)\geqslant 0 \quad \forall u\in\mathscr{U}_{\mathrm{ad}}. \tag{1.23}$$

Proof.

1) (1.23)⇒(1.22).

In (1.23), put $v=(1-\theta)w+\theta u$ where w is any element of $\mathscr{U}_{\mathrm{ad}}$ and $\theta\in]0,1[$; this implies

$$(1-\theta)J'((1-\theta)w+\theta u)\cdot(w-u)\geqslant 0$$

and hence

$$J'((1-\theta)w+\theta u)\cdot(w-u)\geqslant 0\,.$$

Let $\theta\to 1$. We then obtain

$$J'(u)\cdot(w-u)\geqslant 0 \quad \forall w\in\mathscr{U}_{\mathrm{ad}}$$

which is precisely (1.22).

2) (1.22)⇒(1.23).

First assume that the following result, which is important in its own right, is true:

Theorem 1.5. *Assume that the function $v\to J(v):\mathscr{U}\to\mathbb{R}$ is convex and differentiable. Then the derivative $v\to J'(v):\mathscr{U}\to\mathscr{U}'$ is monotone, that is,*

$$(J'(v)-J'(w))\cdot(v-w)\geqslant 0 \quad \forall v,w\in\mathscr{U}\,. \tag{1.24}$$

Since,

$$J'(v)\cdot(v-u)=J'(u)\cdot(v-u)+(J'(v)-J'(u))\cdot(v-u)$$

utilizing (1.24) we obtain

$$J'(v)\cdot(v-u)\geqslant J'(u)\cdot(v-u)$$

and hence (1.22) implies (1.23).

Proof of Theorem 1.5. We have seen that under the hypotheses of Theorem 1.5, (1.17) is true.

Interchanging v and w in (1.17), we also have

$$J(w)-J(v)\geqslant J'(v)\cdot(w-v)\,. \tag{1.25}$$

Finally combining (1.17) and (1.25) we deduce (1.24). □

Remark 1.8. Summarizing: under the hypotheses of Theorem 1.3, we have *three equivalent formulations* of the problem:

(i) the definition of the problem (when the minimum obtains):

$$J(u) = \operatorname*{Inf}_{v\in\mathscr{U}_{\mathrm{ad}}} J(v) \quad (u\in\mathscr{U}_{\mathrm{ad}}),$$

(ii) $$J'(u)\cdot(v-u)\geqslant 0 \ \ \forall v\in\mathscr{U}_{\mathrm{ad}} \quad (u\in\mathscr{U}_{\mathrm{ad}}),$$

(iii) $$J'(v)\cdot(v-u)\geqslant 0 \ \ \forall u\in\mathscr{U}_{\mathrm{ad}} \quad (u\in\mathscr{U}_{\mathrm{ad}}).$$

1.5. Function J being the Sum of a Differentiable and Non-Differentiable Function

If we assume that the function $v\to J(v)$ is coercive, lower semi-continuous in the weak topology and strictly convex, then there exists a $u\in\mathscr{U}_{\mathrm{ad}}$ such that

$$J(u)\leqslant J(v) \ \ \forall u\in\mathscr{U}_{\mathrm{ad}}.$$

In this case it is clear that we cannot apply criteria (ii) and (iii) of Remark 1.8. However these criteria are still applicable to the differentiable part of J. More precisely, we have the following result:

Theorem 1.6. *Consider the function*

$$J(v)=J_1(v)+J_2(v) \tag{1.26}$$

where we assume that the functions $J_i(v)$, $i=1,2$, are continuous, convex, and lower semi-continuous in the weak topology. Further let

$$J(v)\to+\infty \quad \text{as } \|v\|\to+\infty, \qquad v\in\mathscr{U}_{\mathrm{ad}}.$$

We assume that the function $v\to J_1(v)$ is differentiable, but J_2 is not necessarily differentiable. Finally assume that J is strictly convex.

Then the unique element $u\in\mathscr{U}_{\mathrm{ad}}$ such that $J(u)=\operatorname*{Inf}_{v\in\mathscr{U}_{\mathrm{ad}}} J(v)$ is characterized by

$$J_1'(u)\cdot(v-u)+J_2(v)-J_2(u)\geqslant 0 \ \ \forall u\in\mathscr{U}_{\mathrm{ad}}. \tag{1.27}$$

Proof. 1. Let $u\in\mathscr{U}_{\mathrm{ad}}$ be such that $J(u)\leqslant J(v)$ $\forall v\in\mathscr{U}_{\mathrm{ad}}$. Then

$$\forall v\in\mathscr{U}_{\mathrm{ad}}, \qquad \theta\in\,]0,1[,$$

$$J(u)\leqslant J((1-\theta)u+\theta v)=J_1((1-\theta)u+\theta v)+J_2((1-\theta)u+\theta v),$$

and since J_2 is convex, we deduce

$$J_1(u)+J_2(u)\leqslant J_1((1-\theta)u+\theta v)+(1-\theta)J_2(u)+\theta J_2(v)$$

whence

$$\frac{1}{\theta}[J_1((1-\theta)u+\theta v)-J_1(u)]+J_2(v)-J_2(u)\geqslant 0\,.$$

Letting $\theta\to 0$ we obtain (1.27).

2. Let us suppose that (1.27) is satisfied. In a general manner

$$\begin{aligned}J(v)-J(u)&=J_1(v)-J_1(u)+J_2(v)-J_2(u)\\&=J_1'(u)\cdot(v-u)+J_2(v)-J_2(u)+J_1(v)-J_1(u)-J_1'(u)\cdot(v-u)\end{aligned}$$

and hence from (1.17)

$$J(v)-J(u)\geqslant J_1'(u)\cdot(v-u)+J_2(v)-J_2(u),\qquad v\in\mathscr{U}_{\mathrm{ad}}\,.$$

Therefore if (1.27) is true, $J(u)=\inf_{v\in\mathscr{U}_{\mathrm{ad}}} J(v)$. □

Remark 1.9. Putting $J_2=0$, it is clear that Theorem 1.6 contains Theorems 1.3 and 1.4.

Remark 1.10. Suppose that J is of the form

$$J(v)=J_0(v)+J_1(v)+J_2(v)\,, \tag{1.28}$$

where the J_i's satisfy the same hypothese as in Theorem 1.6 and the functions J_0 and J_1 are differentiable. Then the unique element u such that $J(u)=\inf_{v\in\mathscr{U}_{\mathrm{ad}}} J(v)$ is characterized by one of the following equivalent conditions:

$$J_0'(u)\cdot(v-u)+J_1'(u)\cdot(v-u)+J_2(v)-J_2(u)\geqslant 0\quad\forall\, v\in\mathscr{U}_{\mathrm{ad}}\,, \tag{1.29}$$

$$J_0'(u)\cdot(v-u)+J_1'(v)\cdot(v-u)+J_2(v)-J_2(u)\geqslant 0\quad\forall\, v\in\mathscr{U}_{\mathrm{ad}}\,. \tag{1.30}$$

Remark 1.11. In case we only have existence of the minimizing element u of $\mathscr{U}_{\mathrm{ad}}$ but not necessarily uniqueness, any one of the variational inequalities we have obtained characterizes the set of elements in $\mathscr{U}_{\mathrm{ad}}$ determining the minimum.

In applications to control problems an element $u\in\mathscr{U}_{\mathrm{ad}}$ which determines the minimum is termed "optimal control".

Remark 1.12. All the preceeding results, without any change in their proofs are true when $\mathscr{U}$ is a reflexive Banach space.

1.6. The Convexity Hypothesis on $\mathscr{U}_{ad}$

So far we have assumed that $\mathscr{U}_{\mathrm{ad}}$ is *convex*. The following development shows how we may obtain a (simple) necessary condition for extremality in case $\mathscr{U}_{\mathrm{ad}}$ is assumed to be only *closed* in $\mathscr{U}$.

Definition 1.1. Let $\mathscr{U}_{ad}$ be closed and let $u \in \mathscr{U}_{ad}$. Define,

$$\mathscr{C}(\mathscr{U}_{ad}; u) = \{w \mid w \in \mathscr{U}; \text{ there exists } u_n \in \mathscr{U}_{ad} \text{ and } \lambda > 0, \text{ such that } u_n \to u \text{ and } \lambda_n(u_n - u) \to w \text{ in } \mathscr{U}\}. \tag{1.31}$$

It may be easily verified that

$$\mathscr{C}(\mathscr{U}_{ad}; u) \text{ is a closed cone with its vertex at } \{0\}. \tag{1.32}$$

We then have

Theorem 1.7. *Let $v \to J(v)$ be a function which is differentiable and let u be an element (assumed to exist) of $\mathscr{U}_{ad}$ such that $J(u) \leqslant J(v)$ $\forall v \in \mathscr{U}_{ad}$. Then*

$$J'(v) \cdot w \geqslant 0 \quad \forall w \in \mathscr{C}(\mathscr{U}_{ad}; u). \tag{1.33}$$

Before verifying this result, let us demonstrate that if $\mathscr{U}_{ad}$ is convex then (1.33) is equivalent to (1.22).

We first show that (1.33) implies (1.22). Let $v \in \mathscr{U}_{ad}$; since $\mathscr{U}_{ad}$ is convex,

$$u_n = (1 - \theta_n)v + \theta_n u \in \mathscr{U}_{ad}, \qquad \theta_n \in]0, 1[, \theta_n \to 1$$

and hence $\dfrac{1}{1-\theta_n}(u_n - u) = v - u$. Therefore $w = v - u \in \mathscr{C}(\mathscr{U}_{ad}; u)$ and therefore (1.33) implies (1.22).

Conversely, suppose (1.22) is true and if $w \in \mathscr{C}(\mathscr{U}_{ad}; u)$, we have,

$$J'(u) \cdot (u_n - u) \geqslant 0, \quad \text{and hence } \lambda_n J'(u) \cdot (u_n - u) \geqslant 0, \qquad \lambda_n > 0,$$

i.e. $J'(u) \cdot \lambda_n(u_n - u) \geqslant 0$. Passing to the limit we obtain (1.33).

Proof of Theorem 1.7. Let $u \in \mathscr{U}_{ad}$ be such that $J(u) \leqslant J(v)$ $\forall v \in \mathscr{U}_{ad}$ and let $w \in \mathscr{C}(\mathscr{U}_{ad}; u)$. Utilizing (1.31), we obtain

$$J(u) \leqslant J(u_n) = J(u + u_n - u) = J(u) + J'(u) \cdot (u_n - u) + \|u_n - u\| O(1),$$

and hence

$$J'(u) \cdot (u_n - u) + \|u_n - u\| O(1) \geqslant 0.$$

Therefore

$$J'(u) \cdot \lambda_n(u_n - u) + \lambda_n \|u_n - u\| O(1) \geqslant 0,$$

whence passing to the limit in n we obtain (1.33). ☐

Remark 1.13. Conditions of the type (1.33) are "first order necessary conditions", in the terminology of Calculus of Variations.

1.7. Orientation

Our point of departure was a problem in minimization: "*search for* $\operatorname{Inf}_{v\in\mathcal{U}_{ad}} J(v)$". We reduced the problem to the study of variational inequalities of the type

$$J'(u)\cdot(v-u)\geqslant 0 \quad \forall v\in\mathcal{U}_{ad}.$$

We may then ask the question that if $J'(u)$ is replaced by an operator, (not necessarily linear) $u\to A(u)$, what should the properties of A be such that the variational inequality

$$A(u)(v-u)\geqslant 0 \quad \forall v\in\mathcal{U}_{ad}$$

admits at least one solution in $\mathcal{U}_{ad}$. This type of problem has given rise to numerous developments, an example of which (section 2 following) is given below. For other references, the bibliographic notes may be consulted.

In the following we shall study a few examples and in particular the case, important for optimal control, where J is not coercive.

2. A Direct Solution of Certain Variational Inequalities

2.1. Problem Statement

According to the guidelines presented in 1.7, we shall consider problem (1.13) *a priori*. More precisely:

Let $\pi(u,v)$ be a given bilinear form on $\mathcal{U}$ which is not necessarily symmetric; we search for a $u\in\mathcal{U}_{ad}$ which satisfies

$$\pi(u,v-u)\geqslant L(v-u) \quad \forall v\in\mathcal{U}_{ad}, \tag{2.1}$$

(where L is a given linear form on $\mathcal{U}$).

It should be noted that when π is not symmetric, (2.1) does not correspond to a problem in the Calculus of Variations, at least not posed in the form (usual) of a search for extrema of a functional. In spite of this, inequalities of the type (2.1) are referred to as variational inequalities.

2.2. An Existence and Uniqueness Theorem

Theorem 2.1. *Assume that the not necessarily symmetric bilinear form* $\pi(u,v)$ *satisfies*

$$\pi(v_1-v_2,v_1-v_2)\geqslant c\|v_1-v_2\|^2, \qquad c>0, \quad \forall v_1,v_2\in\mathcal{U}_{ad}. \tag{2.2}$$

Then there exists a unique u *in* $\mathcal{U}_{ad}$ *which satisfies* (2.1).

Proof of Uniqueness. Suppose that there exists two solutions u_1 and u_2 in $\mathscr{U}_{ad}$. We then have

$$\pi(u_1, v-u_1) \geqslant L(v-u_1) \quad \forall v \in \mathscr{U}_{ad}, \tag{2.3}$$

$$\pi(u_2, v-u_2) \geqslant L(v-u_2) \quad \forall v \in \mathscr{U}_{ad}. \tag{2.4}$$

Putting $v=u_2$ (resp. $v=u_1$) in (2.3) (resp. (2.4)) and adding (2.3) and (2.4) we obtain

$$-\pi(u_1-u_2, u_1-u_2) \geqslant 0$$

and hence from (2.2) $\pi(u_1-u_2, u_1-u_2)=0$ and thus $u_1=u_2$. □

Proof of Existence. 1. We shall apply Theorem 1.2 with $\pi(u,v)=(v,u)_{\mathscr{U}}$, the scalar product on $\mathscr{U}$. Hence for given g in $\mathscr{U}$, there exists

$$w=S(g) \in \mathscr{U}_{ad} \quad \text{unique} \tag{2.5}$$

such that

$$(w, v-w)_{\mathscr{U}} \geqslant (g, v-w)_{\mathscr{U}} - \rho[\pi(g, v-w) - L(v-w)] \quad \forall v \in \mathscr{U}_{ad} \tag{2.6}$$

where $\rho>0$ is any fixed number.

If u is a fixed point in $\mathscr{U}_{ad}$ of the map $g \to S(g)$ so defined, then u satisfies (2.1).

The theorem will be proved if we can show that

$$\begin{gathered} \text{we may choose } \rho>0 \text{ such that} \\ g \to S(g) \text{ is a contraction on } \mathscr{U}_{ad}. \end{gathered} \tag{2.7}$$

2. The form $v \to \pi(g,v)$ being linear and continuous on $\mathscr{U}$ may be written as

$$\pi(g,v)=(Bg,v)_{\mathscr{U}}, \quad B \in \mathscr{L}(\mathscr{U};\mathscr{U}). \tag{2.8}$$

Thus (2.6) is equivalent to

$$(w, v-w)_{\mathscr{U}} \geqslant (g-\rho Bg, v-w)_{\mathscr{U}} + \rho L(v-w) \quad \forall v \in \mathscr{U}_{ad}. \tag{2.9}$$

Let $w_1=S(g_1)$; (2.9) applied to g_1, w_1 gives us:

$$(w_1, v-w_1)_{\mathscr{U}} \geqslant (g_1-\rho Bg_1, v-w)_{\mathscr{U}} + \rho L(v-w_1) \quad \forall v \in \mathscr{U}_{ad}. \tag{2.10}$$

Putting $v=w_1$ (resp. $v=w$) in (2.9) (resp. (2.10)) and adding, we obtain

$$-\|w-w_1\|_{\mathscr{U}}^2 \geqslant -((I-\rho B)(g-g_1), w-w_1)_{\mathscr{U}}$$

whence

$$\|w-w_1\|_{\mathscr{U}} \leqslant \|(I-\rho B)(g-g_1)\|_{\mathscr{U}}. \tag{2.11}$$

But

$$\|(I-\rho B)g\|^2 = \|g\|_{\mathscr{U}}^2 + \rho^2\|Bg\|_{\mathscr{U}}^2 - 2\rho(Bg,g)_{\mathscr{U}}.$$

Hence noting from (2.8) that $(Bg,g)_{\mathscr{U}}=\pi(g,g)$, and replacing g by $g-g_1$:

$$\|(I-\rho B)(g-g_1)\|_{\mathscr{U}}^2 \leqslant (1+\rho^2\|B\|^2)\|g-g_1\|_{\mathscr{U}}^2 - 2\rho\,\pi(g-g_1,g-g_1).$$

Applying the above inequality with $g, g_1 \in \mathscr{U}_{\text{ad}}$, we deduce by virtue of (2.2)

$$\|(I-\rho B)(g-g_1)\|_{\mathscr{U}}^2 \leqslant (1+\rho^2\|B\|^2-2c\rho)\|g-g_1\|^2$$

which when combined with (2.11) finally gives

$$\|S(g)-S(g_1)\|_{\mathscr{U}} \leqslant (1+\rho^2\|B\|^2-2c\rho)^{\frac{1}{2}}\|g-g_1\|_{\mathscr{U}}, \quad g, g_1 \in \mathscr{U}_{\text{ad}}. \tag{2.12}$$

Choosing $\rho>0$ such that $1+\rho^2\|B\|^2-2c\rho<1$, we obtain the desired result. □

Remark 2.1. More generally, let $b(v,v)$ be a symmetric bilinear form on $\mathscr{U}$ which is coercive.

For given g in $\mathscr{U}$ and $\rho>0$, let $w=S_b(g)$ be the solution in $\mathscr{U}_{\text{ad}}$ of

$$b(w,v-w) \geqslant b(g,v-w)-\rho[\pi(g,v-w)-L(v-w)] \quad \forall v \in \mathscr{U}_{\text{ad}}. \tag{2.13}$$

Then for an appropriate $\rho>0$, the map $g \to S_b(g)$ is a contraction on $\mathscr{U}_{\text{ad}}$—whose fixed point is the required u; (in the proof of the preceeding theorem, we took

$$b(u,v)=(u,v)_{\mathscr{U}}).$$

In this way, a constructive procedure for the approximation of u is obtained, if for example, the method of successive approximation is employed to solve (2.13). For this purpose it is sufficient to identify a bilinear form $b(u,v)$ for which one can effectively solve (possibly in an approximate fashion) the problem

$$b(u,v-w) \geqslant M(v-u) \quad \forall v \in U_{\text{ad}}.$$

Remark 2.2. Remarks analogous to Remarks 1.5 and 1.6 may be made here.

In particular if $\mathscr{U}_{\text{ad}}=\mathscr{U}$, Theorem 2.1 demonstrates

> if $\pi(u,v)$ is a continuous bilinear form on $\mathscr{U}$
> which is not necessarily symmetric and which satisfies
> $\pi(v,v) \geqslant c\|v\|^2$, $c>0$, $\forall v \in \mathscr{U}$, (2.14)
> then there exists a unique $u \in \mathscr{U}$ such that
> $\pi(u,v)=L(v)$ $\forall v \in \mathscr{U}$.

This result is generally known as the Lax-Milgram Lemma, cf. Lax-Milgram [1], I. M. Visik [1].

3. Examples

3.1. Function Spaces on Ω

Let Ω be an open set in $\mathbb{R}^n$ with boundary Γ.

Throughout we shall assume that Ω is a bounded, open set with boundary Γ which is a C^∞-manifold of dimension $(n-1)$. Locally, Ω is totally on one side of Γ. These hypotheses may be somewhat relaxed in the case of certain examples considered in this book, but we shall not insist on this point.

The following notation will be used:

$\mathscr{D}(\Omega)$ = space of infinitely differentiable functions φ on Ω, with values in R and with compact support (support depending on φ); the topology on $\mathscr{D}(\Omega)$ is the inductive limit topology of L. Schwartz [1]. A "pseudo-topology" on $\mathscr{D}(\Omega)$ is defined in the following way: a sequence $u_n \in \mathscr{D}(\Omega)$ tends to zero if

(i) the φ_n's have their support in a fixed compact set in Ω,

(ii) φ_n and all its derivatives tend towards zero uniformly on Ω.

$\mathscr{D}'(\Omega)$ = space of distributions on Ω = space of continuous linear forms on $\mathscr{D}(\Omega)$; if $f \in \mathscr{D}'(\Omega)$, its derivative $\dfrac{\partial f}{\partial x_i}$ is the unique element of $\mathscr{D}'(\Omega)$ defined by

$$\left\langle \frac{\partial f}{\partial x_i}, \varphi \right\rangle = -\left\langle f, \frac{\partial \varphi}{\partial x_i} \right\rangle \quad \forall \varphi \in \mathscr{D}(\Omega),$$

where $\langle \cdot, \cdot \rangle$ denotes the scalar product between $D'(\Omega)$ and $D(\Omega)$. Generally we shall write distributions as functions; if $f \in \mathscr{D}'(\Omega)$, $\varphi \in \mathscr{D}(\Omega)$, we shall therfore write

$$\langle f, \varphi \rangle = \int_\Omega f(x)\varphi(x)\,dx .$$

For example, if $a \in \Omega$, the Dirac mass $+1$ concentrated at a is denoted by $\delta(x-a)$ and defined by

$$\langle \delta(x-a), \varphi \rangle = \int_\Omega \delta(x-a)\varphi(x)\,dx = \varphi(a).$$

For $1 \leqslant p \leqslant \infty$, we define:

$L^p(\Omega)$ = space (equivalence class) of functions, measurable on Ω such that

$$\|f\|_{L^p(\Omega)} = \left(\int_\Omega |f(x)|^p dx \right)^{1/p} < \infty \quad \text{if } p < \infty$$

and

$$\|f\|_{L^\infty(\Omega)} = \text{ess. sup}\,|f(x)| < \infty \quad \text{if } p = \infty .$$

In particular, $L^2(\Omega)$ $(p=2)$ is a Hilbert space which will be identified with its dual. [Care must be taken not to identify other Hilbert spaces of functions on Ω with their duals by means of the same identification.]

For every $f \in L^p(\Omega)$, we associate the distribution

$$\varphi \xrightarrow{\tilde{f}} \langle f, \varphi \rangle = \int_\Omega f(x)\varphi(x)\,dx\,.$$

We thus obtain a linear continuous map $f \to \tilde{f}: L^p(\Omega) \to \mathscr{D}'(\Omega)$ which is *injective*. We can identify f with $\tilde{f}$ which we shall do; making the same identification for $\mathscr{D}(\Omega)$, we obtain:

$$\mathscr{D}(\Omega) \subset L^p(\Omega) \subset \mathscr{D}'(\Omega)\,. \tag{3.1}$$

In view of (3.1) we may define the distributions $\dfrac{\partial f}{\partial x_i}$ for all $f \in L^p(\Omega)$. More generally, for $f \in \mathscr{D}'(\Omega)$, we form:

$$q = \{q_1, \ldots, q_n\}, \quad |q| = q_1 + \cdots + q_n,$$

$$D^q = D_1^{q_1} \ldots D_n^{q_n}, \qquad D_i = \frac{\partial}{\partial x_i},$$

$$\langle D^q f, \varphi \rangle = (-1)^{|q|} \langle f, D^q \varphi \rangle \quad \forall \varphi \in \mathscr{D}(\Omega). \tag{3.2}$$

We may then define

$$H^m(\Omega) = \{v \mid v \in L^2(\Omega),\ D^q v \in L^2(\Omega) \quad \forall q, |q| \leqslant m\}\,. \tag{3.3}$$

The space $H^m(\Omega)$ is the *Sobolev space of order* m on Ω. For $u, v \in H^m(\Omega)$, we define

$$(u,v)_{H^m(\Omega)} = \sum_{|q| \leqslant m} (D^q u, D^q v)_{L^2(\Omega)}\,. \tag{3.4}$$

It may be easily verified (by utilizing the continuity of the map $f \to D^q f: \mathscr{D}'(\Omega) \to \mathscr{D}'(\Omega)$, $\mathscr{D}'(\Omega)$ being endowed with the weak topology of the dual of $\mathscr{D}(\Omega)$—cf. L. Schwartz [1], J. Horvath [1], K. Yosida [1]) that

Theorem 3.1. *The space $H^m(\Omega)$ endowed with the scalar product* (3.4) *is a Hilbert space.*

3.2. Function Spaces on Γ

In an analogous manner we define the spaces

$$\mathscr{D}(\Gamma), \qquad L^q(\Gamma), \qquad \mathscr{D}'(\Gamma), \qquad H^m(\Gamma)$$

on the manifold Γ, endowed with the measure of $d\Gamma$ of the surface (induced by dx). If Γ is compact, $\mathscr{D}(\Gamma)$ coincides with the space of infinitely differentiable functions on Γ (without any condition on the supports!). The derivatives are defined by local coordinates of the type which allow $H^m(\Omega)$ to be defined by local coordinates.

3.3. Subspaces of $H^m(\Omega)$

For purposes of application one of the most useful results is the trace theorem on $H^m(\Omega)$. For each $u \in H^m(\Omega)$ we may associate the trace of u on Γ as well as that of its normal derivative $\frac{\partial^k u}{\partial n^k}$,[5] for $1 \leqslant k \leqslant m-1$ and in this way characterize the image of $H^m(\Omega)$ by the map

$$u \to u|_\Gamma, \left.\frac{\partial u}{\partial n}\right|_\Gamma, \ldots, \left.\frac{\partial u^{m-1}}{\partial n^{m-1}}\right|_\Gamma. \tag{3.5}$$

This characterization requires *Sobolev spaces of non-integral order* and it is therefore essential to introduce such spaces.

Let us first consider the space $H^m(\Omega)$ with $\Omega = \mathbb{R}^n$. Utilizing the Fourier Transform

$$\mathscr{F}: f \to \mathscr{F} f,\ \mathscr{F} f(\xi) = \int_{\mathbb{R}^n} \exp(2\pi i x \cdot \xi) f(x) dx, \tag{3.6}$$

(where $x \cdot \xi = x_1 \xi_1 + \cdots + x_n \xi_n$) an isomorphism between $L^2(\mathbb{R}^n)$ and itself may be established, with

$$\|\mathscr{F} f\|_{L^2(\mathbb{R}^n)} = \|f\|_{L^2(\mathbb{R}^n)}.$$

The inverse $\mathscr{F}^{-1}$ of $\mathscr{F}$ is given by

$$\mathscr{F}^{-1} f(x) = \int_{\mathbb{R}^n} \exp(2\pi i x \cdot \xi) f(\xi) d\xi.^{6} \tag{3.7}$$

The maps $\mathscr{F}$ and $\mathscr{F}^{-1}$ may be extended, by continuity, to the space of tempered distributions (L. Schwartz [2]).

We then verify that

$$\mathscr{F}(D^q v) = (2\pi i)^{|q|} \xi_1^{q_1} \ldots \xi_n^{q_n} \mathscr{F} v \quad \forall v \in L^2(\mathbb{R}^n). \tag{3.8}$$

---

5 $\frac{\partial}{\partial n}$ = normal derivative at Γ, directed towards the exterior of Ω (to fix ideas)

6 In (3.6) and (3.7), if $f \in L^2(\mathbb{R}^n)$, the integrals are limits in the sense of $L^2(\mathbb{R}^n)$ of functions

$$\int_{|\xi| \leqslant M} \exp(\pm 2\pi i x \xi) f(\xi) d\xi, \quad M \to \infty.$$

From these remarks we obtain

$$H^m(\mathbb{R}^n)=\{v|\xi^q \mathscr{F} v\in L^2(\mathbb{R}^n)\ \forall |q|\leq m\} \tag{3.9}$$

(where $\xi^q=\xi_1^{q_1}\dots\xi_n^{q_n}$),
and without difficulty it may be verified that (3.9) may be replaced by the equivalent condition:

$$H^m(\mathbb{R}^n)=\{v|(1+|\xi|^2)^{m/2}\mathscr{F} v\in L^2(\mathbb{R}^n)\}\,. \tag{3.10}$$

If $u,v\in H^m(\mathbb{R}^n)$, the bilinear form

$$((1+|\xi|^2)^{m/2}\mathscr{F} u,\ (1+|\xi|^2)^{m/2}\mathscr{F} v)_{L^2(R^n)} \tag{3.11}$$

defines a scalar product which is equivalent to that defined in (3.4).

But if $H^m(\mathbb{R}^n)$ is defined as in (3.10), it does not matter whether m is an integer or positive. We may then define for any $s\in\mathbb{R}$

$$H^s(\mathbb{R}^n)=\{v|(1+|\xi|^2)^{s/2}\mathscr{F} v\in L^2(\mathbb{R}^n)\}\,, \tag{3.12}$$

which is a Hilbert space when endowed with the scalar product:

$$((u,v))_{H^s}=((1+|\xi|^2)^{s/2}\mathscr{F} u,(1+|\xi|^2)^{s/2}\mathscr{F} v)_{L^2(\mathbb{R}^n)}. \tag{3.13}$$

This is equivalent (but not identical) to (3.4) if $s=m=\text{integer}>0$.

It may be shown (cf. Lions-Magenes [1], Chapter 1) that $H^s(\mathbb{R}^n)$ is of a local type: if $\varphi\in\mathscr{D}(\mathbb{R}^n)$, the map $u\to\varphi u$ is a linear continuous map from $H^s(\mathbb{R}^n)$ into $H^s(\mathbb{R}^n)$.

This allows us to define $H^s(\Gamma)$ by means of local coordinates.

Example 3.1. Let Ω be the disc $|x|<1$ in $\mathbb{R}^2$; Γ then is the circle $|x|=1$, if $x_1=\cos\theta$, $x_2=\sin\theta$, $0\leqslant\theta\leqslant 2\pi$. Then

$$\left.\begin{aligned} &f\in H^s(\Gamma) \quad \text{is equivalent to}\\ &f=\sum_{n=-\infty}^{+\infty} f_n e^{in\theta},\\ &\sum_{n=-\infty}^{+\infty}(1+n^2)^s|f_n|^2<\infty. \end{aligned}\right\} \tag{3.14}$$

If we identify $L^2(\Omega)$ (resp. $L^2(\Gamma)$) with its dual, we may identify $H^s(\mathbb{R}^n)$ (resp. $H^s(\Gamma)$) with $H^{-s}(\mathbb{R}^n)$ (resp. $H^{-s}(\Gamma)$) by means of the "same identification".

In general, if we denote by X' the dual of X, we obtain

$$(H^s(\mathbb{R}^n))'=H^{-s}(\mathbb{R}^n),\ \ \forall s\in\mathbb{R}, \tag{3.15}$$

and

$$(H^s(\Gamma))'=H^{-s}(\Gamma),\quad \forall s\in\mathbb{R}. \tag{3.16}$$

We are now in a position to state the trace theorem (for the proof we refer to Lions-Magenes [1]):

Theorem 3.2. *For any $u \in H^m(\Omega)$, we may define in a unique manner its traces*

$$u|_\Gamma, \left.\frac{\partial u}{\partial n}\right|_\Gamma, \ldots, \left.\frac{\partial^{m-1} u}{\partial n^{m-1}}\right|_\Gamma .$$

We have

$$\left.\frac{\partial^k n}{\partial n^k}\right|_\Gamma \in H^{m-k-\frac{1}{2}}(\Gamma), \quad 0 \leqslant k \leqslant m-1, \tag{3.17}$$

the map $u \to \left\{ \left.\frac{\partial^k u}{\partial n^k}\right|_\Gamma, 0 \leqslant k \leqslant m-1 \right\}$ *being a linear, continuous and surjective map of* $H^m(\Omega)$ *onto* $\prod_{k=0}^{m-1} H^{m-k-\frac{1}{2}}(\Gamma)$. *We also have the following result:*

Theorem 3.3. *The kernel of the map* (3.5) (*i.e. the space of* $u \in H^m(\Omega)$ *such that* $\left.\frac{\partial^k u}{\partial n^k}\right|_\Gamma = 0$, $0 \leqslant k \leqslant m-1$) *coincides with the closure of* $\mathscr{D}(\Omega)$ *in* $H^m(\Omega)$.

We denote this subspace by $H_0^m(\Omega)$. Thus

$$H_0^m(\Omega) = \left[u \,|\, u \in H^m(\Omega), \left.\frac{\partial^k u}{\partial n^k}\right|_\Gamma = 0, 0 \leqslant k \leqslant m-1 \right]. \tag{3.18}$$

Since $\mathscr{D}(\Omega)$ is dense in $H_0^m(\Omega)$, the dual of $H_0^m(\Omega)$ may be identified with a subspace of $\mathscr{D}'(\Omega)$;

$$H^{-m}(\Omega) = (H_0^m(\Omega))'; \tag{3.19}$$

then

$$H_0^m(\Omega) \subset L^2(\Omega) \subset H^{-m}(\Omega). \tag{3.20}$$

3.4. Examples of Boundary Value Problems

Let a_{ij}, $i,j = 1, \ldots, n$, be given functions on Ω with the properties

$$a_{ij} \in L^\infty(\Omega) \quad \text{(with real values)}. \tag{3.21}$$

$$\sum_{i,j=1}^{n} a_{ij}(x)\xi_i\xi_j \geqslant \alpha(\xi_1^2 + \cdots + \xi_n^2), \quad \alpha > 0, \ \forall \xi \in \mathbb{R}^n, \tag{3.22}$$

almost everywhere on Ω.

Let $a_0 \in L^\infty(\Omega)$ with

$$a_0(x) \geqslant \alpha > 0 \quad \text{almost everywhere on } \Omega. \tag{3.23}$$

For $u, v \in H^1(\Omega)$, set

$$\pi(u,v) = \sum_{i,j=1}^{n} \int_\Omega a_{ij} \frac{\partial u}{\partial x_j} \frac{\partial v}{\partial x_i} dx + \int_\Omega a_0 u v \, dx. \tag{3.24}$$

We have thus defined a linear form on $H^1(\Omega)$ and by virtue of (3.22) and (3.23)

$$\pi(v,v) \geqslant \alpha \|v\|^2_{H^1(\Omega)} \quad \forall v \in H^1(\Omega). \tag{3.25}$$

Let us choose (according to the notation used in the general theory of section 2),

$$\begin{aligned} &\mathscr{U} = \text{closed subspace of } H^1(\Omega) \text{ (endowed with the induced norm)} \\ &\quad \text{such that } \mathscr{U} \supset H_0^1(\Omega) \text{ (with the possibility of equality),} \end{aligned} \tag{3.26}$$

and let

$$L = \text{continuous linear form on } \mathscr{U}. \tag{3.27}$$

From Theorem 2.1 (2.14) we obtain:

$$\left.\begin{aligned} &\text{there exists a unique } u \in \mathscr{U} \text{ such that} \\ &\pi(u,v) = L(v), \ \forall v \in \mathscr{U}. \end{aligned}\right\} \tag{3.28}$$

Example 3.2. Equation (3.28) is equivalent to

$$\left.\begin{aligned} &Au = f, \\ &u \in H_0^1(\Omega) \end{aligned}\right\} \tag{3.29}$$

where

$$Av = -\sum_{i,j=1}^{n} \frac{\partial}{\partial x_i}\left(a_{ij}(x) \frac{\partial v}{\partial x_j}\right) + a_0(x) v. \tag{3.30}$$

In fact if $\mathscr{U} = H^1(\Omega)$, equation (3.28) is equivalent to

$$\pi(u, \varphi) = L(\varphi) \quad \forall \varphi \in \mathscr{D}(\Omega),$$

which is equivalent to (3.29), provided we use the definition of the derivative in the distribution sense.

Problem (3.29) is the *Dirichlet Problem.*

Example 3.3. $\mathscr{U} = H^1(\Omega)$; coefficients a_{ij}, a_0 regular in $\bar{\Omega}$ (for example, $a_{ij} \in C^k(\bar{\Omega})$, $a_0 \in C^0(\bar{\Omega})$, where $C^k(\bar{\Omega})$ = space of functions which are k-times continuously differentiable in $\bar{\Omega}$); let us choose $L(v)$ to be of the form

$$\left.\begin{aligned} L(v) &= \int_\Omega f v\,dx + \int_\Gamma g v\,d\Gamma, \\ f &\in L^2(\Omega), \quad g \in H^{-\frac{1}{2}}(\Gamma). \end{aligned}\right\} \tag{3.31}$$

We note that (3.31) defines a continuous linear form on $H^1(\Omega)=\mathcal{U}$, since from Theorem 3.2 the map $v \to v|_\Gamma$ is a continuous map from $H^1(\Omega) \to H^{\frac{1}{2}}(\Gamma)$.

It remains to interpret (3.28).

Consider first $v=\varphi \in \mathcal{D}(\Omega)$. Then

$$Au=f \quad \text{in } \Omega. \tag{3.32}$$

We now reason in a formal manner[7]. Let us multiply both sides of (3.32) by v and apply Green's formula[8]:

$$\int_\Omega (Au)v\,dx = -\int_\Gamma \left(\frac{\partial u}{\partial \nu_A}\right) v\,d\Gamma + \pi(u,v) = \int_\Omega f v\,dx$$

where

$$\left.\begin{aligned} &\frac{\partial u}{\partial \nu_A} = \sum_{i,j} a_{ij} \frac{\partial u}{\partial x_j} \cos(n, x_j) \text{ on } \Gamma, \\ &\cos(n, x_j) = i\text{-th direction cosine of } n,\ n \text{ being} \\ &\text{the normal at } \Gamma \text{ exterior to } \Omega. \end{aligned}\right\} \tag{3.33}$$

But, from (3.28)

$$\pi(u,v) = \int_\Omega f v\,dx + \int_\Gamma g v\,d\Gamma,$$

whence

$$\int_\Gamma \left(-\frac{\partial u}{\partial \nu_A} + g\right) v\,d\Gamma = 0 \quad \forall v \in H^1(\Omega);$$

whence

$$\frac{\partial u}{\partial \nu_A} = g.$$

We have thus demonstrated—provided the application of Green's formula can be justified—the existence and uniqueness of $u \in H^1(\Omega)$ satisfying

$$\left.\begin{aligned} Au &= f \quad \text{in } \Omega, \\ \frac{\partial u}{\partial \nu_A} &= g \quad \text{on } \Gamma. \end{aligned}\right\} \tag{3.34}$$

This is the *Neumann problem* relative to A.

---

[7] This may be entirely justified; cf. Lions-Magenes [1], Chapter 2.

[8] This is what needs to be justified.

Remark 3.1. It may be shown (cf. Lions-Magenes [1], Chapter 2) that for $u\in H^1(\Omega)$ such that $Au\in L^2(\Omega)$, $\frac{\partial u}{\partial \nu_A}$ may be defined uniquely on Γ. Moreover $\frac{\partial u}{\partial \nu_A}\in H^{-\frac{1}{2}}(\Gamma)$.

Example 3.4. $\mathscr{U}=H^1(\Omega)$, *coefficients* a_{ij} *discontinuous.*

Suppose that $\Omega=\Omega_1\cup\Omega_2\cup S$ (cf. Fig. 1); let a_{ij}^1 (resp. a_{ij}^2) be functions which are regular on $\bar{\Omega}_1$ (resp. $\bar{\Omega}_2$), which do not coincide on S, and let

$$a_{ij}=a_{ij}^k \text{ in } \Omega_k, \qquad k=1,2.$$

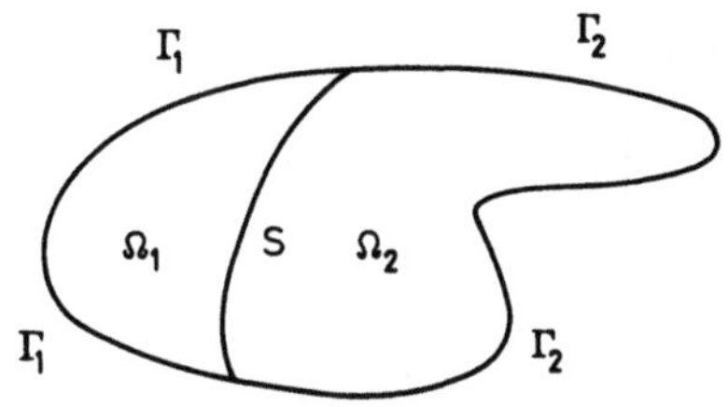

Fig. 1

Let us take

$$L(v)=\int_{\Omega_1} f_1 v\,dx+\int_{\Omega_1} f_2 v\,dx, \qquad f_k\in L^2(\Omega_k), \qquad k=1,2.$$

We again have to interpret (3.28).

Let us first take successively $v=\varphi\in\mathscr{D}(\Omega_1)$ and $v=\varphi\in\mathscr{D}(\Omega_2)$. We find that u satisfies:

$$A^k u=f_k \quad \text{in } \Omega_k, \qquad k=1,2,$$

where

$$A^k w=-\sum_{i,j=1}^{n}\frac{\partial}{\partial x_i}\left(a_{ij}^k\frac{\partial w}{\partial x_j}\right)+a_0^k w, \qquad k=1,2. \tag{3.35}$$

Put $u=u_k$ in Ω_k.

Then taking $v\in H^1(\Omega)$ to be identically zero in $\bar{\Omega}_2$ and in the neighborhood of S in Ω_1, we see as in example 3.3, that

$$\frac{\partial u_1}{\partial \nu_{A^1}}=0 \quad \text{on } \Gamma_2. \tag{3.36}$$

Similarly

$$\frac{\partial u_2}{\partial \nu_{A^2}}=0 \quad \text{on } \Gamma_2. \tag{3.37}$$

Now take $v\in\mathscr{D}(\Omega)$ to be non-zero on S. From the equation $A^k u_k = f_k$ we deduce that:

$$\int_{\Omega_1}(A^1u_1)v\,dx+\int_{\Omega_2}(A^2u_2)v\,dx=\int_{\Omega_1}f_1v\,dx+\int_{\Omega_2}f_2v\,dx.$$

Applying Green's Formula to the left hand side, we obtain:

$$-\int_S\frac{\partial u_1}{\partial\nu_{A^1}}v\,dS-\int_S\frac{\partial u_2}{\partial\nu_{A^2}}v\,ds+\pi(u,v)=L(v),$$

where $\dfrac{\partial}{\partial\nu_{A^k}}$ is taken on S, oriented by the normal on S directed towards the exterior of Ω_k. Thus

$$\int_S\left(\frac{\partial u_1}{\partial\nu_{A^1}}+\frac{\partial u_2}{\partial\nu_{A^2}}\right)v\,dS=0$$

and hence

$$\frac{\partial u_1}{\partial\nu_{A^1}}+\frac{\partial u}{\partial\nu_{A^2}}=0\quad\text{on } S. \tag{3.38}$$

In summary, we have obtained $u_1\in H^1(\Omega_1)$, $u_2\in H^2(\Omega_2)$ satisfying

$$\left.\begin{aligned}A^1u_1&=f_1\quad\text{in }\Omega_1,\\A^2u_2&=f_2\quad\text{in }\Omega_2,\end{aligned}\right\} \tag{3.39}$$

and satisfying the conditions (3.36) and (3.37) on Γ_1 and Γ_2 respectively. On S we have the conditions of transmission

$$\left.\begin{aligned}&u_1=u_2\quad\text{on } S,\\&\text{and (3.38)}.\end{aligned}\right\} \tag{3.40}$$

Example 3.5. Let Ω be as shown in Fig. 2. Let

$$\mathscr{U}=\{u\mid u\in H^1(\Omega),\quad u|_\Gamma=0\}.$$

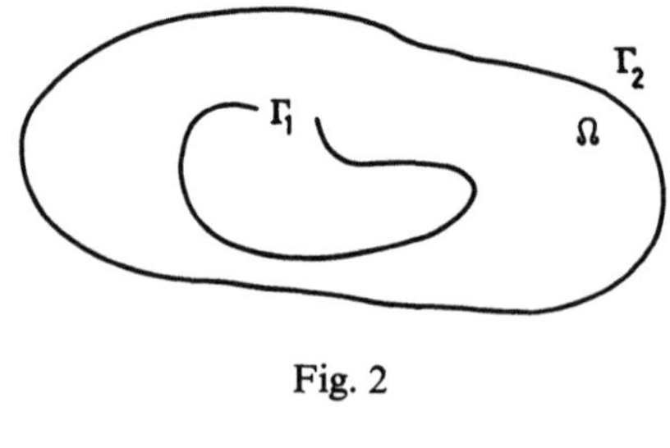

Fig. 2

Let $\pi(u,v)$ be as in the preceeding examples and let

$$L(v)=\int_{\Omega} f v\,dx+\int_{\Gamma_2} g v\,d\Gamma_2, \qquad g\in H^{-\frac{1}{2}}(\Gamma_2), \qquad f\in L^2(\Omega).$$

For this problem (3.28) may be interpreted to obtain:

$$\left.\begin{aligned} Au&=f \quad \text{in } \Omega,\\ u&=0 \quad \text{on } \Gamma_1,\\ \frac{\partial u}{\partial \nu_A}&=g \quad \text{on } \Gamma_2.\end{aligned}\right\} \tag{3.41}$$

Problem (3.41) is a mixed problem: Dirichlet conditions on one part of the boundary and Neumann conditions on the rest of the boundary.

3.5. Unilateral Boundary Value Problems (I)

So far, we have applied the general theory of section 2 to the case where there are "no constraints", i.e. with $\mathscr{U}_{\text{ad}}=\mathscr{U}$.

We shall take $\pi(u,v)$ to be the same as in 3.4 but $\mathscr{U}_{\text{ad}}$ will now be taken as a closed, convex subset of $\mathscr{U}$.

We first consider the case where $\mathscr{U}=H^1(\Omega)$ and

$$\mathscr{U}_{\text{ad}}=\{v|v\in\mathscr{U}, v|_{\Gamma}\geqslant 0\}. \tag{3.42}$$

Lemma 3.1. *The set $\mathscr{U}_{\text{ad}}$ defined by* (3.42) *is a closed, convex cone of $\mathscr{U}=H^1(\Omega)$ with vertex $\{0\}$.*

Proof. $\mathscr{U}_{\text{ad}}$ is clearly a convex cone with vertex $\{0\}$. It suffices to prove that $\mathscr{U}_{\text{ad}}$ is closed. But this is an immediate consequence of the continuity (Theorem 3.2) of the map $v\to v|_{\Gamma}$ of $H^1(\Omega)\to H^{\frac{1}{2}}(\Gamma)$. □

Let us choose L as in example 3.3. Since remark 1.6 is applicable, it is seen that there exists a $u\in\mathscr{U}_{\text{ad}}$ such that

$$\pi(u,v)\geqslant L(v) \quad \forall v\in\mathscr{U}_{\text{ad}}, \tag{3.43}$$

$$\pi(u,u)=L(u). \tag{3.44}$$

It remains to interpret the problem defined by (3.43), (3.44). Let $v=\pm\varphi$, $\varphi\in\mathscr{D}(\Omega)$ (because then $v\in\mathscr{U}_{\text{ad}}$). We then have $\pi(u,\varphi)=L(\varphi)$ $\forall\varphi\in\mathscr{D}(\Omega)$ and hence

$$Au=f \quad \text{in } \Omega. \tag{3.45}$$

Multiplying (3.45) by $v \in \mathscr{U}_{\mathrm{ad}}$ and applying Green's formula (in a formal manner, which may be justified as in Example 3.3), we obtain

$$-\int_{\Gamma} \frac{\partial u}{\partial v_A} v \, d\Gamma + \pi(u, v) = \int_{\Omega} f v \, dx$$

and from (3.43), we deduce

$$\int_{\Gamma} \left(\frac{\partial u}{\partial v_A} - g \right) v \, d\Gamma \geqslant 0 . \tag{3.46}$$

In (3.46), $v \in \mathscr{U}_{\mathrm{ad}}$ and hence $v \geqslant 0$ on Γ and therefore (3.46) is equivalent to

$$\frac{\partial u}{\partial v_A} - g \geqslant 0 \quad \text{on } \Gamma \quad (\text{in the sense of } H^{-\frac{1}{2}}(\Gamma)) .^{9} \tag{3.47}$$

From (3.44) and (3.46) we deduce

$$\int_{\Gamma} \left(\frac{\partial u}{\partial v_A} - g \right) u \, d\Gamma = 0 ,$$

which combined with (3.47) and the fact that $u \geqslant 0$ on Γ shows that

$$u \left(\frac{\partial u}{\partial v_A} - g \right) = 0 \quad \text{on } \Gamma . \tag{3.48}$$

Conclusion: We have shown the existence of a unique $u \in H^1(\Omega)$ satisfying (3.45) and the boundary conditions:

$$\left. \begin{aligned} & u \geqslant 0 \quad \text{on } \Gamma , \\ & \frac{\partial u}{\partial v_A} - g \geqslant 0 \quad \text{on } \Gamma , \\ & u \left(\frac{\partial u}{\partial v_A} - g \right) = 0 \quad \text{on } \Gamma . \end{aligned} \right\} \tag{3.49}$$

Remark 3.2. The boundary conditions (3.49) are referred to as unilateral boundary conditions.

From the last conditions in (3.49), we see that there exists a subset Γ_0 of Γ on which $u = 0$ and hence $\dfrac{\partial u}{\partial v_A} - g = 0$ on $\Gamma - \Gamma_0$; but Γ_0

---

[9] According to L. Schwartz [2], all distributions $\geqslant 0$ is a measure $\geqslant 0$. Therefore $\dfrac{\partial u}{\partial v_A} - g$ is a measure on Γ, $\in H^{-\frac{1}{2}}(\Gamma)$ and $\geqslant 0$.

naturally is not given a priori. Hence we are dealing with a problem which has a certain analogy with problems having a free boundary (cf. also 3.6 in the sequel).

Remark 3.3. We remark that although A is a linear operator, the problem (3.45), (3.49) is non-linear. The following is a general remark: if the constraints are not defined by a vector space, the problem is automatically non-linear.

3.6. Unilateral Boundary Value Problems (II)

Let $\pi(u,v)$ be as in 3.5, $\mathcal{U}=H^1(\Omega)$ and

$$\mathcal{U}_{\text{ad}}=\{v|v\in\mathcal{U},\, v\geqslant 0 \text{ a.e. in } \Omega\}. \tag{3.50}$$

We may easily verify that $\mathcal{U}_{\text{ad}}$ is a convex cone with its vertex at $\{0\}$. Choosing L as in 3.5, we again have existence and uniqueness of a $u\in\mathcal{U}_{\text{ad}}$ such that (3.43), (3.44) obtain.

In this case the interpretation of the problem is however more delicate (we refer the reader to H. Brezis-G. Stampacchia [1] for a deeper study of the problem). We may divide Ω into two regions

$$\begin{aligned}\Omega_0&=\{x|x\in\Omega,\, u(x)=0\},\\ \Omega_+&=\{x|x\in\Omega;\, u(x)>0\}.\end{aligned} \tag{3.51}$$

These regions are shown schematically in Fig. 3. Let us stress the fact that Ω_0 is not given a priori and also that in general we ignore the question of the "degree of regularity" of the surface S, which is the interface between Ω_0 and Ω_+.

Let $\varphi\in\mathcal{D}(\Omega)$ have compact support in Ω_+. For ε sufficiently small the functions $v=u\pm\varepsilon\varphi$ are in $\mathcal{U}_{\text{ad}}$ (since $u>0$ on Ω_+). Now from (3.43) we obtain $(u,\varphi)=L(\varphi)$ and hence we already have

$$\begin{aligned}u&=0 \quad \text{in } \Omega_0,\\ Au&=f \quad \text{in } \Omega_+, \qquad u>0.\end{aligned} \tag{3.52}$$

Now, as in 3.5, on Γ_+ (cf. Fig. 3 for the definition of Γ_+) we have,

$$\frac{\partial u}{\partial \nu_A}-g\geqslant 0, \qquad u\left(\frac{\partial u}{\partial \nu_A}-g\right)=0. \tag{3.53}$$

Since $u\in H^1(\Omega)$, we have

$$u=0 \quad \text{on } S \text{ (limit on } S \text{ taken in } \Omega_+). \tag{3.54}$$

Finally, it can be shown that in an appropriate sense

$$\left.\frac{\partial u}{\partial \nu_A}\right|_S = 0 \quad \text{(limit taken on } \Omega_+\text{)}. \tag{3.55}$$

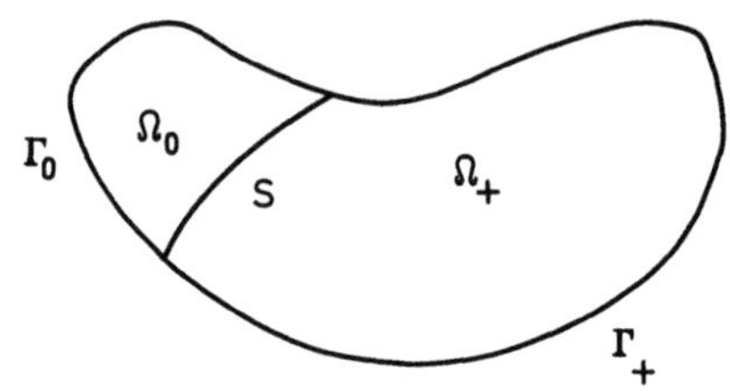

Fig. 3

The unilateral problem (3.52), (3.53), (3.54), (3.55) is a problem with free boundary; the surface S (which forms part of the unknown) is the free surface.

Remark 3.4. The fundamental difficulty of this problem resides in the study of the regularity of the solution u; we refer the reader to Brezis-Stampacchia, loc. cit.

A large number of problems of this type is open. In particular problems of this type which we shall meet in later chapters remain unsolved.

3.7. Unilateral Boundary Value Problems (III)

Let $\pi(u,v)$ be as in 3.5, 3.6 preceeding and let $\mathscr{U}=H_0^1(\Omega)$. Further, let

$$\mathscr{U}_{\text{ad}}=\{v \mid |\operatorname{grad} v(x)| \leqslant 1 \text{ a.e. in } \Omega\}. \tag{3.56}$$

It may immediately be verified that $\mathscr{U}_{\text{ad}}$ is a closed, bounded convex set in $H_0^1(\Omega)$. Let

$$L(v) = \int_\Omega f v\,dx, \quad f \in H^{-1}(\Omega). \tag{3.57}$$

Then there exists a unique $u \in \mathscr{U}_{\text{ad}}$ such that

$$\pi(u, v-u) \geqslant L(v-u) = \int_\Omega f(v-u)\,dx, \quad v \in \mathscr{U}_{\text{ad}}. \tag{3.58}$$

Let us interpret this problem formally. We again divide (formally) Ω into two regions:

$$\Omega_1 = \{x \mid |\operatorname{grad} u(x)| = 1\}.$$
$$\Omega_2 = \{x \mid |\operatorname{grad} u(x)| < 1\}.$$

Then in Ω_2, we again have

$$A u = f, \tag{3.59}$$

the fundamental difficulty being the study of the regularity of the solution on the "interface" separating Ω_1 and Ω_2 (cf. H. Brezis-G. Stampacchia [1]).

3.8. Unilateral Boundary Value Problems; Case of Systems

All the preceeding examples may be extended to the case of systems. Let us confine ourselves to a very simple example.

Let

$$\mathscr{U} = H^1(\Omega) \times H^1(\Omega); \tag{3.60}$$

hence every $u \in \mathscr{U}$ is of the form

$$u = \{u_1, u_2\}, \quad u_i \in H^1(\Omega).$$

In $\mathscr{U}$ consider the set

$$\mathscr{U}_{\mathrm{ad}} = \{v \mid v = \{v_1, v_2\}, v_1 \geqslant 0 \text{ on } \Gamma, v_2 \geqslant 0 \text{ on } \Gamma\}. \tag{3.61}$$

Naturally, as in Lemma 3.1, $\mathscr{U}_{\mathrm{ad}}$ is a closed convex cone in $\mathscr{U}$ with vertex at $\{0\}$.

Let

$$\begin{aligned}\pi(u,v) = &\int_\Omega \operatorname{grad} u_1 \cdot \operatorname{grad} v_1 \, dx + \int_\Omega \operatorname{grad} u_2 \cdot \operatorname{grad} v_2 \, dx \\ &+ \int_\Omega (u_1 v_1 + u_2 v_2 + u_1 v_2 - u_2 v_1) \, dx.\end{aligned} \tag{3.62}$$

$$\begin{aligned}&L(v) = \int_\Omega f_1 v_1 \, dx + \int_\Omega f_2 v_2 \, dx + \int_\Gamma g_1 v_1 \, d\Gamma + \int_\Gamma g_2 v_2 \, d\Gamma, \\ &f_i \in L^2(\Omega), \quad g_i \in H^{\frac{1}{2}}(\Gamma), \quad i = 1,2.\end{aligned} \tag{3.63}$$

Since

$$\pi(v,v) = \|v_1\|^2_{H^1(\Omega)} + \|v_2\|^2_{H^1(\Omega)},$$

the general theory is applicable and there exists a unique $u \in \mathscr{U}_{\mathrm{ad}}$ satisfying (3.43), (3.44).

The problem is interpreted as in 3.5. We obtain:

$$\begin{aligned} -\Delta u_1 + u_1 - u_2 &= f_1 \quad \text{in } \Omega,^{10} \\ -\Delta u_2 + u_2 + u_1 &= f_2 \quad \text{in } \Omega, \end{aligned} \tag{3.64}$$

together with the unilateral boundary conditions:

$$u_i \geqslant 0 \quad \text{on } \Gamma,$$

$$\frac{\partial u_i}{\partial n} - g_i \geqslant 0 \quad \text{on } \Gamma, \quad u_i\left(\frac{\partial u_i}{\partial n} - g_i\right) = 0 \quad \text{on } \Gamma, \quad i=1,2. \tag{3.65}$$

3.9. Elliptic Operators of Order Greater than Two

The elliptic operators used in the preceeding examples were of second order. We now give a simple example of an elliptic operator of order 4 and of the associated unilateral boundary value problem.

Let us define

$$\mathscr{U} = \{v \mid v \in L^2(\Omega),\ \Delta v \in L^2(\Omega)\} \tag{3.66}$$

(Δv being taken in the distribution sense in Ω). For $u, v \in \mathscr{U}$, set

$$(u,v)_{\mathscr{U}} = (u,v)_{L^2(\Omega)} + (\Delta u, \Delta v)_{L^2(\Omega)};$$

it may be easily verified that with this scalar product $\mathscr{U}$ is a Hilbert space.

It may be shown (cf. Lions-Magenes [1], Chapter 2) that

$$\left.\begin{aligned} &\text{for each } v \in \mathscr{U}, \text{ we may define in a unique fashion} \\ &v|_\Gamma, \quad \left.\frac{\partial v}{\partial n}\right|_\Gamma; \\ &v|_\Gamma \in H^{-\frac{1}{2}}(\Gamma), \quad \left.\frac{\partial v}{\partial n}\right|_\Gamma \in H^{-\frac{3}{2}}(\Gamma), \\ &\text{and the map } v \to \left\{v|_\Gamma, \left.\frac{\partial v}{\partial n}\right|_\Gamma\right\} \text{ is a continuous map from} \\ &\mathscr{U} \to H^{-\frac{1}{2}}(\Gamma) \times H^{-\frac{3}{2}}(\Gamma). \end{aligned}\right\} \tag{3.67}$$

We may define

$$\mathscr{U}_{\text{ad}} = \left\{v \mid v|_\Gamma \geqslant 0 \text{ on } \Gamma, \left.\frac{\partial v}{\partial n}\right|_\Gamma \geqslant 0 \text{ on } \Gamma\right\}. \tag{3.68}$$

---

[10] $\Delta = \sum_{i=1}^{n} \frac{\partial^2}{\partial x_i^2}$.

Remark 3.5. The conditions $v|_\Gamma \geqslant 0$ and $\left.\frac{\partial v}{\partial n}\right|_\Gamma \geqslant 0$ are taken in the sense of $H^{-\frac{1}{2}}(\Gamma)$ and $H^{-\frac{3}{2}}(\Gamma)$ respectively. It therefore results that $v|_\Gamma$ and $\left.\frac{\partial v}{\partial n}\right|_\Gamma$ are positive measures.

By virtue of (3.67), $\mathscr{U}_{\text{ad}}$ is a closed convex set in $\mathscr{U}$ with vertex 0.

Let us now choose

$$\pi(u,v) = \int_\Omega a_1(x)\Delta u \Delta v\, dx + \int_\Omega a_0 u v\, dx,$$

$$a_0, a_1 \in L^\infty(\Omega), \qquad a_0(x) \geqslant \alpha > 0, \qquad a_1(x) \geqslant \alpha, \quad \text{a.e. in } \Omega. \tag{3.69}$$

Clearly

$$\pi(v,v) \geqslant \alpha \|v\|_{\mathscr{U}}^2.$$

Further, let

$$L(v) = \int_\Omega f v\, dx - \int_\Omega g_2 v\, d\Gamma + \int_\Omega g_1 \frac{\partial v}{\partial n}\, d\Gamma, \tag{3.70}$$

where

$$f \in L^2(\Omega), \qquad g_1 \in H^{\frac{3}{2}}(\Gamma), \qquad g_2 \in H^{\frac{1}{2}}(\Gamma). \tag{3.71}$$

By virtue of (3.67) the linear form $v \to L(v)$ is continuous on $\mathscr{U}$.

Therefore there exists a unique $u \in \mathscr{U}_{\text{ad}}$ satisfying (3.43), (3.44). It remains to interpret the problem—which we do only formally[11].

We first take $v = \pm\varphi$, $\varphi \in \mathscr{D}(\Omega)$ which then is in $\mathscr{U}_{\text{ad}}$; from (3.43) it follows

$$\Delta(a_1 \Delta u) + a_0 u = f \quad \text{in } \Omega. \tag{3.72}$$

If we now assume that a_1 is sufficiently regular in $\bar{\Omega}$, then it may be shown that $u \in \mathscr{U}$ and (3.72) permits us to define in a unique manner:

$$\begin{aligned} &\Delta u|_\Gamma \in H^{-\frac{5}{2}}(\Gamma), \\ &\left.\frac{\partial \Delta u}{\partial n}\right|_\Gamma \in H^{-\frac{7}{2}}(\Gamma). \end{aligned} \tag{3.73}$$

Multiplying both sides of (3.72) by $v \in \mathscr{U}_{\text{ad}}$ and applying (formally) Green's formula, we obtain

$$\int_\Gamma \frac{\partial}{\partial n}(a_1 \Delta u)\cdot v\, d\Gamma - \int_\Gamma a_1 \Delta u \cdot \frac{\partial v}{\partial n}\, d\Gamma + \pi(u,v) = \int_\Omega f v\, dx.$$

[11] We do not attempt to justify the formal calculations which follow.

Utilizing (3.43) and (3.70), the above is equivalent to

$$\int_\Gamma (a_1 \Delta u - g_1) \frac{\partial v}{\partial n} \, d\Gamma - \int_\Gamma \left(\frac{\partial}{\partial n} (a_1 \Delta u) - g_2 \right) v \, d\Gamma \geqslant 0. \tag{3.74}$$

From (3.44), amongst other things, we see that the expression appearing in (3.74) is zero if we replace v by u.

In conclusion, we have the following unilateral conditions on Γ:

$$\begin{aligned} &u \geqslant 0, \quad \frac{\partial}{\partial n}(a_1 \Delta u) - g_2 \leq 0, \quad u\left(\frac{\partial}{\partial n}(a_1 \Delta u) - g_2\right) = 0, \\ &\frac{\partial u}{\partial n} \geqslant 0, \quad a_1 \Delta u - g_1 \geqslant 0, \quad \frac{\partial u}{\partial n}(a_1 \Delta u - g_1) = 0. \end{aligned} \tag{3.75}$$

Remark 3.6. As we have previously observed, the preceeding calculations are formal. In fact, for example $u \in H^{-\frac{1}{2}}(\Gamma)$ and $\frac{\partial}{\partial n}(a_1 \Delta u) \in H^{-\frac{7}{2}}(\Gamma)$ and the product $u\left(\frac{\partial}{\partial n}(a_1 \Delta u)\right)$ is not even defined. In order to obtain a rigorous interpretation, Remark 3.5 must be certainly utilized.

3.10. Non-differentiable Functionals

We now present an example of the application of Theorem 1.6.

Let Ω be a bounded open set in $\mathbb{R}^n$ and $\mathscr{U} = H_0^1(\Omega)$. We consider

$$J(v) = J_1(v) + J_2(v) \tag{3.76}$$

where

$$J_1(v) = \alpha \int_\Omega |\operatorname{grad} v|^2 dx - 2 \int_\Omega f v \, dx, \quad f \text{ given in } H^{-1}(\Omega), \quad \alpha \geqslant 0, \tag{3.77}$$

$$J_2(v) = 2\beta \int_\Omega |\operatorname{grad} v| dx, \quad \beta > 0. \tag{3.78}$$

It may be easily verified that the conditions for the application of Theorem 1.6 are met[12].

Let us also note that the functional $v \to J_2(v)$ is not differentiable.

Let $\mathscr{U}_{\text{ad}} = \mathscr{U}$ (case of "no constraints").

Then

$$\operatorname{Inf}_{v \in \mathscr{U}} \cdot J(v) = J(u), \quad u \text{ unique},$$

[12] $\int_\Omega v^2 dx \leqslant c \int_\Omega |\operatorname{grad} v|^2 dx \quad \forall v \in H_0^1(\Omega)$.

the element u being characterized by:

$$J_1'(u)\cdot(v-u)+J_2(v)-J_2(u)\geqslant 0 \quad \forall v\in\mathscr{U}\,. \tag{3.79}$$

To simplify the expressions, set

$$\pi(u,v)=\alpha\int_\Omega \operatorname{grad} u\cdot\operatorname{grad} v\,dx. \tag{3.80}$$

(3.79) is then equivalent to

$$\pi(u,v-u)-\int_\Omega f(v-u)\,dx+\beta\int_\Omega |\operatorname{grad} v|\,dx-\beta\int_\Omega |\operatorname{grad} u|\,dx\geqslant 0\,.$$

Set,

$$X(u,v)=\pi(u,v)-\int_\Omega fv\,dx+\beta\int_\Omega |\operatorname{grad} v|\,dx\,. \tag{3.81}$$

(3.79) is further equivalent to

$$X(u,v)\geqslant X(u,u) \quad \forall v\in\mathscr{U}\,. \tag{3.82}$$

Replacing v by λv, $\lambda\geqslant 0$, we deduce

$$\lambda X(u,v)\geqslant X(u,u) \quad \forall v\in\mathscr{U},\ \forall\lambda\geqslant 0\,. \tag{3.83}$$

Set $\lambda=0$ to obtain $X(u,u)\leqslant 0$ and hence (3.83) is only possible if $X(u,v)\geqslant 0$, $\forall v\in\mathscr{U}$. Hence $X(u,u)\geqslant 0$ and

$$\left.\begin{array}{l}\textit{the solution } u \textit{ of the problem } \operatorname{Inf} J(v)=J(u),\ J\\ \textit{being given by } (3.76), (3.77), (3.78) \textit{ is characterized by}\\ X(u,v)\geqslant 0 \quad \forall v\in\mathscr{U},\\ X(u,u)=0,\\ \textit{where } X(u,v) \textit{ is given by } (3.81) \textit{ and } (3.80).\end{array}\right\} \tag{3.84}$$

Let us now verify

Proposition 3.1. *A necessary and sufficient condition for the solution of* (3.84) *to be identically zero is*

$$\int_\Omega fv\,dx\leqslant\beta\int_\Omega |\operatorname{grad} v|\,dx \quad \forall v\in H_0^1(\Omega)=\mathscr{U}\,. \tag{3.85}$$

Proof. We may write conditions (3.84) as

$$\left.\begin{array}{l}\pi(u,v)-\beta\int_\Omega |\operatorname{grad} v|\,dx\geqslant\int_\Omega fv\,dx \quad \forall v\in\mathscr{U}\,,\\ \pi(u,u)+\beta\int_\Omega |\operatorname{grad} u|\,dx=\int_\Omega fu\,dx\,.\end{array}\right. \tag{3.86}$$

If we put $u=0$, the first condition of (3.86) reduces to (3.85).

If (3.85) holds, applying condition (3.85) with $v-u$ the second condition of (3.86) reduces to $\pi(u,u)\leqslant 0$ and hence $u=0$. □

Hence if condition (3.85) is not in effect, $u\neq 0$, $J(u)<0$ and hence u is characterized by (3.86). □

4. A Comparison Theorem

4.1. General Results

We consider a bilinear form $\pi(u,v)$, not necessarily symmetric, on the Hilbert space $\mathcal{U}$ which satisfies[13]

$$\pi(v,v) \geqslant c\|v\|^2 \quad \forall v \in \mathcal{U}, \qquad c>0. \tag{4.1}$$

We are given

$L=$ a continuous linear form on $\mathcal{U}$, and two closed, convex subsets $\mathcal{U}_{\text{ad}}$ and $\mathcal{U}^*_{\text{ad}}$ of $\mathcal{U}$.

From Theorem 2.1 it follows that there exists a unique u (resp. u^*) in $\mathcal{U}_{\text{ad}}$ (resp. $\mathcal{U}^*_{\text{ad}}$) such that

$$\pi(u,v-u) \geqslant L(v-u) \quad \forall v \in \mathcal{U}_{\text{ad}}, \tag{4.2}$$

(resp.

$$\pi(u^*,v-u^*) \geqslant L(v-u^*) \quad \forall v \in \mathcal{U}_{\text{ad}}). \tag{4.3}$$

We then have the following (examples of which are given in Section 4.2 below):

Theorem 4.1. *Let* (4.1) *hold.*

We assume that we may find $w \in \mathcal{U}_{\text{ad}}$, $w^* \in \mathcal{U}^*_{\text{ad}}$, *such that*

$$w+w^*=u+u^*. \tag{4.4}$$

$$\pi(w-u^*, w-u)=0. \tag{4.5}$$

Then

$$w=u, \qquad w^*=u^*. \tag{4.6}$$

Proof. Let us take $v=w$ in (4.2) and $v=w^*$ in (4.3) and add the two inequalities. By virtue of (4.4),

$$\pi(u,w-u)+\pi(u^*,w^*-u^*) \geqslant 0.$$

But since from (4.4) $w^*-u^*=-(w-u)$,

$$\pi(u-w,w-u)+\pi(w,w-u)-\pi(u^*,w-u) \geqslant 0$$

and hence

$$-\pi(w-u,w-u)+\pi(w-u^*,w-u) \geqslant 0.$$

Using (4.5) this reduces to $-\pi(w-u,w-u) \geqslant 0$ whence $w=u$ according to (4.1). Therefore $w^*=u^*$. □

---

[13] To simplify. We may generalize as in (2.2) and the inequality on $\mathcal{U}^*_{\text{ad}}$.

4.2. An Application

Consider the problem of section (3.5). Then u is the solution of (3.49). Let Γ_1 be a measurable subset of Γ; let us define

$$\mathscr{U}_{ad}(\Gamma_1)=\{v|v\in H^1(\Gamma),\ v=0 \text{ on } \Gamma_1\}.^{14} \tag{4.7}$$

In this manner we define a closed subspace of $\mathscr{U}$. If we take

$$\mathscr{U}^*_{ad}=\mathscr{U}_{ad}(\Gamma),$$

the corresponding solution u^* is

$$u^*=u_{\Gamma_1},$$

where

$$\left.\begin{aligned} A u_{\Gamma_1} &= f, \\ u_{\Gamma_1} &= 0 \quad \text{on } \Gamma_1, \\ \frac{\partial u_{\Gamma_1}}{\partial \nu_A} - g &= 0 \quad \text{on } \Gamma-\Gamma_1. \end{aligned}\right\} \tag{4.8}$$

By applying Theorem 4.1 we shall prove

Theorem 4.2. *If u is a solution of* (3.49) *and u_{Γ_1} a solution of* (4.8), *we have*

$$u=\sup_{\Gamma_1} u_{\Gamma_1}, \tag{4.9}$$

where Γ_1 varies over the family of measurable subsets of Γ.

Proof. 1. Since from (3.49) u is of the form $u=u_{\Gamma_1}$ for an appropriate Γ_1, we obtain

$$\operatorname*{Sup}_{\Gamma_1} u_{\Gamma_1}\geqslant u. \tag{4.10}$$

Hence it is sufficient to prove the inequality (4.10) in the opposite sense.

2. Therefore let Γ_1 be fixed and let $\mathscr{U}^*=\mathscr{U}_{ad}(\Gamma_1)$ and $u_{\Gamma_1}=u^*$. It is sufficient to prove that

$$u^*\leqslant u \quad \text{almost everywhere in } \Omega. \tag{4.11}$$

Let us define

$$w=\sup(u,u^*), \qquad w^*=\inf(u,u^*). \tag{4.12}$$

Clearly (4.4) is satisfied[15]. Let us verify (4.5); setting $u-u^*=\psi$, we obtain,

$$\begin{aligned}\pi(w-u^*,w-u) &= \pi(\sup(u-u^*,0),\ \sup(0,u^*-u)) \\ &= -\pi(\sup(\psi,0),\ \inf(\psi,0)).\end{aligned}$$

---

[14] We have $\mathscr{U}_{ad}(\Gamma_1)$ is a proper subset of Γ_1 if and only if Γ_1 is of capacity >0.

[15] Let us note that if $u,v\in H^1(\Omega)$, then $\sup(u,v)$ and $\inf(u,v)\in H^1(\Omega)$ (the same property however is not true in $H^m(\Omega)$ when $m\geqslant 2$).

If ψ_n be a sequence of functions in $C^1(\bar{\Omega})$ such that $\psi_n \to \psi$ in $H^1(\Omega)$[16] then

$$\pi(\sup(\psi,0),\inf(\psi,0)) = \lim_n \pi(\sup(\psi_n,0),\inf(\psi_n,0))$$

and

$$\pi(\sup(\psi_n,0),\inf(\psi_n,0)) = 0. \quad \text{Therefore (4.5) obtains}.$$

Thus Theorem 4.1 is applicable giving us $w = u$, i.e. (4.11). □

Remark 4.1. The preceeding proof is analogous to the proof of the maximum principle for elliptic equations. In fact, let f be a given element of $H^1(\Omega)$ with

$$f \geqslant 0, \tag{4.13}$$

and u be the solution in $H_0^1(\Omega)$ of

$$\pi(u,v) = \int_\Omega f v\, dx, \quad \forall v \in H_0^1(\Omega), \tag{4.14}$$

π being given as above. We have

$$u \geqslant 0. \tag{4.15}$$

In fact, let $w = \sup(u,0)$; we may verify $w \in H_0^1(\Omega)$[17] and $\pi(w, v-u) = 0$ (as in the preceeding proof); (4.14) gives us

$$\pi(u, w-u) = \int_\Omega f(w-u)\, dx \geqslant 0 \quad \text{by virtue of (4.13)}$$

and hence

$$\pi(u, w-u) - \pi(w, w-u) \geqslant 0.$$

Therefore $-\pi(w-u, w-u) \geqslant 0$ and hence $w = u$, from which we obtain (4.15).

For more precise results in the sense of the preceeding remark, cf. G. Stampacchia [2].

5. Non Coercive Forms

5.1. Convexity of the Set of Solutions

Let us consider the functional $v \to J(v)$ which is continuous, convex and lower semi-continuous in the weak topology. Let us denote by X

---

[16] A sequence of this type exists. Cf. for example Lions-Magenes [1], Chapter 1.

[17] As in the case of proof of Theorem 4.2, this is a special property of Sobolev spaces of order one.

the set of elements $u \in \mathscr{U}_{\text{ad}}$ such that

$$J(u) = \inf_{v \in \mathscr{U}_{\text{ad}}} J(v). \tag{5.1}$$

The set X may be empty. However if $\mathscr{U}_{\text{ad}}$ is also bounded then X is non-empty (or else also in the coercive case, cf. for example Theorem 1.6). In any case

Theorem 5.1. *X is a closed, convex subset of $\mathscr{U}_{\text{ad}}$.*

Proof. X is clearly a closed subset of $\mathscr{U}_{\text{ad}}$. To prove convexity, if $u_1, u_2 \in X$, then

$$J(u_k) \leqslant J(v) \quad \forall v \in \mathscr{U}_{\text{ad}}, \qquad k = 1, 2.$$

If $\theta \in]0,1[$,

$$J((1-\theta)u_1 + \theta v_2) \leq (1-\theta)J(u_1) + \theta J(u_2) \leq J(v) \quad \forall v \in \mathscr{U}_{\text{ad}},$$

and hence $(1-\theta)u_1 + \theta u_2 \in X$. □

This property is valid for variational inequalities which do not necessarily correspond to variational problems (cf. 2.1). For example

Theorem 5.2. *Let $\pi(u,v)$ be a continuous, not necessarily symmetric bilinear form on $\mathscr{U}$ satisfying*

$$\pi(v,v) \geqslant 0 \quad \forall v \in \mathscr{U}. \tag{5.2}$$

Then the set X of solutions of

$$u \in \mathscr{U}_{\text{ad}}, \qquad \pi(u, v-u) \geqslant L(v-u) \quad \forall v \in \mathscr{U}_{\text{ad}} \tag{5.3}$$

is a closed, convex subset of $\mathscr{U}$.

Proof. Indeed, (5.3) is equivalent to (cf. 1.4 and Remark 2.3)

$$\pi(v, v-u) \geqslant L(v-u) \quad \forall v \in \mathscr{U}_{\text{ad}}, \tag{5.4}$$

and it may be immediately verified that if u_1, u_2 satisfy (5.4) so does $u = (1-\theta)u_1 + \theta u_2$, $\theta \in]0,1[$. □

Remark 5.1. We shall see that if $\mathscr{U}_{\text{ad}}$ is also bounded, then the set X of Theorem 5.2 is non-empty (this is obvious if π is also symmetric).

5.2. Approximation Theorem

Let us consider the set X of Theorem 5.2, which we assume to be non-empty.

Let

$$\begin{aligned}&b(u,v) \text{ be a continuous bi-linear form on } \mathscr{U}\text{, which may be}\\&\text{symmetric or not; let } b(u,v) \text{ be coercive:}\\&b(v,v)\geqslant\beta\|v\|^2,\quad \beta>0,\ \forall v\in\mathscr{U},\end{aligned}\tag{5.5}$$

and let

$$M \text{ be a continuous linear form on } \mathscr{U}. \tag{5.6}$$

From Theorem 2.1 we know that there exists a unique element u_0 such that

$$u_0\in X,\qquad b(u_0,v-u_0)\geqslant M(v-u_o)\ \ \forall v\in X. \tag{5.7}$$

Example 5.1. Consider the case where $b(u,v)=(u,v)_{\mathscr{U}}$ and $M(v)=(g,v)_{\mathscr{U}}$. Then (cf. Example 1.1, section 1.2)

$$u_0=\text{projection of } g \text{ on } X. \tag{5.8}$$

Our objective now is to present a constructive procedure to obtain u_0 (without assuming X is known!).

The essential idea of this procedure is to replace the form $\pi(u,v)$ by the "regularized" form

$$\pi_\varepsilon(u,v)=\pi(u,v)+\varepsilon b(u,v),\qquad \varepsilon>0. \tag{5.9}$$

From (5.2) and (5.5) we have

$$\pi_\varepsilon(v,v)\geqslant\varepsilon_\beta\|v\|^2_{\mathscr{U}}, \tag{5.10}$$

and hence from Theorem 2.1, by setting

$$L_\varepsilon=L+\varepsilon M \tag{5.11}$$

there exists a unique $u_\varepsilon\in\mathscr{U}_{\text{ad}}$ such that

$$\pi_\varepsilon(u_\varepsilon,v-u_\varepsilon)\geqslant L_\varepsilon(v-u_\varepsilon)\ \ \forall v\in\mathscr{U}_{\text{ad}}. \tag{5.12}$$

We thus have

Theorem 5.3. *Suppose that conditions* (5.2) *and* (5.3) *are satisfied. Fror each* $\varepsilon>0$, *let* u_ε *be the corresponding solution of* (5.12) *and let* u_0 *be the solution of* (5.7), *it being assumed that the set* X *of solutions of* (5.3) *is non-empty. Then*

$$u_\varepsilon\to u_0\quad \text{in } \mathscr{U} \text{ as } \varepsilon\to 0. \tag{5.13}$$

Proof. 1. Setting $v=u_0$ in (5.12), $v=u_\varepsilon$, $u=u_0$ in (5.3) and adding, we obtain

$$-\pi(u-u_0,u_\varepsilon-u_0)+b(u_\varepsilon,u_0-u_\varepsilon)\geqslant\varepsilon M(u_0-u_\varepsilon).$$

Hence from (5.2) we may deduce

$$b(u_\varepsilon,u_0-u_\varepsilon)\geqslant M(u_0-u_\varepsilon). \tag{5.14}$$

Using (5.5) we may now deduce from (5.14) that

$$\beta\|u_\varepsilon\|^2 \leqslant c_1 + c_2\|u_\varepsilon\|_{\mathscr{U}}, \qquad (c_i = \text{constant})$$

and hence

$$\|u_\varepsilon\|_{\mathscr{U}} \leqslant c_3 . \tag{5.15}$$

2. We may then extract a subsequence from the sequence u_ε, which we also denote by u_ε, such that $u_\varepsilon \to w$ in $\mathscr{U}$ weakly. Since $\mathscr{U}_{\text{ad}}$ is weakly closed, $w \in \mathscr{U}_{\text{ad}}$. It is easily verified that $v \to \pi(v,v)$ is lower semi-continuous in the weak topology of $\mathscr{U}$. Then, in the limit (5.12) gives

$$\pi(w, v-w) \geqslant L(v-w) \quad \forall v \in \mathscr{U}_{\text{ad}} . \tag{5.16}$$

This shows that $w \in X$.[18]

Passing to the limit in (5.14) ($b(v,v)$ being weakly lower semi-continuous), we get

$$b(w, u_0 - w) \geqslant M(u_0 - w) . \tag{5.17}$$

But since $w \in X$ we may take $v = w$ in (5.7) and add to (5.17). This gives us,

$$-b(w-u_0, w-u_0) \geqslant 0$$

hence $w = u_0$ and hence without extracting a subsequence $u_\varepsilon \to u_0$ in $\mathscr{U}$ weakly.

3. It remains to show that $u_\varepsilon \to u_0$ in $\mathscr{U}$ strongly, that is,

$$b(u_\varepsilon - u_0, u_\varepsilon - u_0) \to 0 . \tag{5.18}$$

From (5.14),

$$b(u_\varepsilon - u_0, u_\varepsilon - u_0) \leqslant M(u_\varepsilon - u_0) - b(u_0, u_\varepsilon - u_0)$$

whence (5.18). □

Remark 5.2. If $\mathscr{U}_{\text{ad}}$ is bounded, then $X \neq \emptyset$.

Remark 5.3. X may turn out to be empty if π is non coercive, $\mathscr{U}_{\text{ad}}$ is unbounded and L is of a particular form. Cf. Lions-Stampacchia [1].

Remark 5.4. By using the procedure outlined in (5.12) with b and M chosen as in Example 5.1, we may obtain a u_ε which approximates the projection of g on X (without knowledge of the set X).

Remark 5.5. Let $\mathscr{H}$ be a Hilbert space and let $\mathscr{C} \in \mathscr{L}(\mathscr{U}; \mathscr{H})$; let us suppose that

$$\pi(u,v) = (\mathscr{C}u, \mathscr{C}v)_{\mathscr{H}} \quad \text{and} \quad L(v) = (h, \mathscr{C}v)_{\mathscr{H}}, \qquad h \in \mathscr{H} .$$

[18] But this does not imply that $X \neq \emptyset$; we have utilized this hypothesis.

If X is non-empty, we have:

$$X=(u+\operatorname{Ker}\mathscr{C})\cap\mathscr{U}_{\text{ad}} \tag{5.19}$$

where u is any element of X and where

$$\operatorname{Ker}\mathscr{C}=\text{Kernel of } \mathscr{C}=\{v|v\in\mathscr{U},\mathscr{C}v=0\}.$$

((5.19) is easily verified, since X is the set of all elements where $J(v) =\|\mathscr{C}v\|^2_{\mathscr{H}}-2(h,\mathscr{C}v)$ attains its minimum and since $J(v+z)=J(v)$ if $z\in\operatorname{Ker}\mathscr{C}$.)

Notes

Theorem 2.1 is due to G. Stampacchia [1], the proof given in the text being that of Lions-Stampacchia [1].

Inequalities of the type (2.1) have been studied and their scope generalized by several authors:

(i) For the elliptic or stationary case (where time does not intervene), these problems were introduced in mechanics by Prager and Signorini and the equations of elasticity have been studied by G. Fichera [1]. A study of the non-coercive case has been made by Lions-Stampacchia [1], [2]. The case where $\pi(u,v)$ is not a bi-linear form has been studied by Hartman-Stampacchia [1], Browder [1], Brezis [1], Minty [1] (cf. also the bibliography of these works).

(ii) For evolution inequalities—problems which have not been considered in this book—in the case of "*parabolic inequalities*" cf. Lions-Stampacchia [1], Brezis [1], [2] and in the case of "*hyperbolic inequalities*" cf. Brezis-Lions [1].

In the following chapters we shall often have the occasion to consider new types of "*variational inequalities*".

In the formulation (1.27) the term $J_1(u)(v-u)$ may be replaced by $\pi(u,v-u)$, π as in Theorem 2.1 (cf. Theorem 2.2 of Lions-Stampacchia [1]) or even with π "*monotone in u*" (cf. Brezis [1], Theorem 24).

For the topology of $\mathscr{D}(\Omega)$ and $\mathscr{D}'(\Omega)$, other than the book of L. Schwartz [1], the book of J. Horvath [1] may be consulted.

Sections 3.1 and 3.3 indicate some of the fundamental properties of Sobolev spaces (Sobolev [1]). For proofs (and various addtional properties) cf. Lions-Magenes [1], Chapter 1. Other Sobolev spaces and properties of these spaces will be indicated in various parts of this book.

Clearly, our objective is not to study *usual* boundary value problems. It is for this reason that we have presented examples of boundary value problems which are a necessary minimum for studying the following chapters. For usual elliptic boundary value problems, one may consult for example Lions-Magenes, loc. cit. Chapter 2 and the bibliography of this work.

Unilateral boundary value problems are presented in Section 3.5 and in the following sections. As we shall see, the theory of optimal control of systems governed by partial differential equations leads to a number of problems which is of a very similar nature to those presented here. The remarks given in Section 3.6 and 3.7 are formal. For a deeper study of these problems, cf. H. Brezis-G. Stampacchia [1]. Problems of the type given in Section 3.7 are encountered in mechanics (elasto-plasticity); cf. B. D. Avin [1], Lanchon-Duvaut [1]. In Section 3.8 and 3.9 we present examples of unilateral problems for systems of equations and for operators of order greater than two. These examples were chosen for their simplicity. Clearly an infinite number of variations on these problems are possible. Unilateral problems connected with quasi-elliptic operators etc. may equally well be considered: A study of such problems will not be undertaken in this book.

The example considered in Section 3.10 occurs in visco-plasticity; cf. P. P. Mosolov-V. P. Miasnikov [1] for further considerations.

A problem with origin in physics (filters) leading to a similar question has been studied by Berkovitz-Pollard [1].

Theorems 4.1 and 4.2 are due to Y. Haugazeau [1], [2] where other applications may be found.

Theorem 5.3 is due to Stampacchia and the author, cf. Lions-Stampacchia [1]. This approximation procedure is also applicable to problems where $\pi(u,v)$ is not linear in u; cf. Brezis-Sibony [1], Browder [1].

Other general results (duality, Lagrange multiplier) are given in Chapter 3, Sections 12 and 13. For this we refer the reader to the work of J. J. Moreau and R. T. Rockafellar mentioned in the bibliography. Somewhat similar problems (with different technical details) arise in convex approximation—cf. P. J. Laurent [1], [2] (and the bibliography of these papers).

Necessary inditions of optimality have now been obtained in a very general setting. For this the reader may consult A. Yu. Dubovicki and A. A. Miljutin [1], Gamkrelidze [1] Halkin [2], Halkin-Neustadt [1], Neustadt [1], [2], Pschenichny [3], Varaiya [2].

CHAPTER II

Control of Systems Governed by Elliptic Partial Differential Equations

1. Control of Elliptic Variational Problems

1.1. Problem Statement

Let V and H be two Hilbert Spaces on $\mathbb{R}$.[1] The norm on V (resp. H) will be denoted by $\| \ \|$ (resp. $| \ |$) and the corresponding scalar products will be denoted by $((\,,))$ (resp. $(\,,)$). The index V or H ($\| \ \|_V$, etc.) will be added in case there is ambiguity. Assume

$$\begin{aligned}&V\subset H, \text{ the injection of } V \text{ into } H \text{ is continuous}^2,\\ &\text{and } V \text{ is dense in } H.\end{aligned} \tag{1.1}$$

H will be identified with its dual space. If V' denotes the dual space, H may be identified with a subspace of V' and we may write

$$V\subset H\subset V'. \tag{1.2}$$

In (1.2) V is dense in H and H is dense in V' and the corresponding injections are continuous[3].

Now let

$$\begin{aligned}&a(u,v) \text{ be a continuous, coercive, bilinear form on } V, \text{ that is,}\\ &a(v,v)\geqslant\alpha\|v\|^2 \quad \forall\, v\in V, \qquad \alpha>0\end{aligned} \tag{1.3}$$

(the bilinear form $a(u,v)$ is not necessarily symmetric).

Finally let L be a continuous linear form on V. By virtue of (1.2),

$$L(v)=(f,v) \qquad f\in V', \tag{1.4}$$

where (f,v) denotes the scalar product between $f\in V'$ and $v\in V$ (the scalar product coincides with the scalar product in H if $f\in H$).

---

[1] Extension to the complex case is immediate.

[2] Hence there exists a constant c such that $|v|\leqslant c\,\|v\|$.

[3] In case $V=H=V'=$ space of finite dimension, $a(u,v)=(A\,u,v)$ $A=$ matrix, coercivity is equivalent to $A+A^*\geqslant 2\alpha$ (Identity).

Identifying,

$$\mathscr{U} = V,$$

$$\pi(u,v) = a(u,v),$$

L given by (1.4), we may apply Theorem 2.1 and Remark 2.2 to obtain,

Theorem 1.1. *Assume that* (1.3) *is satisfied. For given f in V', there exists a unique $y \in V$ such that:*

$$a(y,\psi) = (f,\psi) \quad \forall \psi \in V. \tag{1.5}$$

The change of notation in (1.5) is justified in the sequel. We may interpret (1.5) in the following way: the form $v \to a(u,v)$ being linear and continuous may be written as:

$$a(u,v) = (Au,v), \qquad Au \in V' \tag{1.6}$$

which defines

$$A \in \mathscr{L}(V,V'). \tag{1.7}$$

Hence (1.5) is equivalent to

$$Ay = f. \tag{1.8}$$

Examples of problems which can be solved by using Theorem 1.1 have already been given in Chapter 1, sections 3, 4 and other examples will be given in section 2 of this chapter.

We may now formulate the first control problem. The Hilbert space $\mathscr{U}$ being the space of controls (as in Chapter 1) is given and an operator B with

$$B \in \mathscr{L}(\mathscr{U}; V') \tag{1.9}$$

is given.

A *system* (physical, mechanical etc.) which is governed by the operator A as in (1.7) is given. For a control $u \in \mathscr{U}$, the state of the system y is given by the solution of

$$Ay = f + Bu, \qquad y \in V.$$

Clearly y depends on u and hence we write $y(u)$. Thus:

$$Ay(u) = f + Bu, \qquad y(u) \in V, \tag{1.10}$$

an equation which defines $y(u)$ uniquely by virtue of Theorem 1.1.

We are also given an *observation equation*

$$z(u) = Cy(u), \tag{1.11}$$

where $C \in \mathscr{L}(V; \mathscr{H})$, $\mathscr{H}$ being a Hilbert space.

Finally we are given

$$\begin{aligned} &N \in \mathcal{L}(\mathcal{U};\mathcal{U}), \qquad N \text{ hermitian, positive definite;} \\ &(N u, u)_{\mathcal{U}} \geqslant \nu \|u\|_{\mathcal{U}}^2, \qquad \nu > 0. \end{aligned} \tag{1.12}$$

With every control $u \in \mathcal{U}$ we associate the cost:

$$J(u) = \|C y(u) - z_d\|_{\mathcal{H}}^2 + (N u, u)_{\mathcal{U}}, \tag{1.13}$$

where z_d is a given element in $\mathcal{H}$.

Let

$$\mathcal{U}_{\mathrm{ad}} = \text{closed, convex subset of } \mathcal{U}. \tag{1.14}$$

The control problem then is:

$$\text{Find } \inf_{v \in \mathcal{U}_{\mathrm{ad}}} J(v). \tag{1.15}$$

Remark 1.1. We shall also study the case where $N = 0$.

1.2. First Remarks on the Control Problem

According to (1.10) the map $u \to y(u)$ is an *affine* map of $\mathcal{U} \to V$. Let us therefore write $J(u)$ in the form

$$J(u) = \| C(y(u) - y(0)) + C y(0) - z_d\|_{\mathcal{H}}^2 + (N u, u)_{\mathcal{U}}.$$

If we set

$$\pi(u, v) = (C(y(u) - y(0)), C(y(v) - y(0)))_{\mathcal{H}} + (N u, v)_{\mathcal{U}}, \tag{1.16}$$

$$L(v) = (z_d - C y(0), C(y(v) - y(0)))_{\mathcal{H}} \tag{1.17}$$

the form $\pi(u, v)$ is a continuous bilinear form on $\mathcal{U}$ and we have,

$$J(v) = \pi(v, v) - 2 L(v) + \| z_d - C y(0)\|_{\mathcal{H}}^2. \tag{1.18}$$

Since $\|z_d - C y(0)\|_{\mathcal{H}}^2$ is clearly $\geqslant 0$, from (1.12) we have,

$$\pi(v, v) \geqslant \nu \|v\|_{\mathcal{U}}^2 \quad \forall v \in \mathcal{U}. \tag{1.19}$$

Therefore we have reduced the problem to a form where Theorem 1.1 of Chapter 1 can be applied.

Thus we have,

Theorem 1.2. *Let us suppose that the hypotheses* (1.3) *and* (1.12) *hold and that the state of the system is given by* (1.10). *Then there exists a unique element* $u \in \mathcal{U}_{\mathrm{ad}}$ *such that*

$$J(u) = \inf_{v \in \mathcal{U}_{\mathrm{ad}}} J(v). \tag{1.20}$$

Definition 1.1. The element $u \in \mathcal{U}_{ad}$ *for which* $J(v)$ *attains its minimum is termed the optimal control.*

The Case $N=0$.

If $N=0$, from (1.19) one can in general only conclude[4]

$$\pi(v,v) \geqslant 0 \tag{1.21}$$

and hence one may apply Theorem 5.2 and Remark 5.2 of Chapter 1. We thus have,

Theorem 1.3. *Let us suppose that* (1.3) *holds and* $N=0$. *We also assume that* $\mathcal{U}_{ad}$ *is bounded. Then there exists a non-empty subset* X *of* $\mathcal{U}_{ad}$ *such that*

$$J(u) = \operatorname*{Inf}_{v \in \mathcal{U}_{ad}} J(v) \quad \forall u \in X. \tag{1.22}$$

The set X *is closed and convex.*

Definition 1.2. The set of $u \in X$ *is termed the set of optimal controls.*

Remark 1.2. Theorem 5.3 *of Chapter* 1 *provides a method of approximation for obtaining a particular element* $u_0 \in X$ *(for example the projection of* $g \in \mathcal{U}$ *on* X*).*

Remark 1.3. The preceeding is nothing else but an immediate application of the theory developed in Chapter 1. The only difference lies in the fact that in Chapter 1 the function $J(v)$ is given directly as a function of v whereas here the function is given in terms of the intermediary of two operators:

(i) the operator $u \to y(u)$ which transforms the control into the state (it is this fact which motivated the change in notation in 1.5);

(ii) the operator $y(u) \xrightarrow{C} Cy(u)$ which transforms the state into an observation.

Orientation. The problems that now remain to be solved are the following:

(i) To write down a set of equations which define the optimal control u (the case of Theorem 1.2) or the set of optimal controls (the case of Theorem 1.3);

(ii) Study the set of equations in order to extract the maximum amount of information regarding the optimal control (or the set X).

Naturally, step (ii) which is far more difficult and interesting than step (i), can be carried out only in certain concrete cases.

---

[4] It may turn out that $\|Cy(v) - y(0)\|_{\mathcal{H}}^2$ is coercive.

1.3. The Set of Inequalities Defining the Optimal Control

If u is an optimal control (cf. Theorem 1.3, Chapter 1), then

$$J'(u)\cdot(v-u)\geqslant 0 \quad \forall v\in\mathscr{U}_{\text{ad}} \tag{1.23}$$

and conversely.

Since A is an isomorphism from V onto V' (according to Theorem 1.1), we may write

$$y(u)=A^{-1}(f+Bu)$$

whence

$$y'(u)\cdot\psi=A^{-1}B\psi\,,$$

and hence

$$y'(u)\cdot(v-u)=A^{-1}B(v-u)=y(v)-y(u)\,.$$

Therefore (after division by 2), (1.23) is equivalent to

$$\big(Cy(u)-z_d,\,C(y(v)-y(u))\big)_{\mathscr{H}}+(Nu,v-u)_{\mathscr{U}}\geqslant 0 \quad \forall v\in\mathscr{U}_{\text{ad}}. \tag{1.24}$$

Let us introduce the dual $\mathscr{H}'$ of $\mathscr{H}$[5] and set:

$$\Lambda=\Lambda_{\mathscr{H}}=\text{canonical isomorphism of } \mathscr{H} \text{ onto } \mathscr{H}'. \tag{1.25}$$

Then denoting $C^*\in\mathscr{L}(\mathscr{H}';V')$ as the adjoint of C, for φ and ψ in V, we have

$$(C^*\Lambda C\psi,\varphi)\,(\text{duality } V',V)=(\Lambda C\psi,C\varphi)\,(\text{duality } \mathscr{H}',\mathscr{H})=(C\psi,C\varphi)_{\mathscr{H}}\,.$$

Therefore, (1.24) is equivalent to

$$\big(C^*\Lambda(Cy(u)-z_d),\,y(v)-y(u)\big)_{\mathscr{H}}+(Nu,v-u)_{\mathscr{U}}\geqslant 0 \quad \forall v\in\mathscr{U}_{\text{ad}}. \tag{1.26}$$

We shall now transform (1.26) by utilizing the *adjoint state.*

Let $A^*\in\mathscr{L}(V;V')$ be the adjoint of A; it is associated with a bilinear form

$$(A^*\varphi,\psi)=(\varphi,A\psi)=a(\psi,\varphi),\ \varphi,\ \psi\in V,$$

and A^* is an isomorphism of V onto V'. For a control $v\in\mathscr{U}$, the *adjoint state* $p(v)\in V$ is defined by:

$$A^*p(v)=C^*\Lambda(Cy(v)-z_d)\,. \tag{1.27}$$

---

[5] Naturally we may identity $\mathscr{H}'$ with $\mathscr{H}$, but in the case where $\mathscr{H}\subset V'$ this leads to difficulties since H has already been identified with its dual.

Let us now transform (1.26); from (1.27) we deduce

$$\begin{aligned}(A^* p(u), y(v)-y(u)) &= (C^* \Lambda(C y(u)-z_d), y(v)-y(u))\\ &= (\text{from the definition of the adjoint } A^*)\,(p(u), A y(v)-A y(u))\\ &= (\text{from } (1.10))\,(p(u), B(v-u)) = (B^* p(u), v-u)_{\mathscr{U}},\end{aligned}$$

where $B^* \in \mathscr{L}(V;\mathscr{U}')$ is the adjoint of B, $\mathscr{U}'$ being the dual of $\mathscr{U}$.

If we now set:

$$\Lambda_{\mathscr{U}} = \text{canonical isomorphism of } \mathscr{U} \text{ onto } \mathscr{U}', \tag{1.28}$$

we have for $v \in \mathscr{U}$, $\psi \in V$:

$$\begin{aligned}(\psi, B v)\,(\text{duality } V, V') &= (B^* \psi, v)\,(\text{duality } \mathscr{U}', \mathscr{U})\\ &= (\Lambda_{\mathscr{U}}^{-1} B^* \psi, v)_{\mathscr{U}}.\end{aligned}$$

Hence (1.26) is equivalent to

$$(\Lambda_{\mathscr{U}}^{-1} B^* p(u) + N u, v-u)_{\mathscr{U}} \geqslant 0 \quad \forall v \in \mathscr{U}_{\text{ad}}. \tag{1.29}$$

We also have,

$(B^* p(u) + \Lambda_{\mathscr{U}} N u, v-u) \geqslant 0$ (the bracket here denoting the scalar product between $\mathscr{U}'$ and $\mathscr{U}$), $\forall v \in \mathscr{U}_{\text{ad}}$.[6] (1.29)

The results may now be summarized in

Theorem 1.4. *Assume that* (1.3) *holds. The cost function being given by* (1.3), *a necessary and sufficient condition for u to be an optimal control is that the following equations and inequalities be satisfied:*

$$A y(u) = f + B u, \tag{1.10}$$

$$A^* p(u) = C^* \Lambda(C y(u) - z_d), \tag{1.30}$$

$$u \in \mathscr{U}_{\text{ad}}, \quad (\Lambda_{\mathscr{U}}^{-1} B^* p(u) + N u, v-u)_{\mathscr{U}} \geqslant 0 \quad \forall v \in \mathscr{U}_{\text{ad}}. \tag{1.29}$$

If N satisfies (1.12), *then the optimal control is unique (i.e. the system of equations* (1.10), (1.30), (1.29) *admits a unique solution).*

If $N=0$ *and* $\mathscr{U}_{\text{ad}}$ *is bounded then the system of equations* (1.10), (1.30), (1.29) *admits at least one solution. The set of solutions corresponds to a family of optimal controls forming a closed, convex subset X of* $\mathscr{U}_{\text{ad}}$. □

Remark 1.4. (1.29) *may be written in the form (equivalent)*

$$(\Lambda_{\mathscr{U}}^{-1} B^* p(u) + N u, u)_{\mathscr{U}} = \inf_{v \in \mathscr{U}_{\text{ad}}} (\Lambda_{\mathscr{U}}^{-1} B^* p(u) + N u, v)_{\mathscr{U}}. \tag{1.31}$$

The objective of the following sections is to study the system of equations (1.10), (1.30), (1.29).

---

[6] We have implicity shown that $\frac{1}{2} J'(u) = B^* p(u) + \Lambda_{\mathscr{U}} N u$.

Remark 1.5. Assume that $\mathscr{U}_{ad}$ = *closed, convex cone with vertex at* $\{0\}$. *Then* (*cf*. *Remark* 1.6, *Chapter* 1) (1.29) *may be equivalently written as*

$$
\begin{aligned}
&u \in \mathscr{U}_{ad}, \\
&(\Lambda_{\mathscr{U}}^{-1} B^* p(u) + N u, v)_{\mathscr{U}} \geqslant 0 \quad \forall v \in \mathscr{U}_{ad}, \\
&(\Lambda_{\mathscr{U}}^{-1} B^* p(u) + N u, u)_{\mathscr{U}} = 0 .
\end{aligned}
\tag{1.32}
$$

2. First Applications

2.1. System Governed by the Dirichlet Problem; Distributed Control

We begin with the simplest case (without being trivial). We set

$V = H_0^1(\Omega)$, $H = L^2(\Omega)$ (in the notation of Chapter 1,[7] section 3), (2.1)

$$
a(\varphi, \psi) = \sum \int_\Omega a_{ij} \frac{\partial \varphi}{\partial x_j} \frac{\partial \psi}{\partial x_i} dx + \int_\Omega a_0 \varphi \, dx \tag{2.2}
$$

where (cf. Chapter 1, sections 3 and 4)

$$
\left.
\begin{aligned}
&a_{ij}, a_0 \in L^\infty(\Omega), \\
&\sum_{i,j=1}^{n} a_{ij}(x) \xi_i \xi_j \geqslant \alpha(\xi_1^2 + \cdots + \xi_n^2), \quad \alpha > 0, \quad \text{almost everywhere in } \Omega, \\
&a_0(x) \geqslant \alpha \quad \text{almost everywhere.}
\end{aligned}
\right\}
\tag{2.3}
$$

Therefore the operator A is the *second order elliptic operator*

$$
A\varphi = -\sum \frac{\partial}{\partial x_j}\left(a_{ij} \frac{\partial}{\partial x_j} \varphi\right) + a_0 \varphi . \tag{2.4}
$$

We choose

$$
\mathscr{U} = H (= L^2(\Omega)), \quad \text{and hence } \Lambda_{\mathscr{U}} = \text{identity operator.} \tag{2.5}
$$

We say that the control is distributed (in Ω).[8] Further, we set

B = identity operator, (2.6)

C = injection of V into H ($\mathscr{H} = H$ and therefore Λ = identity). (2.7)

Remark 2.1. The hypothesis (2.7) *signifies that we observe* $y(u)$ *on all of* Ω, *which is not practically realistic. Later we shall study the case*

---

[7] In this case $V' = H^{-1}(\Omega)$.

[8] In contrast to the case where the control is a function defined on the boundary (or on a subset of the boundary) (we then say boundary control)—a case, which we shall see is far more interesting both from a theoretical and practical viewpoint.

where we observe $y(u)$ on a subset of Ω or on Γ. This case is also mathematically more interesting.

The state $y(u)$, therefore, is given by the solution of the Dirichlet Problem

$$\left.\begin{aligned} &A\,y(u)=f+u,\\ &y(u)\in H_0^1(\Omega) \quad (\text{hence } y(u)=0 \text{ on } \Gamma), \end{aligned}\right\} \tag{2.8}$$

and we wish to find

$$\inf_{v\in\mathscr{U}_{\mathrm{ad}}}\left[\int_\Omega (y(v)-z_d)^2\,dx+(N\,v,v)\right]^9 \tag{2.9}$$

where $\mathscr{U}_{\mathrm{ad}}$ is a closed, convex subset of $\mathscr{U}$.

Remark 2.2. (On the notation) *The solution* $y(u)$ of (2.8) *is a function on Ω, given by*

$$x \xrightarrow{y(u)} y(x;u)\,.$$

The expression $J(v)$ may thus be written as

$$J(v)=\int_\Omega (y(x;v)-z_d(x))^2\,dx+\int_\Omega (N\,v)(x)\,v(x)\,dx\,.$$

By virtue of (2.3), we may apply Theorem 1.4, and equations (1.10), (1.30) (1.29) of this theorem become in this case

$$\left.\begin{aligned} A\,y(u)&=f+u \quad &&\text{in } \Omega, \quad y(u)=0 \quad \text{on } \Gamma,\\ A^*\,p(u)&=y(u)-z_d \quad &&\text{in } \Omega, \quad p(u)=0 \quad \text{on } \Gamma,\\ &\int_\Omega (p(u)+N\,u)(v-u)\,dx\geqslant 0 \quad \forall\, v\in\mathscr{U}_{\mathrm{ad}},\\ &&& u\in\mathscr{U}_{\mathrm{ad}}. \end{aligned}\right\} \tag{2.10}$$

Example 2.1. Case without any constraints: $\mathscr{U}_{\mathrm{ad}}=\mathscr{U}$.

In this case the last condition in (2.10) reduces to

$$p(u)+N\,u=0\,.$$

We may then eliminate $u(u=-N^{-1}p)$ and the optimal control is given by the following rule:

(i) The following system of partial differential equations needs to be solved[10]:

---

[9] $(N\,v,v)=(N\,v,v)_{\mathscr{U}}$ since $\mathscr{U}=H$.

[10] We note that the word "system" is used in two different senses (both allowed due to common usage): a) physical system which we observe: the state $y(u)$ of the system, b) system of partial differential equations.

$$\left.\begin{aligned} A y+N^{-1} p&=f && \text{in } \Omega, \\ A^* p-y&=-z_d && \text{in } \Omega, \\ y=0, \quad p&=0 && \text{on } \Gamma; \end{aligned}\right\} \tag{2.11}$$

(ii) then

$$u=-N^{-1} p. \tag{2.12}$$

Remark 2.3. Regularity of the Optimal Control.

Let us suppose that the coefficients of A are regular in $\bar{\Omega}$. Then by utilizing the regularity of solutions of elliptic problems (cf. for example Lions-Magenes [1], Chapter 2) we may show that the solution p of the equation

$$A^* p=y-z_d, \quad p|_\Gamma=0$$

is such that

$$p \in H^2(\Omega). \tag{2.13}$$

If we further assume that N^{-1} *maps* $H^2(\Omega)$ *into itself*, we see that the *optimal control* u *belongs* to $H^2(\Omega)$ (whereas the search for u was originally in all of $L^2(\Omega)$). We shall see that this regularity of the optimal control is not generally true.

Example 2.2. Let us now consider the case where

$$\mathscr{U}_{\text{ad}}=\{v \mid v \geqslant 0 \text{ almost everywhere in } \Omega\}. \tag{2.14}$$

According to Remark 1.5, we obtain

$$\left.\begin{aligned} u&\geqslant 0 && \text{almost everywhere in } \Omega, \\ p(u)+N u&\geqslant 0 && \text{almost everywhere in } \Omega, \\ u(p(u)+N u)&=0 && \text{almost everywhere in } \Omega. \end{aligned}\right\} \tag{2.15}$$

u may be eliminated from the above equations in the following way:

$$\left.\begin{aligned} A y-f&\geqslant 0 && \text{in } \Omega, \\ A^* p-y&=-z_d && \text{in } \Omega, \\ p+N(A y-f)&\geqslant 0 && \text{in } \Omega, \\ (A y-f)[p+N(A y-f)]&=0 && \text{in } \Omega, \\ y|_\Gamma=0, \quad p|_\Gamma&=0, \end{aligned}\right\} \tag{2.16}$$

and then $u=A y-f$.

Clearly from (2.15) either $u=0$ or $p+N u=0$ or both u and $p+N u$ equal to zero. In the last case, we have simultaneously $A y=f$ and

$y = z_d$ and hence $Az_d = f$. Hence, assuming (which evidently is the general case) that

$$Az_d \neq f \quad \text{almost everywhere in } \Omega,$$

only two cases are possible (Ω_i being defined up to a set of measure zero):

(i) $u=0$ in Ω_0 and $p>0$,

(ii) $u>0$ in Ω_1 and $p+Nu=0$.

To simplify further assume that

$$N = \nu I, \qquad \nu > 0. \tag{2.17}$$

Hence from (i) and (ii) above, we deduce:

$$u = -\frac{1}{\nu} \inf(0, p). \tag{2.18}$$

From the above the following conclusion obtains:

the optimal control is given by (2.18), where p is furnished by the solution of the boundary value problem (non-linear),

$$\left.\begin{aligned} Ay + \frac{1}{\nu}\inf(0,p) &= f, \\ A^* p - y &= z_d, \\ y|_\Gamma = 0, \quad p|_\Gamma &= 0. \end{aligned}\right\} \tag{2.19}$$

$$u \in H_0^1(\Omega) \quad \text{(regularity result on the optimal control)}. \tag{2.20}$$

Example 2.3. Let us consider the case where

$$\mathcal{U}_{\text{ad}} = \{v \mid \xi_0(x) \leqslant v(x) \leqslant \xi_1(x), \text{ almost everywhere in } \Omega, \; \xi_0, \xi_1 \text{ given functions in } L^\infty(\Omega)\}. \tag{2.21}$$

It may be verified without difficulty that

$\mathcal{U}_{\text{ad}}$ defined by (2.21) is a closed, bounded, convex set in $\mathcal{U} = L^2(\Omega)$. (2.22)

From the third condition in (2.10) (cf. section 2.5 for the general case) we may easily deduce the following local conditions

$$p(x;u) + (Nu)(x)(\xi - u(x)) \geqslant 0 \;\; \forall \xi \in [\xi_0(x), \xi_1(x)], \text{ almost everywhere in } \Omega. \tag{2.23}$$

To simplify matters assume that (2.17) obtains. From (2.23) we then deduce

(i) $p(x;u)+\nu u(x)>0, \quad u(x)=\xi_0(x),$

(ii) $p(x;u)+\nu u(x)<0, \quad u(x)=\xi_1(x),$

(iii) $p(x;u)+\nu u(x)=0, \quad u(x)=-\frac{1}{\nu}p(x;u)$ (clearly!).

Conclusion: In a general manner, set for $g\in H_0^1(\Omega)$:

$$\Phi(\xi_0,\xi_1,\nu)g=\left\{\begin{array}{ll} -\frac{1}{\nu}g(x) & \text{if } \xi_0(x)\leqslant -\frac{1}{\nu}g(x)\leqslant\xi_1(x), \\ \xi_0 & \text{if } -\frac{1}{\nu}g(x)<\xi_0(x), \\ \xi_1 & \text{if } -\frac{1}{\nu}g(x)>\xi_1(x). \end{array}\right\} \tag{2.24}$$

Then the optimal control is given by the solution of the non-linear problem

$$\left.\begin{array}{r} Ay-\Phi(\xi_0,\xi_1,\nu)p=f, \\ A^*p-y=-z_d, \\ y|_\Gamma=0, \quad p|_\Gamma=0 \end{array}\right\} \tag{2.25}$$

and

$$u=\Phi(\xi_0,\xi_1,\nu)p. \tag{2.26}$$

In this case the operator (non-linear) Φ maps $H_0^1(\Omega)$ into $L^2(\Omega)$ and (in general) there are no results on the regularity of the optimal control. However if $\xi_i\in L^\infty(\Omega)\cap H^1(\Omega)$, $i=0,1$, then $\Phi(\xi_0,\xi_1,\nu)p\in H^1(\Omega)$.

2.2. The Case with No Constraints

The situation examined in Example 2.1 of the preceeding Section 2.1 is quite general: if we assume

$$\mathscr{U}_{\text{ad}}=\mathscr{U}$$

condition (1.29) reduces to

$$\Lambda_{\mathscr{U}}^{-1}B^*p(u)+Nu=0$$

and eliminating u we obtain the following rule:

(i) one solves the boundary value problem (for a system of operator equations)

$$\left.\begin{aligned} Ay+BN^{-1}\Lambda_{\mathscr{U}}^{-1}B^*p&=f,\\ A^*p-C^*Cy&=-C^*z_d;\end{aligned}\right\} \tag{2.27}$$

(ii) then the optimal control is given by

$$u=-N^{-1}B^*p. \tag{2.28}$$

(The regularity of u—cf. Remark 2.3—depends on particular situations).

2.3. System Governed by a Neumann Problem; Distributed Control

We now take (cf. Chapter 1, Section 3)

$$V=H^1(\Omega), \qquad H=L^2(\Omega) \tag{2.29}$$

the form $a(\varphi,\psi)$ being given by (2.2) (together with (2.3)).

In this case care must be taken: if Ω is an open, bounded set in $\mathbb{R}^n$, the space $\mathscr{D}(\Omega)$ is not dense in V (the closure of $\mathscr{D}(\Omega)$ in $H^1(\Omega)$ is the proper subspace $H_0^1(\Omega)$; hence $H=L^2(\Omega)$ being already identified with its dual, V' cannot be identified with a subspace of $\mathscr{D}(\Omega)$).

Two typical continuous linear forms on V are[11]:

$$L_1(\psi)=\int_\Omega h\psi\,dx, \qquad h\in L^2(\Omega), \tag{2.30}$$

$$L_2(\psi)=\int_\Gamma g\psi\,d\Gamma, \qquad g\in H^{-\frac{1}{2}}(\Gamma) \tag{2.31}$$

(in (2.31) we utilize the results of Chapter 1, Section 3) (clearly linear combinations of L_1 and L_2 can also be considered).

The isomorphism A of V onto V' must be carefully interpreted under the variational form: for a continuous linear form f on V, there exists a unique y such that

$$a(y,\psi)=f(\psi)^{12} \quad \forall\psi\in V. \tag{2.32}$$

Let us first assume that (as in section 2.1):

$\mathscr{U}=L^2(\Omega)$,

$B=$ identity map,

$C=$ injection map of V into H.

---

[11] Cf. Lions-Magenes [1], Chapter 1 for a far more precise study of this question. What follows is sufficient for our purpose.

[12] $f(\psi)=$ value of f at $\psi=(f,\psi)$.

In (1.10), let us choose f as,

$$f(\psi)=\int_\Omega f_1 \psi \, dx+\int_\Gamma g \psi \, d\Gamma, \quad f_1 \in L^2(\Omega), \quad g \in H^{-\frac{1}{2}}(\Gamma). \tag{2.33}$$

The state of the system $y(u)$ is then given by the solution of

$$a(y(u),\psi)=f(\psi)+\int_\Omega u \psi \, dx \quad \forall \psi \in H^1(\Omega), \tag{2.34}$$

which may be interpreted as (cf. Chapter 1, section 3) (Neumann Problem)

$$\left.\begin{aligned} A\, y(u)&=f_1+u \quad \text{in } \Omega, \\ \frac{\partial y}{\partial \nu_A}(u)&=g \quad \text{on } \Gamma. \end{aligned}\right\} \tag{2.35}$$

The cost function $J(v)$ is the same as given in (2.9). The adjoint state is given by the solution to the adjoint Neumann problem,

$$\left.\begin{aligned} A^* p(u)&=y(u)-z_d \quad \text{in } \Omega, \\ \frac{\partial p}{\partial \nu_{A^*}}(u)&=0 \quad \text{on } \Gamma. \end{aligned}\right\} \tag{2.36}$$

The optimal control u is given by the solution of (2.35), (2.36) together with

$$\int_\Omega (p(u)+N u)(v-u)\, dx \geq 0 \quad \forall v \in \mathcal{U}_{ad}, \; u \in \mathcal{U}_{ad}. \tag{2.37}$$

Examples analogous to that given in section 2.1 can be constructed—it is sufficient to replace the Dirichlet conditions by Neumann Conditions.

2.4. System Governed by a Neumann Problem; Boundary Control

Let us now assume that

$$\mathcal{U}=H^{-\frac{1}{2}}(\Gamma); \tag{2.38}$$

$$\left.\begin{aligned} &\text{for every } u \in \mathcal{U}, \; Bu \in V' \text{ is given by} \\ &(Bu,\psi) = \int_\Gamma u \psi \, d\Gamma, \end{aligned}\right. \tag{2.39}$$

C = injection map of V into H,[13] f is given as in section 2.3.

[13] Control is exercised through the boundary, but the observation is still "distributed in Ω". Later on, we shall examine the case where observation is also on the boundary.

Then the state of the system $y(u)$ is given by the solution of

$$a(y(u),\psi)=f(\psi)+\int_\Gamma u\psi\,d\Gamma, \tag{2.40}$$

which may be interpreted as,

$$\left.\begin{aligned} &A\,y(u)=f_1,\\ &\frac{\partial y}{\partial \nu_A}(u)=g+u. \end{aligned}\right\} \tag{2.41}$$

The cost function is the same as given in (2.9).

The adjoint state is again given by (2.36).

The optimal control u is given by the solution of (2.41), (2.36) together with

$$(\Lambda_{\mathscr{U}}^{-1}B^* p(u)+N u, v-u)_{\mathscr{U}}\geqslant 0 \quad \forall\, v\in\mathscr{U}_{\text{ad}},\ u\in\mathscr{U}_{\text{ad}}. \tag{2.42}$$

In order to interpret (2.42)[14], we have to explicitly calculate $\Lambda_{\mathscr{U}}^{-1}B^*$. For this purpose we shall use the Laplace-Beltrami operator $-\Delta_\Gamma$ on Γ, whose powers $(-\Delta_\Gamma)^\alpha$, $\alpha\in\mathbb{C}$ may be defined. We choose as scalar product on $\mathscr{U}$[15]—a choice which is perfectly valid, cf. Lions-Magenes [1], Chapters 1 and 2,

$$(u,v)_{\mathscr{U}}=\int_\Gamma(-\Delta_\Gamma)^{-\frac{1}{4}}u\cdot(-\Delta_\Gamma)^{-\frac{1}{4}}v\,dx=\int_\Gamma(-\Delta_\Gamma)^{-\frac{1}{2}}u\cdot v\,dx.$$

With this choice of scalar product,

$$\Lambda_{\mathscr{U}}^{-1}B^*\psi=(-\Delta_\Gamma)^{+\frac{1}{2}}(\psi|_\Gamma), \qquad \psi\in H^1(\Omega).\text{[16]} \tag{2.43}$$

Indeed,

$$\begin{aligned}(\Lambda_{\mathscr{U}}^{-1}B^*\psi,u)_{\mathscr{U}}&=(\psi,Bu)=\int_\Gamma\psi u\,d\Gamma=\int_\Gamma(-\Delta_\Gamma)^{\frac{1}{2}}(\psi|_\Gamma)(-\Delta_\Gamma)^{-\frac{1}{2}}u\,d\Gamma\\ &=((-\Delta_\Gamma)^{\frac{1}{2}}(\psi|_\Gamma),u)_{\mathscr{U}} \quad \text{whence (2.43).}\end{aligned}$$

(2.42) is then equivalent to

$$((-\Delta_\Gamma)^{\frac{1}{2}}(p(u)|_\Gamma)+N u, v-u)_{\mathscr{U}}\geqslant 0 \quad \forall\, v\in\mathscr{U}_{\text{ad}} \tag{2.44}$$

which is further equivalent to

$$\int_\Gamma(p(u)|_\Gamma+(-\Delta_\Gamma)^{-\frac{1}{2}}N u)(v-u)\,d\Gamma\geqslant 0 \quad \forall\, v\in\mathscr{U}_{\text{ad}}. \tag{2.45}$$

---

[14] The reading of this section is not indispensable for an understanding of the sections which follow.

[15] It is necessary to make a choice right at the beginning since $(N u,u)_{\mathscr{U}}$ appears as a term in the cost function.

[16] In other words, we take $\psi|_\Gamma$ to be an element of $H^{\frac{1}{2}}(\Gamma)$, which is operated on by $(-\Delta_\Gamma)^{\frac{1}{2}}$.

Example 2.4. Let us choose

$$\mathcal{U}_{\text{ad}}=\{v \mid v \geqslant 0 \text{ on } \Gamma,\ v \in H^{-\frac{1}{2}}(\Gamma)\}\,. \tag{2.46}$$

The differential system specifying the optimal control is then,

$$\left.\begin{aligned}
&A\,y(u)=f_1 \quad \text{in } \Omega, \qquad \frac{\partial y}{\partial \nu_A}(u)=g+u \quad \text{on } \Gamma,\\
&A^*\,p(u)=y(u)-z_d \quad \text{in } \Omega, \qquad \frac{\partial p}{\partial \nu_{A^*}}(u)=0 \quad \text{on } \Gamma,\\
&\qquad u \geqslant 0 \quad \text{on } \Gamma,\\
&p(u)+(-\Delta_\Gamma)^{-\frac{1}{2}} N u \geqslant 0 \quad \text{on } \Gamma,\\
&u(p(u)+(-\Delta_\Gamma)^{-\frac{1}{2}} N u)=0 \quad \text{on } \Gamma.
\end{aligned}\right\} \tag{2.47}$$

(Let us remark that it is known that the above boundary value problem—of the type "unilateral problem", cf. Chapter 1—admits a unique solution).

Remark 2.4. The complicated nature of the preceeding example is due to the choice of $\mathcal{U}=H^{-\frac{1}{2}}(\Gamma)$ (which introduces fractional powers of $-\Delta_\Gamma$).

The situation is simplified in the more realistic case where

$$\mathcal{U}=L^2(\Gamma)\,. \tag{2.48}$$

Condition (2.44) is then replaced by

$$\int_\Gamma (p(u)+N\,u)(v-u)\,d\Gamma \geqslant 0 \quad \forall\, v \in \mathcal{U}_{\text{ad}}, \tag{2.49}$$

the other equations remaining unchanged.

Example 2.5. Let us choose

$$\mathcal{U}_{\text{ad}}=\{v \mid v \geqslant 0 \text{ on } \Gamma,\ v \in L^2(\Gamma)\}\,. \tag{2.50}$$

The differential system governing the optimal control is then,

$$\left.\begin{aligned}
&A\,y(u)=f_1 \quad \text{in } \Omega, \qquad \frac{\partial y}{\partial \nu_A}(u)=g+u \quad \text{on } \Gamma,\\
&A^*\,p(u)=y(u)-z_d \quad \text{in } \Omega, \qquad \frac{\partial p}{\partial \nu_{A^*}}=0 \quad \text{on } \Gamma,\\
&\qquad u \geqslant 0 \quad \text{on } \Gamma,\\
&p(u)+N\,u \geqslant 0 \quad \text{on } \Gamma,\\
&u(p(u)+N\,u)=0 \quad \text{on } \Gamma.
\end{aligned}\right\} \tag{2.51}$$

On Γ, there are three possibilities:

1. $u=0, \quad p \doteq 0,$
2. $u>0, \quad p+Nu=0,$
3. $p>0, \quad u=0.$

It would be interesting to study the properties of the three regions of Γ corresponding to the three possibilities above. This appears to be an open problem.

2.5. Local and Global Constraints

Let S be either an open set in $\mathbb{R}^N$ or a manifold in $\mathbb{R}^N$ [17] which is sufficiently differentiable. Let S be endowed with a measure dS and let E be a Hilbert space. Let $L^2(S;E)$ denote the space of (equivalence classes of) measurable functions

$$s \to f(s) \quad \text{of } S \to E$$

such that $\int_S \|f(s)\|_E^2 dS < \infty$. Endowed with the scalar product

$$\int_S (f(s), g(s))_E \, dS$$

this is a Hilbert space (cf. Bourbaki [1]).

In this section we take

$$\mathcal{U} = L^2(S;E). \tag{2.52}$$

We say that the closed, convex subset U_{ad} of $\mathcal{U}$ is defined by *local constraints*[18] if

$$\begin{aligned} \mathcal{U}_{\text{ad}} = \{v \mid v \in \mathcal{U},\ & v(s) \in K \text{ for almost all } s \in S, \\ & \text{where } K = \text{closed, convex subset of } E\}. \end{aligned} \tag{2.53}$$

If $\mathcal{U}_{\text{ad}}$ is not of the form (2.53), we say that $\mathcal{U}_{\text{ad}}$ is defined by *global constraints;* for example,

$$\mathcal{U}_{\text{ad}} = \left\{ v \,\middle|\, \left(\int_S \|v(s)\|^2 dS \right)^{\frac{1}{2}} \leqslant M \right\}. \tag{2.54}$$

or

$$\mathcal{U}_{\text{ad}} = \left\{ v \,\middle|\, \int_S v(s)\, dS \in K,\ K = \text{closed, convex subset of } E \right\} \text{ etc.}^{19}. \tag{2.55}$$

[17] These hypotheses, which may be relaxed are sufficient for our purposes.
[18] One also says that one is concerned with "Hard constraints".
[19] One also says that one is concerned with "Soft constraints".

We shall prove that

Theorem 2.1. *We assume that* (2.53) *holds. Then condition* (1.29) *of Theorem* 1.4 *is equivalent to the following: setting,*

$$\Lambda_{\mathscr{U}}^{-1} B^* p(u) + N u = \zeta \tag{2.56}$$

we have

$$(\zeta(s), k - u(s))_E \geqslant 0 \quad \forall k \in K, \tag{2.57}$$

except on a subset of S of measure zero.

Remark 2.5. Clearly (2.57) *may also be written as*

$$(\zeta(s), u(s))_E = \inf_{k \in K} (\zeta(s), k). \tag{2.58}$$

Proof of Theorem 2.1.
1. (2.57)⇒(1.29).

This is obvious. If $v \in \mathscr{U}_{\mathrm{ad}}$, $v(s) \in K$ almost everywhere, and hence from (2.57)

$$(\zeta(s), v(s) - u(s))_E \geqslant 0 \quad \text{almost everywhere}$$

and therefore integrating over S we obtain (1.29).

2. (1.29)⇒(2.57).

Let S_0 be the subset of S consisting of Lebesgue points for the two functions

$$s \to \zeta(s) \quad \text{and} \quad s \to (\zeta(s), u(s))_E.$$

For a fixed s_0 in S_0, there exists a sequence of neighborhoods $\mathcal{O}_j$ of s_0, with diameter → 0, such that with $|\mathcal{O}_j|$ denoting the measure of $\mathcal{O}_j$, we have

$$\begin{aligned} &\frac{1}{|\mathcal{O}_j|} \int_{\mathcal{O}_j} \zeta(s)\, ds \to \zeta(s_0) \quad \text{in } E, \\ &\frac{1}{|\mathcal{O}_j|} \int_{\mathcal{O}_j} (\zeta(s), u(s))_E\, ds \to (\zeta(s_0), u(s_0))_E. \end{aligned} \tag{2.59}$$

It is known that the complement of S_0 in S is of measure zero. If (2.57) can be verified on S_0 the theorem will be proved.

Let k be an arbitrary but fixed element in K and s_0 be fixed in S_0 (the $\mathcal{O}_j$ being chossen as in (2.59)). Set

$$\begin{aligned} &v_j = \chi_j k + (1 - \chi_j) u \\ &\chi_j = \text{characteristic function of } \mathcal{O}_j. \end{aligned} \tag{2.60}$$

Since $\mathscr{U}_{ad}$ is defined by (2.58), the function v_j defined by (2.60) is an element of $\mathscr{U}_{ad}$ and therefore in (1.29) we may take $v=v_j$. We thus obtain,

$$\int_{\mathcal{O}_j} (\zeta(s), k-u(s))_E dS \geqslant 0,$$

and dividing by $|\mathcal{O}_j|$,

$$\frac{1}{|\mathcal{O}_j|}\int_{\mathcal{O}_j} (\zeta(s), k-u(s))_E dS \geqslant 0. \tag{2.61}$$

Using (2.59) we may pass to the limit in (2.61). We then obtain,

$$(\zeta(s_0), k-u(s_0))_E \geqslant 0$$

and this is true for $\forall s_0 \in S_0$, whence the desired result. □

Remark 2.6. This is a particular case of the preceeding theorem. We used that theorem in Example 2.3.

2.6. System Governed by a Differential System[20]

We consider the simple case of a differential system which we considered in Section 3.8, Chapter 1.

Let the state of the system $y(u)=\{y_1(u), y_2(u)\}$ be defined by

$$\left.\begin{aligned} -\Delta y_1(u)+y_1(u)-y_2(u)&=f_1+u_1,^{21}\\ -\Delta y_2(u)+y_2(u)+y_1(u)&=f_2+u_2,\\ y_1(u)&=0 \quad \text{on } \Gamma,\\ y_2(u)&=0 \quad \text{on } \Gamma, \end{aligned}\right\} \tag{2.62}$$

where

$$u=\{u_1,u_2\}\in\mathscr{U}=L^2(\Omega)\times L^2(\Omega). \tag{2.63}$$

Using the same notation as in the general theory, we have

$$\begin{aligned} &V=H_0^1(\Omega)\times H_0^1(\Omega), \qquad H=L^2(\Omega)\times L^2(\Omega), \qquad \mathscr{U}=H,\\ &A=(\varphi=\{\varphi_1,\varphi_2\})\to A\varphi=\{-\Delta\varphi_1+\varphi_1-\varphi_2, -\Delta\varphi_2+\varphi_2+\varphi_1\};\\ &B=\text{identity map}. \end{aligned}$$

[20] As we have already remarked, the word system is being used in two different senses.

[21] We restrict ourselves to distributed controls.

We choose (as in section 2.1),

$$C = \text{injection map of } V \text{ into } H.$$

For a given $z_d = \{z_{1d}, z_{2d}\}$ in H, we have

$$J(v) = \int_\Omega [(y_1(v) - z_{1d})^2 + (y_2(v) - z_{2d})^2]\, dx + (Nv, v)_{\mathscr{U}}. \tag{2.64}$$

The adjoint state is defined by $A^* p(u) = y(u) - z_d$, which may be written explicitly as

$$\left.\begin{aligned} -\Delta p_1(u) + p_1(u) + p_2(u) &= y_1(u) - z_{1d}, \\ -\Delta p_2(u) + p_2(u) - p_1(u) &= y_2(u) - z_{2d}, \\ p_1(u) &= 0 \quad \text{on } \Gamma, \\ p_2(u) &= 0 \quad \text{on } \Gamma. \end{aligned}\right\} \tag{2.65}$$

The optimal control is given by the solution of the set of equations (2.62), (2.65) together with

$$\int_\Omega [p_1(u)(v_1 - u_1) + p_2(u)(v_2 - u_2)]\, dx + (Nu, v-u)_{\mathscr{U}} \geqslant 0 \quad \forall v \in \mathscr{U}_{\text{ad}}, \quad u \in \mathscr{U}_{\text{ad}}. \tag{2.66}$$

Example 2.6. Let us suppose that N is a diagonal matrix of operators: $N\{u_1, u_2\} = \{N_1 u_1, N_2 u_2\}$ and that

$$\mathscr{U}_{\text{ad}} = \{u \mid u_1 \text{ arbitrary in } L^2(\Omega),\ u_2 \geqslant 0 \text{ almost everywhere in } \Omega\}. \tag{2.67}$$

(Thus there are no constraints on u_1).

Then (2.66) is equivalent to

$$\left.\begin{aligned} p_1(u) + N_1 u_1 &= 0, \\ p_2(u) + N_2 u_2 &\geqslant 0, \qquad u_2 \geqslant 0, \\ u_2(p_2(u) + N_2 u_2) &= 0. \end{aligned}\right\} \tag{2.68}$$

Thus under the hypotheses of (2.67), the optimal control is given by the solution of the system of equations and inequalities,

$$\left.\begin{aligned} -\Delta y_1 + y_1 - y_2 + N_1^{-1} p_1 &= f_1, \\ -\Delta y_2 + y_2 + y_1 - f_2 &\geqslant 0, \\ -\Delta p_1 + p_1 + p_2 - y_1 &= -z_{1d}, \\ -\Delta p_2 + p_2 - p_1 - y_2 &= -z_{2d}, \\ p_2 + N_2(-\Delta y_2 + y_2 + y_1 - f_2) &\geqslant 0, \\ (-\Delta y_2 + y_2 + y_1 - f_2)[p_2 + N_2(-y_2 + y_2 + y_1 - f_2)] &= 0, \\ y_1 = y_2 = p_1 = p_2 &= 0 \quad \text{on } \Gamma. \end{aligned}\right\} \tag{2.69}$$

Further

$$u_1 = -N_1^{-1} p_1, \qquad u_2 = -\Delta y_2 + y_2 + y_1 - f_2. \tag{2.70}$$

Remark 2.7.

It is evident that by modifying
—the boundary conditions,
—the nature of the control (distributed, boundary),
—the nature of the observation,
—the initial differential system,
an infinity of variations on the above problem is possible.

2.7. System Governed by a 4th Order Differential Operator [22]

Let us consider the 4th order differential operator: $\Delta(a_1\Delta\varphi)+a_0\varphi$, which we introduced in Chapter 1, section 3.9. Therefore,

$$V=\{\varphi\,|\,\varphi\in L^2(\Omega),\ \Delta\varphi\in L^2(\Omega)\},$$
$$H=L^2(\Omega).$$

As in section 2.3, the space V' is not identified with a space of distributions on Ω. We define (cf. (3.69), Chapter 1),

$$\left.\begin{aligned}&a(\varphi,\psi)=\int_\Omega a_1\Delta\varphi\,\Delta\psi\,dx+\int_\Omega a_0\varphi\psi\,dx,\\&a_0,a_1\in L^\infty(\Omega),\ a_0(x)\geqslant\alpha>0,\ a_1(x)\geqslant\alpha\quad\text{almost everywhere in }\Omega.\end{aligned}\right\}\tag{2.71}$$

We take,

$$\mathscr{U}=H^{\frac{3}{2}}(\Gamma)\times H^{\frac{1}{2}}(\Gamma),\tag{2.72}$$

and suppose that the state $y=y(u)$ is given by

$$\left.\begin{aligned}&a(y(u),\psi)=\int_\Omega f\psi\,dx+\int_\Gamma u_1\frac{\partial\psi}{\partial n}\,d\Gamma-\int u_2\psi\,d\Gamma,\\&\forall\psi\in V.\end{aligned}\right\}\tag{2.73}$$

This problem admits a unique solution; the state $y(u)$ is given by the solution of the equivalent system

$$\left.\begin{aligned}A\,y(u)&=f\quad\text{in }\Omega,\ \text{where } A\varphi=\Delta(a_1\Delta\varphi)+a_0\varphi,\\a_1\Delta y(u)&=u_1\quad\text{on }\Gamma,\\\frac{\partial}{\partial n}(a_1\Delta y(u))&=u_2\quad\text{on }\Gamma.\end{aligned}\right\}\tag{2.74}$$

[22] The reading of this section is not indispensable for an understanding of the sections which follow.

If we again take

$$C = \text{injection map of } V \text{ into } H,$$

the adjoint state $p(u)$ is defined by

$$a(p(u),\psi) = \int_\Omega (y(u)-z_d)\psi\, dx \quad \forall \psi \in V, \tag{2.75}$$

or equivalently by,

$$\left.\begin{aligned} A\,p(u) &= y(u)-z_d \quad \text{in } \Omega,\\ a_1 \Delta\, p(u) &= 0 \quad \text{on } \Gamma,\\ \frac{\partial}{\partial n}(a_1 \Delta\, p(u)) &= 0 \quad \text{on } \Gamma. \end{aligned}\right\} \tag{2.76}$$

Let us recapitulate that (1.29) was transformed to the equivalent form (1.23) by utilizing the adjoint state. In case there are a number of boundary conditions, it is simpler to transform (1.23) directly rather than calculate $\Lambda_{\mathscr{U}}^{-1} B^*$ separately. This of course amounts to calculating $\Lambda_{\mathscr{U}}^{-1} B^*$ implicitly. In the present case, condition (1.23) may be written as

$$\int_\Omega (y(u)-z_d)(y(v)-y(u))\,dx + (N u, v-u)_{\mathscr{U}} \geqslant 0. \tag{2.77}$$

Multiplying the first equation in (2.76) by $y(v)-y(u)$ and applying Green's formula, we obtain,

$$\int_\Omega (y(u)-z_d)(y(v)-y(u))\,dx = \int_\Gamma \frac{\partial p}{\partial n}(a_1 \Delta\, y(v) - a_1 \Delta\, y(u))\,d\Gamma$$
$$- \int_\Gamma p\left(\frac{\partial}{\partial n}(a_1 \Delta\, y(v)) - \frac{\partial}{\partial n}(a_1 \Delta\, y(u))\right) d\Gamma$$

and taking into account the boundary conditions in (2.74), the right hand side becomes

$$= \int_\Gamma \frac{\partial p}{\partial n}(v_1 - u_1)\,d\Gamma - \int_\Gamma p(v_2-u_2)\,d\Gamma.$$

Therefore (2.77) is equivalent to

$$\int_\Gamma \frac{\partial p}{\partial n}(v_1 - u_1)\,d\Gamma - \int_\Gamma p(v_2-u_2)\,d\Gamma + (N u, v-u)_{\mathscr{U}} \geqslant 0. \tag{2.78}$$

Let us now suppose that $N=\{N_1,N_2\}$ is a diagonal matrix of operators and the scalar product on $\mathcal{U}$ (cf. section 2.4) is chosen as

$$(u,v)_{\mathcal{U}} = \int_\Gamma (-\Delta_\Gamma)^{\frac{3}{2}} u_1 \cdot v_1 \, d\Gamma + \int_\Gamma (-\Delta_\Gamma)^{\frac{1}{2}} u_2 \cdot v_2 \, d\Gamma . \tag{2.79}$$

Then (2.78) is equivalent to

$$\int_\Gamma \left(\frac{\partial p}{\partial n} + (-\Delta_\Gamma)^{\frac{3}{2}} u_1\right)(v_1 - u_1)\, d\Gamma + \int_\Gamma (-p + (-\Delta_\Gamma)^{\frac{1}{2}} u_2)(v_2 - u_2)\, d\Gamma \geqslant 0, \tag{2.80}$$

$\forall\, v \in \mathcal{U}_{\text{ad}}$.

Conclusion

The optimal control is given by the solution of (2.74), (2.76) *and* (2.80).

2.8. Orientation

Our efforts now need to be concentrated on

(i) the case where *observation can be made only on the boundary*, a case which has not been considered as yet;

(ii) other cases where *control is exercised through the boundary;* we have already studied two examples (sections 2.4 and 2.8), but these do not cover the case

$$A\, y(u) = f \quad \text{in } \Omega,$$
$$y(u)|_\Gamma = u \quad \text{on } \Gamma .$$

We shall see presently that the study of this case requires some additional developments.

Before attacking these problems, we shall examine a case where $N=0$, but $\mathcal{U}_{\text{ad}}$ is not necessarily a bounded set.

3. A Family of Examples with $N=0$ and $\mathcal{U}_{\text{ad}}$ Arbitrary

3.1. General Case

We place ourselves within the framework of section 1 with

$$\mathcal{U} = V', \qquad B = \text{identity map}, \tag{3.1}$$

$$C = \text{injection map of } V \text{ into } V \text{ (hence } \mathcal{H} = V). \tag{3.2}$$

We therefore have to utilize

$$\Lambda = \text{canonical isomorphism of } V \text{ onto } V'. \tag{3.3}$$

The state of the system is thus given by

$$A\,y(u)=f+u \tag{3.4}$$

and the observation is $z(u)=y(u)$ in V.[23]

The cost function is given by (since $N=0$),

$$J(v)=\|y(v)-z_d\|_V^2,\quad z_d \text{ being a given element in } V. \tag{3.5}$$

We then have,

Theorem 3.1. *Under the hypotheses* (3.1)—(3.5) *and hypothesis* (1.3) *being satisfied, there exists a unique element* u (optimal control) *such that*

$$J(u)=\inf_{v\in\mathscr{U}_{ad}} J(v).$$

The element u *is given by the solution of the following problem: equation* (3.4) *adjoined to the equation and inequality*

$$A^*p(u)=\Lambda(y(u)-z_d), \tag{3.6}$$

$$(p(u),v-u)\geqslant 0\quad \forall v\in\mathscr{U}_{ad}, \text{ the bracket denoting the scalar product between } V \text{ and } V'. \tag{3.7}$$

Proof. In this case the bi-linear form (1.16) reduces to

$$\pi(u,v)=(y(u)-y(0),y(v)-y(0))_V=(A^{-1}u,A^{-1}v)_V$$

and therefore

$$\pi(v,v)=\|A^{-1}v\|_V^2\geqslant c\|v\|_{V'}^2$$

(since A is an isomorphism of V onto V'). Hence the general theory is applicable also in the case $N=0$. ☐

3.2. Application (I)

In section (3.1) let us take

$V=H^1(\Omega)$,

$A=$ differential operator as given in section 2.1.

The problem then is the following: the state $y(u)$ being given by

$$\begin{aligned} A\,y(u)&=f+u \quad \text{in } \Omega,\\ y(u)|_\Gamma&=0, \end{aligned} \tag{3.8}$$

---

[23] Since $V\subset H$, observing $y(u)$ in V is more restrictive than observing $y(u)$ in H (which was the case in the examples of section 2).

find the optimal control which minimizes

$$J(v) = \int_\Omega \left[(y(v)-z_d)^2 + \sum_{i=1}^{n} \left(\frac{\partial}{\partial x_i} (y(v)-z_d) \right)^2 \right] dx.^{24} \tag{3.9}$$

The operator Λ is equal to $-\Delta+I$ and hence the adjoint state is given by

$$\left. \begin{aligned} A^* p(u) &= (-\Delta+I)(y(u)-z_d), \\ p(u)|_\Gamma &= 0 \end{aligned} \right\} \tag{3.10}$$

and (3.7) may be written as

$$\int_\Omega p(u)(v-u)\,dx \geqslant 0 \quad \forall v \in \mathscr{U}_{\text{ad}}. \tag{3.11}$$

Example 3.1. Let us take $\mathscr{U} = \mathscr{U}_{\text{ad}}$ (the case where there are no constraints). Then from (3.11) we obtain $p(u)=0$ and $y(u)=z_d$. By setting

$$u = A z_d - f$$

this result becomes clear from the start.

Example 3.2. Let us take

$$\mathscr{U}_{\text{ad}} = \{ v \mid v \in H^{-1}(\Omega),\ v \geqslant 0 \text{ in } \Omega \}.^{25}$$

Therefore, as we have already seen in section 2, (3.11) reduces to

$$\left. \begin{aligned} &u \geqslant 0, \\ &p(u) \geqslant 0, \quad u \cdot p(u) = 0 \quad \text{in } \Omega. \end{aligned} \right\} \tag{3.12}$$

The optimal control is given by the solution of the following boundary value problem (of "unilateral" type):

$$\left. \begin{aligned} A y - f &\geqslant 0, \\ A^* p - (-\Delta+I) y &= \Delta z_d - f, \\ p &\geqslant 0, \\ (A y - f) p &= 0 \quad \text{in } \Omega, \\ y|_\Gamma = 0, &\quad p|_\Gamma = 0, \end{aligned} \right\} \tag{3.13}$$

and therefore

$$u = A y - f.$$

---

[24] The norm in V is chosen equal to

$$\left\{ \int_\Omega \left[\psi^2 + \sum_{i=1}^{n} \left(\frac{\partial \psi}{\partial x_i} \right)^2 \right] dx \right\}^{\frac{1}{2}}.$$

[25] Hence v is a measure $\geqslant 0$.

Here again, it would be interesting to obtain more information about the various regions of Ω where $u=0, p>0; u>0, p=0; u=0, p=0$. This appears to be an open problem.

3.3. Application (II)

Let us now reconsider section 3.1 with $V=H^1(\Omega)$ normed by

$$\left\{\int_\Omega \left[\psi^2 + \sum_{i=1}^n \left(\frac{\partial \psi}{\partial x_i}\right)^2\right] dx\right\}^{\frac{1}{2}}.$$

The canonical isomorphism of V onto V' is given in the following way:

$$(\Lambda \varphi, \psi) = (\varphi, \psi)_V \quad \forall \varphi, \psi \in V.$$

The state $y(u)$ is given by the variational problem,

$$a(y(u), \psi) = (f+u, \psi)_V \quad \forall \psi \in H^1(\Omega); \tag{3.14}$$

the adjoint state by

$$a^*(p(u), \psi) = (y(u) - z_d, \psi)_V \quad \forall \psi \in H^1(\Omega) \tag{3.15}$$

(where $a^*(\varphi, \psi) = a(\psi, \varphi)$ $\forall \varphi, \psi \in V$), and the optimal control is given by the solution of (3.14), (3.15) and

$$(p(u), v-u) \geqslant 0 \quad \forall v \in \mathscr{U}_{ad}. \tag{3.16}$$

Example 3.3.

$$\mathscr{U}_{ad} = \left\{v \,|\, (v, \psi) = \int_\Gamma v_1 \psi \, d\Gamma, \; v_1 \in H^{-\frac{1}{2}}(\Gamma)\right\}. \tag{3.17}$$

In this manner we define a closed linear subspace of $H^{-\frac{1}{2}}(\Gamma)$, since v_1 being given in $H^{-\frac{1}{2}}(\Gamma)$, the form $v \in V'$ defined by $(v, \psi) = \int_\Gamma v_1 \psi \, d\Gamma$ has a norm defined by

$$\|v\|_{V'} = \sup_{\psi \in V} \frac{\left|\int_\Gamma v_1 \psi \, d\Gamma\right|}{\|\psi\|_V}.$$

But

$$V = H^1(\Omega) = H_0^1(\Omega) \oplus H_\Delta^1(\Omega),$$
$$H_\Delta^1(\Omega) = \{v \,|\, v \in H_\Delta^1(\Omega), \; (-\Delta + 1) v = 0\},$$

and

$$\|v\|_{V'} = \sup_{\psi \in H^1(\Omega)} \frac{\left|\int_\Gamma v_1 \psi \, d\Gamma\right|}{\|\psi\|_V}. \tag{3.18}$$

In other words, $\psi \to \psi|_\Gamma$ is an *isomorphism* of $H^1_\Delta(\Omega)$ onto $H^{\frac{1}{2}}(\Gamma)$. This isomorphism is of a type which permits us to define a norm *equivalent* to $\|v_1\|_{H^{-\frac{1}{2}}(\Gamma)}$ from which the desired result follows.

But, with $\mathcal{U}_{\text{ad}}$ defined by (3.17), (3.16) reduces to

$$p(u)=0 \quad \text{on } \Gamma.$$

Hence the optimal control is given by the solution of the following system of equations:

$$\left.\begin{aligned} &Ay=f,\\ &a^*(p,\psi)=(y-z_d,\psi)_V \quad \forall \psi\in H^1(\Omega),\\ &p|_\Gamma=0, \end{aligned}\right\} \tag{3.19}$$

together with the condition

$$u=\frac{\partial y}{\partial \nu_A} \quad \text{on } \Gamma. \tag{3.20}$$

The system of equations (3.19) may be formally interpreted as follows:

$$\left.\begin{aligned} &Ay=f,\\ &A^*p=(-\Delta+I)(y-z_d),\\ &p|_\Gamma=0, \quad \frac{\partial p}{\partial \nu_{A^*}}=\frac{\partial}{\partial n}(y-z_d). \end{aligned}\right\} \tag{3.21}$$

Example 3.4. Let us take $\mathcal{U}_{\text{ad}}$ to be a subset of the set defined by (3.17):

$$\mathcal{U}_{\text{ad}}=\left\{v \,\middle|\, (v,\psi)=\int_\Gamma v_1\psi\, d\Gamma,\ v_1\in H^{-\frac{1}{2}}(\Gamma),\ v_1\geqslant 0 \text{ on } \Gamma\right\}. \tag{3.22}$$

Then (3.16) reduces to

$$p\geqslant 0 \quad \text{on } \Gamma, \qquad u.p.=0 \quad \text{on } \Gamma \text{ (identifying } u \text{ with } u_1). \tag{3.23}$$

The adjoint state is defined as in (3.19).

4. Observation on the Boundary

4.1. System Governed by a Dirichlet Problem (I)

We again consider the problem studied in section 2.1. The state of the system is thus given by

$$\left.\begin{aligned} &Ay(u)=f+u \quad \text{in } \Omega,\\ &y(u)|_\Gamma=0. \end{aligned}\right\} \tag{4.1}$$

We assume that the coefficients of A are sufficiently regular such that (regularity theorem)

$$y(u)\in H^2(\Omega). \tag{4.2}$$

We may therefore define

$$\left.\frac{\partial y}{\partial n}\right|_\Gamma,$$

and from Chapter 1, section 3, $\left.\frac{\partial y}{\partial n}\right|_\Gamma \in H^{\frac{1}{2}}(\Gamma)$.

In order to avoid as much as possible the use of spaces H^S, S not an integer, we may consider in particular that $\left.\frac{\partial y}{\partial n}\right|_\Gamma \in L^2(\Gamma)$ and we may take the observation to be:

$$\begin{aligned} z(u) &= M\frac{\partial y}{\partial \nu_A}(u), \\ M &= \text{given function in } L^\infty(\Gamma).^{26} \end{aligned} \tag{4.3}$$

Example. If $M=1$ (respectively, $M=$ Characteristic function of $\Gamma_0\subset\Gamma$), (4.3) signifies that we observe $\frac{\partial y}{\partial \nu_A}$ on Γ (respectively Γ_0).

The cost function is given by

$$J(v) = \int_\Gamma \left(M\frac{\partial y}{\partial \nu_A}(v)-z_d\right)^2 d\Gamma + (Nv,v)_{\mathscr{U}}, \tag{4.4}$$

z_d being a given element in $L^2(\Gamma)$.

It is clear that Theorem 1.2 is applicable in this case. Hence, there exists a unique optimal control u, which is characterized by

$$J'(u)\cdot(v-u)\geqslant 0 \quad \forall\, v\in\mathscr{U}_{\text{ad}}.$$

The above condition when explicitly calculated for this case gives us

$$\int_\Gamma \left(M\frac{\partial y}{\partial \nu_A}(u)-z_d\right)\left(M\frac{\partial y}{\partial \nu_A}(v)-M\frac{\partial y}{\partial \nu_A}(u)\right)d\Gamma + (Nu,v-u)_{\mathscr{U}}\geqslant 0, \tag{4.5}$$

$\forall\, v\in\mathscr{U}_{\text{ad}}$.

---

[26] Since $y(u)|_\Gamma$ is known and equal to zero, the data on $\frac{\partial y}{\partial \nu_A}$ is equivalent to the data on $\frac{\partial y}{\partial n}$. If $M\in L^\infty(\Gamma)$, $M\frac{\partial y}{\partial \nu_A}\in L^2(\Gamma)$.

In order to transform (4.5), we first proceed in a formal manner[27].

Let us introduce the adjoint state by means of[28]

$$\left.\begin{aligned} &A^* p(u)=0\,, \\ &p(u)|_\Gamma = -M\left(M\frac{\partial y}{\partial \nu_A}(u)-z_d\right). \end{aligned}\right\} \tag{4.6}$$

Now, multiplying the first equation in (4.6) by $y(v)-y(u)$ and applying Green's Formula, we obtain

$$0=-\int_\Gamma \frac{\partial p}{\partial \nu_A}(u)(y(v)-y(u))\,d\Gamma+\int_\Gamma p(u)\left(\frac{\partial y}{\partial \nu_A}(v)-\frac{\partial y}{\partial \nu_A}(u)\right)d\Gamma$$
$$+\int_\Gamma p(u)(v-u)\,dx.$$

Since $y|_\Gamma=0$, utilizing the second condition in (4.6), gives us,

$$\int_\Gamma\left(M\frac{\partial y}{\partial \nu_A}(u)-z_d\right)\left(M\frac{\partial y}{\partial \nu_A}(v)-M\frac{\partial y}{\partial \nu_A}(u)\right)d\Gamma=\int_\Omega p(u)(v-u)\,dx \tag{4.7}$$

and hence (4.5) is equivalent to

$$\int_\Omega (p(u)+N u)(v-u)\,dx \geqslant 0 \quad \forall v\in\mathscr{U}_{\mathrm{ad}}. \tag{4.8}$$

It remains to solve (4.6). Let us note, that in general, if (4.6) admits a solution, the solution is not in $H^1(\Omega)$ (since otherwise $p(u)|_\Gamma$ will be in $H^{\frac{1}{2}}(\Gamma)$, which is generally not the case).

The solution of problems of type (4.6)[29] therefore requires some additional results which will be given in the following sections[30].

4.2. Some Results on Non-homogeneous Dirichlet Problems

As in section 4.1, let A be a second-order elliptic operator with coefficients sufficiently regular in $\bar{\Omega}$.

---

[27] This will be later justified.

[28] It is precisely this point that needs to be justified.

[29] Non-homogeneous Dirichlet Problem in the terminology of Lions-Magenes [1].

[30] Here we shall be concerned with very special cases of techniques utilized in Lions-Magenes [1].

Let f (respectively g) be functions or distributions on Ω (respectively Γ) (the precise hypotheses will be indicated later). We search for y (in an appropriate class to be indicated) which satisfies

$$\begin{aligned} Ay &= f \quad \text{in } \Omega, \\ y|_\Gamma &= g. \end{aligned} \tag{4.9}$$

Let ψ be a "test function" in $\bar{\Omega}$ (the properties of ψ will be indicated later). Multiplying the first equation in (4.9) by ψ and utilizing Green's Formula, we obtain,

$$\int_\Omega (Ay)\psi\,dx = \int_\Omega f\psi\,dx = -\int_\Gamma \frac{\partial y}{\partial \nu_A}\psi\,d\Gamma + \int_\Gamma y\frac{\partial \psi}{\partial \nu_{A^*}}\,d\Gamma + \int_\Omega y(A^*\psi)\,dx, \tag{4.10}$$

from which, using the second equation in (4.9) we deduce,

$$\int_\Omega y(A^*\psi)\,dx = \int_\Omega f\psi\,dx - \int_\Gamma g\frac{\partial \psi}{\partial \nu_{A^*}}\,d\Gamma + \int_\Gamma \frac{\partial y}{\partial \nu_A}\psi\,d\Gamma. \tag{4.11}$$

It is then natural to impose on ψ the condition

$$\psi|_\Gamma = 0. \tag{4.12}$$

Hence we have,

$$\int_\Omega y(A^*\psi)\,dx = \int_\Omega f\psi\,dx - \int_\Gamma g\frac{\partial \psi}{\partial \nu_{A^*}}\,d\Gamma. \tag{4.13}$$

Conversely, if y satisfies (4.13) for "all" functions ψ satisfying (4.12), then (4.9) obtains.

The idea therefore is to utilize (4.13) *in order to define a function* y *which by definition is a "weak" solution of* (4.9).

Our point of departure for putting this programme into effect will be results which we have already utilized.

Let φ be a given element in $L^2(\Omega)$. The solution of

$$\begin{aligned} A^*\psi &= \varphi, \\ \psi|_\Gamma &= 0 \end{aligned} \tag{4.14}$$

is in the space $H^2(\Omega)$ (cf. for example Lions-Magenes [1], Chapter 2).

In other words,

$$A^* \text{ is an isomorphism of } H^2(\Omega)\cap H_0^1(\Omega) \text{ onto } L^2(\Omega). \tag{4.15}$$

By transposing the isomorphism (4.15) we also see,

$$\left.\begin{array}{l}\text{if } \psi\to\mathscr{L}(\psi) \text{ is a continuous linear form on } H^2(\Omega)\cap H_0^1(\Omega),\\ \text{then there exists a unique } y\in L^2(\Omega) \text{ such that}\\ \int_\Omega y(A^*\psi)\,dx=\mathscr{L}(\psi)\ \ \forall\psi\in H^2(\Omega)\cap H_0^1(\Omega).\end{array}\right\} \tag{4.16}$$

Let us now take

$$f\in L^2(\Omega),\qquad g\in H^{-\frac{1}{2}}(\Gamma). \tag{4.17}$$

Then the linear form $\mathscr{L}(\psi)$ defined by

$$\mathscr{L}(\psi)=\int_\Omega f\psi\,dx-\int_\Gamma g\frac{\partial\psi}{\partial\nu_{A^*}}\,d\Gamma \tag{4.18}$$

is continuous on $H^2(\Omega)\cap H_0^1(\Omega)$. We have thus proved.

Theorem 4.1. *For f, g given as in* (4.17), *there exists a unique* $y\in L^2(\Omega)$ *such that* (4.13) *is true.*

By definition, y is a weak solution (in $L^2(\Omega)$) of (4.9).

Remark 4.1. It is easy to see that

$$Ay=f$$

in the sense of $\mathscr{D}'(\Omega)$. To see this it is sufficient to take $\psi\in\mathscr{D}(\Omega)$ in (4.16).

Then $y\in L^2(\Omega)$ and $Ay=f\in L^2(\Omega)$.

It may then be shown (Lions-Magenes [1], Chapter 2) that $y|_\Gamma$ and $\left.\frac{\partial y}{\partial\nu_A}\right|_\Gamma$ have meaning with $y|_\Gamma\in H^{-\frac{1}{2}}(\Gamma)$ and $\left.\frac{\partial y}{\partial\nu_A}\right|_\Gamma\in H^{-\frac{3}{2}}(\Gamma)$.

Remark 4.2. Let us suppose that the operator A is symmetric, that is $A^*=A$, and let us utilize the eigenfunctions of A as a basis to calculate y:

$$\begin{aligned}&Aw_j=\lambda_j w_j,\quad \int_\Omega w_j^2\,dx=1,\quad j=1,2,\dots\infty,\ \lambda_j \text{ increasing},\\ &w_j|_\Gamma=0.\end{aligned} \tag{4.19}$$

Then y may be represented uniquely as,

$$y=\sum_{j=1}^{\infty} y_j w_j,\qquad \sum_{j=1}^{\infty} y_j^2<\infty. \tag{4.20}$$

Setting $\psi=w_j$ in (4.16) (where $A^*=A$), we obtain,

$$\int_\Omega y(\lambda_j w_j)\,dx=\mathscr{L}(w_j)=\int_\Omega f w_j\,dx-\int_\Gamma g\frac{\partial w_j}{\partial\nu_{A^*}}\,d\Gamma.$$

The left hand side is equal to $\lambda_j y_j$ and hence,

$$y_j = \frac{1}{\lambda_i}\left[\int_\Omega f w_j dx - \int_\Gamma g \frac{\partial w_j}{\partial \nu_{A^*}} d\Gamma\right]. \tag{4.21}$$

4.3. System Governed by a Dirichlet Problem (II)

Using the ideas introduced in section 4.2, we are now in a position to justify the formal calculations carried out in section 4.1.

Let us first remark that the adjoint state formally defined in (4.6) is (rigorously) determined as the solution in $L^2(\Omega)$ of

$$\int_\Omega p(u)(A\psi)dx = \int_\Gamma M\left(M\frac{\partial y}{\partial \nu_A}(u) - z_d\right)\frac{\partial \psi}{\partial \nu_A} d\Gamma \quad \forall \psi \in H^2(\Omega) \cap H_0^1(\Omega). \tag{4.22}$$

Replacing ψ by $y(v) - y(u)$ in (4.22) we obtain (4.7). We have thus proved

Theorem 4.2. *The state of the system being given by* (4.1) *and the cost function by* (4.4), *the optimal control u is given by the simultaneous solution of the set of equations* (4.1), (4.6) ((4.6) *being equivalent to* (4.22)) *and the inequality* (4.8).

Remark 4.3. With these ideas in hand, we are in a position to embark on the study of the case where the system is governed by a Dirichlet Problem, but both the control and observation are on the boundary. Section 5 is devoted to this.

4.4. System Governed by a Neumann Problem

We consider the same problem as in section (2.4) with

$$\mathcal{U} = L^2(\Gamma). \tag{4.23}$$

The state is therefore given by,

$$\left.\begin{aligned} &A y(u) = f \quad \text{in } \Omega,\\ &\frac{\partial y}{\partial \nu_A}(u) = u \quad \text{on } \Gamma, \qquad u \in L^2(\Gamma). \end{aligned}\right\} \tag{4.24}$$

Since $y(u) \in H^1(\Omega)$, we may consider $y(u)|_\Gamma$ and if $M \in L^\infty(\Gamma)$ we may observe $M y(u)$. Therefore let the cost function be,

$$J(v) = \int_\Gamma |M y(v) - z_d|^2 d\Gamma + (N v, v)_{\mathcal{U}}. \tag{4.25}$$

The adjoint state is defined by,

$$\left.\begin{aligned} &A^* p(u)=0 \quad \text{in } \Omega, \\ &\frac{\partial p}{\partial \nu_{A^*}}(u)=M(M(y(u)-z_d)) \quad \text{on } \Gamma. \end{aligned}\right\} \tag{4.26}$$

The optimal control u is unique and determined by the simultaneous solution of (4.24), (4.26) and

$$\int_\Gamma (p(u)+N u)(v-u)\,d\Gamma \geqslant 0 \;\; \forall v \in \mathscr{U}_{ad} (u \in \mathscr{U}_{ad}). \tag{4.27}$$

Example 4.1.

$$\left.\begin{aligned} &\mathscr{U}_{ad}=\{v \mid v \text{ has support in } \Gamma_0 \subset \Gamma,\ \xi_0(x) \leqslant v(x) \leqslant \xi_1(x), \\ &\text{for almost all } x \in \Gamma_0 \text{ (for the measure } d\Gamma),\ \xi_i \in L^\infty(\Gamma)\}, \end{aligned}\right\} \tag{4.28}$$

$$M=\text{characteristic function of } \Gamma_1, \quad \Gamma_1 \subset \Gamma, \quad \Gamma_0 \cap \Gamma_1=\emptyset. \tag{4.29}$$

Condition (4.27) reduces to (assuming $N=\nu I$),

$$\begin{aligned} &(p(x;u)+\nu u(x))(\xi-u(x)) \geqslant 0 \;\; \forall \xi \in [\xi_0(x), \xi_1(x)] \\ &\text{for almost all } x \in \Gamma_0. \end{aligned} \tag{4.30}$$

Example 4.2. Case where $N=0$.

Let us consider the preceeding example with $N=0$ and hence $\nu=0$. Then (4.30) reduces to

$$p(x;u)(\xi-u(x)) \geqslant 0 \;\; \forall \xi \in [\xi_0(x), \xi_1(x)], \quad x \in \Gamma_0. \tag{4.31}$$

But we have

Lemma 4.1. *Let us suppose that the boundary Γ of Ω is real analytic and that the coefficients of A are real analytic in $\bar{\Omega}$. Then*

$$p(x;u) \neq 0 \quad \textit{almost everywhere on } \Gamma_0, \textit{ unless } \quad p(u)=0. \tag{4.32}$$

Proof. We know $p(u)$ satisfies

$$\begin{aligned} &A^* p(u)=0 \quad \text{in } \Omega, \\ &\frac{\partial p}{\partial \nu_{A^*}}(u)=0 \quad \text{on } \Gamma-\Gamma_1 \quad (\text{since } M \text{ is the characteristic function of } \Gamma_1). \end{aligned}$$

Hence $\frac{\partial p}{\partial \nu_{A^*}}(u)=0$ on Γ_0. Now if $p(x;u)=0$ on a set $E \subset \Gamma_0$ of measure >0, we would have, $A^* p(u)=0$ in Ω, $p(u)=0$ and $\frac{\partial p}{\partial \nu_{A^*}}=0$ on $E \subset \Gamma$, measure of $E>0$. Under the analyticity assumptions of the lemma, this would imply $p(u) \equiv 0$. □

If $p(u)\equiv 0$, we may find u such that

$$M\,y(u)-z_d=0. \tag{4.33}$$

If therefore $\operatorname*{Inf}_{v\in\mathscr{U}_{ad}} J(v)>0$, then $p(x;u)=0$ and from (4.31) we may deduce

$$\left.\begin{array}{l}\text{if}\quad p(x;u)>0\quad \text{on } \Gamma_0 \text{ then } u(x)=\xi_0(x);\\ \text{if}\quad p(x;u)<0\quad \text{on } \Gamma_0 \text{ then } u(x)=\xi_1(x).\end{array}\right\} \tag{4.34}$$

Remark 4.4. Property (4.34) is the so-called "bang-bang" property. If, for example, $\xi_0=-1$, $\xi_1=1$, the optimal control takes on the values ± 1 only.

Remark 4.5. It would be interesting to obtain information on the regions where u takes on the values $\xi_0(x)$ and $\xi_1(x)$ and in particular on the interface of the two regions, that is on the *switching surface.*

5. Control and Observation on the Boundary. Case of the Dirichlet Problem

5.1. Orientation

In section 4.4 we have studied systems governed by the Neumann problem and with control and observations on the boundary.

We shall now study some analogous examples where the system is governed by a Dirichlet problem. We shall see that this introduces some additional difficulties.

5.2. Boundary Control in $L^2(\Gamma)$

Take

$$\mathscr{U}=L^2(\Gamma). \tag{5.1}$$

According to 4.2 (the notation remaining unchanged), there exists a unique element $y(u)$ in $L^2(\Omega)$ satisfying

$$\left.\begin{array}{ll}A\,y(u)=f & \text{in } \Omega,\\ y(u)|_\Gamma=u & \text{on } \Gamma,\end{array}\right\} \tag{5.2}$$

the precise interpretation given to (5.2) being (cf. (4.13)):

$$\int_\Omega y(u)(A^*\psi)\,dx=\int_\Omega f\psi\,dx-\int_\Gamma u\frac{\partial\psi}{\partial\nu_{A^*}}\,d\Gamma\quad \forall\psi\in H^2(\Omega)\cap H_0^1(\Omega). \tag{5.3}$$

The observation may be taken as

$$\frac{\partial}{\partial \nu_A} y(u)$$

provided we give a precise meaning to this derivative. It may be shown (Lions-Magenes [1], Chapter 2) that if *$u \in L^2(\Gamma)$ then $\frac{\partial}{\partial \nu_A} y(u) \in H^{-1}(\Gamma)$, the mapping $u \to \frac{\partial}{\partial \nu_A} y(u)$ being a continuous affine map of $L^2(\Gamma) \to H^{-1}(\Gamma)$.*

The cost function may be taken as

$$J(v) = \left\| \frac{\partial y}{\partial \nu_A}(u) - z_d \right\|^2_{H^{-1}(\Gamma)} + (Nv, v)_{\mathcal{U}};^{31} \tag{5.4}$$

$N: \mathcal{U} \to \mathcal{U}$ is a positive operator.
The general theory is applicable here; there exists a unique optimal control u which is characterized by

$$\left(\frac{\partial y}{\partial \nu_A}(u) - z_d, \frac{\partial y}{\partial \nu_A}(v) - \frac{\partial y}{\partial \nu_A}(u)\right)_{H^{-1}(\Gamma)} + (Nu, v-u)_{\mathcal{U}} \geqslant 0, \quad \forall v \in \mathcal{U}_{\text{ad}}. \tag{5.5}$$

The adjoint state is given by

$$\left.\begin{aligned} &A^* p(u) = 0, \\ &p(u)|_\Gamma = (-\Delta_\Gamma)^{-1}\left(\frac{\partial y}{\partial \nu_A}(u) - z_d\right); \end{aligned}\right\} \tag{5.6}$$

it should be noted that

$$(-\Delta_\Gamma)^{-1}\left(\frac{\partial y}{\partial \nu_A}(u) - z_d\right) \in H^1(\Gamma)$$

and hence $p(u)$ is a regular solution[32]; this is a general property (since y and p are in duality, this is natural): as the regularity of the state diminishes, the regularity of the adjoint increases.

[31] For $\varphi, \psi \in H^{-1}(\Gamma)$, we choose

$$(\varphi, \psi)_{H^{-1}(\Gamma)} = \int_\Gamma ((-\Delta_\Gamma)^{-1}\varphi)\psi \, d\Gamma .$$

[32] More precisely, $p(u) \in H^{\frac{3}{2}}(\Omega)$—cf. Lions-Magenes [1], Chapter 2.

Utilizing (5.6) we may transform (5.5) into the equivalent condition[33]

$$\int_\Gamma \left(\frac{\partial p}{\partial \nu_{A^*}}(u)+Nu\right)(v-u)\,d\Gamma \geqslant 0 \quad \forall v \in \mathscr{U}_{\mathrm{ad}}. \tag{5.7}$$

Example 5.1. Let us take

$$\mathscr{U}_{\mathrm{ad}}=\{v \mid v\geqslant 0 \text{ on } \Gamma,\ v\in L^2(\Gamma)\}. \tag{5.8}$$

The optimal control is then given by the solution of the problem (of unilateral type)

$$\left.\begin{aligned} &Ay=f, \quad A^*p=0 \quad \text{in } \Omega,\\ &(-\Delta_\Gamma)p=\frac{\partial y}{\partial \nu_A}-z_d, \quad y\geqslant 0, \quad \frac{\partial p}{\partial \nu_{A^*}}+Ny\geqslant 0,\\ &y\left(\frac{\partial p}{\partial \nu_{A^*}}+Ny\right)=0 \quad \text{on } \Gamma, \end{aligned}\right\} \tag{5.9}$$

and

$$u=y|_\Gamma. \tag{5.10}$$

Remark 5.1. If the control is given in $L^2(\Gamma)$, the observation $\dfrac{\partial y}{\partial \nu_A}(u)$ *is not in* $L^2(\Gamma)$ *(but only in* $H^{-1}(\Gamma)$*); this result cannot be improved and therefore the introduction of the norm of* $H^{-1}(\Gamma)$ *in* $J(u)$ *(and of the operator* $(-\Delta_\Gamma)$ *in (5.6)) is indispensable.*

5.3. A "Controllability-Like" Problem

Let Ω be an open set with boundary $\Gamma=\Gamma_1\cup\Gamma_2$, the Γ_i's being sufficiently regular varieties of dimension $(n-1)$.

Let us suppose that the state $y(u)$ is given by,

$$Ay(u)=0 \quad \text{in } \Omega, \tag{5.11}$$

$$y(u)|_\Gamma=u, \quad u\in L^2(\Gamma_1), \tag{5.12}$$

$$y(u)|_{\Gamma_2}=0. \tag{5.13}$$

We may then observe (see section 5.2 preceeding) $\dfrac{\partial y}{\partial \nu_A}(u)|_{\Gamma_2}$ which is an element of $H^{-1}(\Gamma_2)$. We have the following property (a "controllability-like" property—cf. Chapter 3, section 10):

[33] Always using the same method: we multiply the first equation in (5.6) by $y(v)-y(u)$ and apply Green's formula.

Theorem 5.1. *As u ranges over* $L^2(\Gamma_1)$, $\dfrac{\partial y}{\partial \nu_A}(u)$ *ranges over a dense subspace of* $H^{-1}(\Gamma_2)$ *(at least in the case where the coefficients of A are sufficiently regular).*

Proof. Let $\varphi \in H^1(\Gamma_2)$ be such that

$$\int_{\Gamma_2} \frac{\partial y}{\partial \nu_A}(u)\varphi \, d\Gamma_2 = 0 \quad \forall u \in L^2(\Gamma_1). \tag{5.14}$$

It suffices to prove that $\varphi = 0$.

We first introduce z as the solution of

$$\left.\begin{aligned} &A^* z = 0, \\ &z|_{\Gamma_1} = 0, \qquad z|_{\Gamma_2} = \varphi. \end{aligned}\right\} \tag{5.15}$$

Then $\int_\Omega (A^* z) y(u) \, dx = 0$. Applying Green's Formula, we obtain,

$$\int_{\Gamma_2} \varphi \frac{\partial y}{\partial \nu_A}(u) \, d\Gamma_2 = \int_{\Gamma_1} \frac{\partial z}{\partial \nu_{A^*}} y(u) \, d\Gamma_1$$

and hence from (5.14) and (5.12),

$$\int_{\Gamma_1} \frac{\partial z}{\partial \nu_{A^*}} u \, d\Gamma_1 = 0 \quad \forall u \in L^2(\Gamma_1).$$

Therefore

$$\frac{\partial z}{\partial \nu_{A^*}} = 0 \quad \text{on } \Gamma_1, \tag{5.16}$$

which when coupled with (5.15) implies $z = 0$ and hence $\varphi = 0$. □

5.4. Pointwise Control and Observation

If $b \in \Gamma$, we denote by $\delta(\gamma - b)$ the Dirac Mass concentrated at b on Γ,

$$\int_\Gamma \delta(\gamma - b)\varphi(\gamma) \, d\Gamma = \varphi(b) \quad \forall \varphi \in C^0(\Gamma). \tag{5.17}$$

We take

$$\mathscr{U} = \mathbb{R}^M, \quad \text{a finite dimensional space} \tag{5.18}$$

(and hence $\mathscr{U}_{\text{ad}}$ = closed convex subset of $\mathscr{U} = \mathbb{R}^M$).

We denote by $y(u)$ the state given as the solution of

$$A\,y(u)=0 \quad \text{in } \Omega, \tag{5.19}$$

$$y(u)|_{\Gamma}=\sum_{i=1}^{N} u_i\delta(\gamma-b_i), \qquad u=\{u_i\}\in\mathbb{R}^M, \qquad b_i\in\Gamma. \tag{5.20}$$

The problem given by (5.19), (5.20) admits a unique solution. In general, the solution $P(x,b)$ of

$$\left.\begin{aligned} &A_x P(x,b)=0,\\ &P(x,b)|_{\Gamma}=\delta(\gamma-b),^{34} \end{aligned}\right\} \tag{5.21}$$

is termed the Poison Kernel.

Then the solution of (5.19), (5.20) is

$$y(u)=\sum_{i=1}^{N} u_i P(x,b_i). \tag{5.22}$$

If we assume that the coefficients of A and the boundary Γ are sufficiently regular, then the function $y(u)$ is sufficiently regular at all points of $\bar{\Omega}$ excepting at the points b_i, and we may make local observations

$$\frac{\partial y}{\partial \nu_A}(c_i;u), \qquad c_i\in\Gamma, \qquad i=1,2,\dots,M', \tag{5.23}$$

provided $c_i\neq b_i\ \forall\, i,j$.

We therefore introduce the cost function

$$J(v)=\sum_{i=1}^{M'}\left|\frac{\partial y}{\partial \nu_A}(c_i;v)-z_d\right|^2+(N\,v,v)_{\mathcal{U}}. \tag{5.24}$$

(N is a positive definite matrix in $\mathbb{R}^M$).

The mapping $u\to\dfrac{\partial y}{\partial \nu_A}(c_i;u)$ is a continuous mapping of $\mathbb{R}^M\to\mathbb{R}$ and hence the general theory is applicable.

The adjoint state is defined by

$$\begin{aligned} &A^*p(u)=0,\\ &p(u)|_{\Gamma}=\sum_{i=1}^{M'}\left(\frac{\partial y}{\partial \nu_A}(c_i;v)-z_d\right)\delta(\gamma-c_i). \end{aligned} \tag{5.25}$$

---

[34] Given, for example, by "duality", that is by (4.13):
$\int_{\Omega} P(x,b)(A^*\psi)\,dx=-\dfrac{\partial\psi}{\partial\nu_{A^*}}(b)\ \forall\psi$ which are test functions vanishing on Γ such that $\dfrac{\partial\psi}{\partial\nu_{A^*}}\in C^0(\Gamma)$.

Applying Green's Formula, we see that the optimal control is given by the simultaneous solution of (5.19), (5.20), (5.25) and

$$\sum_{i=1}^{M} \frac{\partial p}{\partial v_{A^*}}(b_i; u)(v_i - u_i) + (N u, v-u)_{\mathscr{U}} \geqslant 0 \quad \forall v \in \mathscr{U}_{\mathrm{ad}},$$
$$u \in \mathscr{U}_{\mathrm{ad}}. \tag{5.26}$$

6. Constraints on the State

6.1. Orientation

Up to now, the set $\mathscr{U}_{\mathrm{ad}}$ (of constraints) was specified directly. In many problems the set $\mathscr{U}_{\mathrm{ad}}$ is given indirectly via the intermediary of the state. For example, $\mathscr{U}_{\mathrm{ad}}$ could be the set of all $u \in \mathscr{U}$ such that the observation $C y(u)$ belongs to a closed, convex subset of the space of observations K. The problem then is to obtain equations characterizing the optimal control in a manner such that the set K (and not $\mathscr{U}_{\mathrm{ad}}$) enters the equations directly.

We shall give an example of such a situation.

6.2. Control and Constraints on the Boundary

Let us suppose that the state is given by (the notation being the same as in the preceeding section)

$$A y(u) = 0,$$
$$y(u)|_{\Gamma} = 0, \tag{6.1}$$

with

$$\mathscr{U} = L^2(\Gamma), \qquad u \in \mathscr{U}. \tag{6.2}$$

We have seen (cf. section 5.2) that we may define

$$\frac{\partial y}{\partial v_A}(u) \in H^{-1}(\Gamma). \tag{6.3}$$

Let

$$K = \text{closed, convex subset of } H^{-1}(\Gamma). \tag{6.4}$$

We define,

$$\mathscr{U}_{\mathrm{ad}} = \left\{ v \,|\, v \in L^2(\Gamma) = \mathscr{U} \text{ such that } \frac{\partial y}{\partial v_A}(u) \in K \right\}. \tag{6.5}$$

This definition is valid since the mapping $v \to \dfrac{\partial y}{\partial \nu_A}(v)$ being a linear continuous map of $L^2(\Gamma) \to H^{-1}(\Gamma)$, the set $\mathcal{U}_{\text{ad}}$ defined by (6.5) is closed and convex.

We shall consider the cost function

$$J(v) = \left\| \frac{\partial y}{\partial \nu_A}(v) - z_d \right\|^2_{H^{-1}(\Gamma)} + (N v, v)_{\mathcal{U}},$$
$$N \text{ being a positive definite operator mapping } \mathcal{U} \to \mathcal{U}. \tag{6.6}$$

We shall prove

Theorem 6.1. *The unique optimal control u which minimizes $J(v)$ on $\mathcal{U}_{\text{ad}}$ is given by the solution of the following problem:*

$$A y = 0, \quad A^* q = 0 \quad \text{in } \Omega, \tag{6.7}$$

$$\frac{\partial q}{\partial \nu_{A^*}} = N y \quad \text{on } \Gamma, \tag{6.8}$$

$$\left((-\Delta_\Gamma)^{-1} \left(\frac{\partial y}{\partial \nu_A}(u) - z_d \right) + q, g - \frac{\partial y}{\partial \nu_A}(u) \right) \geqslant 0 \;\; \forall g \in K,^{35}$$
$$\frac{\partial y}{\partial \nu_A}(u) \in K. \tag{6.9}$$

Further $$u = y|_\Gamma. \tag{6.10}$$

(We note that the set K enters equation (6.9) directly as desired.)

Proof. 1. The unique optimal control is determined by the condition

$$J'(u) \cdot (v - u) \geqslant 0 \;\; \forall v \in \mathcal{U}_{\text{ad}},$$

which may be written as,

$$\left(\frac{\partial y}{\partial \nu_A}(u) - z_d, \frac{\partial y}{\partial \nu_A}(v) - \frac{\partial y}{\partial \nu_A}(u) \right)_{H^{-1}(\Gamma)} + (N u, v - u)_{\mathcal{U}} \geqslant 0 \;\; \forall v \in \mathcal{U}_{\text{ad}}. \tag{6.11}$$

In order to transform (6.11) so that the set K enters directly, we define $q(u)$ as the solution

$$A^* q(u) = 0 \quad \text{in } \Omega, \tag{6.12}$$

$$\frac{\partial q}{\partial \nu_{A^*}}(u) = N u \in L^2(\Gamma). \tag{6.13}$$

[35] The bracket denotes the scalar product between $H^{-1}(\Gamma)$ and $H^{-1}(\Gamma)$.

(The function $q(u) \in H^1(\Gamma)$.[36])
Multiplying (6.12) by $y(v)-y(u)$ and applying Green's Formula, we obtain

$$-\int_\Gamma \frac{\partial q}{\partial \nu_{A^*}}(u)(v-u)\,d\Gamma + \int_\Gamma q(u)\left(\frac{\partial y}{\partial \nu_A}(v) - \frac{\partial y}{\partial \nu_A}(u)\right)d\Gamma = 0$$

and hence (6.11) is equivalent to

$$\left(\frac{\partial y}{\partial \nu_A}(u) - z_d, \frac{\partial y}{\partial \nu_A}(v) - \frac{\partial y}{\partial \nu_A}(u)\right)_{H^{-1}(\Gamma)} + \int_\Gamma q(u)\left(\frac{\partial y}{\partial \nu_A}(v) - \frac{\partial y}{\partial \nu_A}(u)\right)d\Gamma \geqslant 0, \tag{6.14}$$

$\forall\, v \in \mathscr{U}_{\mathrm{ad}}$.

But, since

$$(\varphi, \psi)_{H^{-1}(\Gamma)} = \int_\Gamma (-\Delta_\Gamma)^{-1}\varphi \cdot \psi\, d\Gamma,$$

(6.14) is further equivalent to

$$\int_\Gamma \left[(-\Delta_\Gamma)^{-1}\left(\frac{\partial y}{\partial \nu_A}(u) - z_d\right) + q(u)\right]\left(\frac{\partial y}{\partial \nu_A}(v) - \frac{\partial y}{\partial \nu_A}(u)\right)d\Gamma \geqslant 0 \tag{6.15}$$

$\forall\, v \in \mathscr{U}_{\mathrm{ad}}$.

2. But the mapping

$$u \to \frac{\partial y}{\partial \nu_A}(u) \tag{6.16}$$

is an isomorphism of $L^2(\Gamma)$ onto $H^{-1}(\Gamma)$ (cf. Lions-Magenes [1], Chapter 2)[37]. Hence for every $g \in K$, there exists $v \in \mathscr{U}_{\mathrm{ad}}$ such that $\frac{\partial y}{\partial \nu_A}(v) = g$, and hence (6.15) is equivalent to (6.9).

Eliminating *u* from (6.13) we obtain the theorem. □

Example 6.1.

$$K = \{g \mid g \geqslant 0 \text{ on } \Gamma\} \quad (\text{hence } g \text{ is a measure} \geqslant 0 \text{ on } \Gamma). \tag{6.17}$$

---

[36] We can even show (cf. Lions-Magenes [1], Chapter 2) that $q(u) \in H^{\frac{3}{2}}(\Omega)$.

[37] The inverse of the mapping (6.16) is given by the solution of the nonhomogeneous Neumann problem:

$$Az = 0, \quad \frac{\partial z}{\partial \nu_A} = \frac{\partial y}{\partial \nu_A}(u) \to z|_\Gamma = u.$$

We then obtain the following conditions equivalent to (6.9):

$$\left.\begin{aligned} \frac{\partial y}{\partial \nu_A} &\geqslant 0, \\ (-\Delta_\Gamma)^{-1}\left(\frac{\partial y}{\partial \nu_A} - z_d\right) + q|_\Gamma &\geqslant 0, \\ \frac{\partial y}{\partial \nu_A}\left[(-\Delta_\Gamma)^{-1}\left(\frac{\partial y}{\partial \nu_A} - z_d\right) + q|_\Gamma\right] &= 0. \end{aligned}\right\} \tag{6.18}$$

The problem given by (6.7), (6.8), (6.9) is again a problem of unilateral type.

Remark 6.1. The following may be verified: suppose that

$$\mathscr{U}_{\mathrm{ad}} = \left\{ v \,\middle|\, \lambda v + \frac{\partial y}{\partial \nu_A}(v) \in K \right\}, \qquad \lambda > 0 \quad \text{given}, \tag{6.19}$$

$J(v)$ remaining unchanged.

We introduce $p(u)$ as the solution of

$$\left.\begin{aligned} &A^* p(u) = 0 \quad \text{in } \Omega, \\ &\lambda p(u) + \frac{\partial p}{\partial \nu_{A^*}}(u) = \lambda(-\Delta_\Gamma)^{-1}\left(\frac{\partial y}{\partial \nu_A}(u) - z_d\right) - N u \quad \text{on } \Gamma. \end{aligned}\right\} \tag{6.20}$$

Then the optimal control is characterized by (6.1), (6.20) and

$$\int_\Gamma \left(\frac{\partial p}{\partial \nu_{A^*}}(u) + N u\right)\left(g - \left(\frac{\partial y}{\partial \nu_A}(u) + \lambda u\right)\right) d\Gamma \geqslant 0 \quad \forall g \in K \tag{6.21}$$

or by the equivalent condition

$$\int_\Gamma \left[(-\Delta_\Gamma)^{-1}\left(\frac{\partial y}{\partial \nu_A}(u) - z_d\right) - p(u)\right]\left[g - \left(\frac{\partial y}{\partial \nu_A}(u) + \lambda u\right)\right] d\Gamma \geqslant 0, \quad \forall g \in K \tag{6.22}$$

(together evidently with

$$\frac{\partial y}{\partial \nu_A}(u) + \lambda u \in K).$$

7. Existence Results for Optimal Controls

7.1. Orientation

In all problems considered up to now, the state of the system was an affine function of the control.

We shall now study by means of examples problems where the state of the system is a non-linear function of control.

Existence theorems for optimal control which were trivial up to now become far less elementary; techniques similar to those used for non-linear boundary value problems have to be used. This is the objective of this section. The following section is devoted to obtaining "first order necessary conditions" of optimality.

7.2. Distributed Control

Let Ω be as in the preceeding section and let A be a second order elliptic operator.

We set

$$\mathscr{U}_+ = \{v | v \in L^\infty(\Omega), v \geqslant 0 \text{ almost everywhere in } \Omega\}, \tag{7.1}$$

$\mathscr{U}_{ad} \subset \mathscr{U}_+$, $\mathscr{U}_{ad}$ = convex set in $L^\infty(\Omega)$, which is closed in the weak topology of the dual of $L^1(\Omega)$[38]—with the possibility of equality $\mathscr{U}_{ad} = \mathscr{U}_+$. (7.2)

For each $v \in \mathscr{U}_+$, we denote by $y(v)$ the solution of

$$A\,y(v) + v\,y(v) = f \quad \text{in } \Omega \tag{7.3}$$

f given, for example, in $L^2(\Omega)$, with

$$y(v)|_\Gamma = 0. \tag{7.4}$$

Remark 7.1. Problem (7.3), (7.4) admits a unique solution: apply Theorem 1.1 with $V = H_0^1(\Omega)$ and

$$a(\varphi, \psi) = \sum_{i,j=1}^{n} \int_\Omega a_{ij} \frac{\partial \varphi}{\partial x_j} \frac{\partial \psi}{\partial x_i} dx + \int_\Omega a_0 \varphi \psi\, dx + \int_\Omega v \varphi \psi\, dx; \tag{7.5}$$

---

[38] In other words, it is the "weak-star topology": if $v_n \in \mathscr{U}_{ad}$ and if

$$\int_\Omega v_n h\, dx \to \int_\Omega v h\, dx \quad \forall h \in L^1(\Omega),$$

then $v \in \mathscr{U}_{ad}$. Note that $\mathscr{U}_+$ has this property.

we have (under the hypotheses of (2.3)):

$$a(\psi,\psi) \geqslant \alpha \|\psi\|^2_{H^1(\Omega)} \tag{7.6}$$

by virtue of the fact that $v \in \mathscr{U}_+$.

Remark 7.2. The solution $y(v)$ belongs to $H_0^1(\Omega)$. If the coefficients and the boundary Γ are sufficiently regular, we even have

$$y(v) \in H^2(\Omega) \tag{7.7}$$

(since from (7.3) we deduce: $A\,y(v) = f - v\,y(v) \in L^2(\Omega)$). In any case the mapping $v \to y(v)$ is non-linear.

Remark 7.3. Everything that follows in this section may be adapted without any modifications to the following situations:
— to boundary conditions other than that given by (7.4), as long as they give rise to a well-posed problem;
— to the case where A is an elliptic operator of order > 2.
Let us take as cost function

$$J(v) = \int_\Omega |y(v) - z_d|^2\,dx + \nu \|v\|_{L^\infty(\Omega)}, \qquad \nu > 0, \qquad z_d \in L^2(\Omega), \tag{7.8}$$

or

$$J(v) = \int_\Gamma \left| \frac{\partial y}{\partial \nu_A}(v) - z_d \right|^2 d\Gamma + \nu \|v\|_{L^\infty(\Omega)}.^{39} \tag{7.9}$$

We shall prove

Theorem 7.1. *Let us assume that* (2.3) *and* (7.2) *hold. If the cost function is defined by* (7.8) *or* (7.9)[40], *there exists at least one optimal control* $u \in \mathscr{U}_{ad}$, *that is, there exists at least one* $\mathscr{U}_{ad}$ *such that*

$$J(u) \leqslant J(v) \quad \forall\, v \in \mathscr{U}_{ad}.$$

Proof. Let v_n be a minimizing sequence, that is $J(v_n) \to \inf_{v \in \mathscr{U}_{ad}} J(v)$, and let $y(v_n) = y_n$. From (7.6) (since $v_n \in \mathscr{U}_+$),

$$\|y_n\|_{H^1(\Omega)} \leqslant \text{constant} \tag{7.10}$$

and from the form of $J(v)$, we have,

$$\|v_n\|_{L^\infty(\Omega)} \leqslant \text{constant}. \tag{7.11}$$

[39] Provided (7.7) is applicable.

[40] Other choices are clearly possible.

Hence, we may extract a subsequence, still denoted by $\{y_n, v_n\}$ such that,

$$y_n \to y \quad \text{in } H^1(\Omega) \quad \text{weakly (or in } H_0^1(\Omega) \text{ weakly)}, \tag{7.12}$$

$$v_n \to v \quad \text{in } L^\infty(\Omega) \quad \text{in the weak topology of the dual of } L^1(\Omega). \tag{7.13}$$

According to (7.2), $v \in \mathscr{U}_{ad}$. The essential point is to show $y = y(v)$. But $A y_n + v_n y_n = f$ and from (7.12) $A y_n \to A y$ in $H^1(\Omega)$. If we can show that

$$v_n y_n \to v y \quad \text{in } \mathscr{D}'(\Omega), \tag{7.14}$$

we would have

$$A y + y v = f, \qquad y \in H_0^1(\Omega)$$

and $y = y(v)$.

To prove (7.14) we have to use compactness arguments. It is known (a variant of Ascoli's theorem due to Rellich—cf. for example Lions-Magenes [1], Chapter 1) that the injection map of $H_0^1(\Omega)$ into $L^2(\Omega)$ is compact if Ω is bounded (and the injection map of $H^1(\Omega)$ into $L^2(\Omega)$ is compact if Ω is bounded with a sufficiently regular boundary). Then (7.12) implies

$$y_n \to y \quad \text{in } L^2(\Omega) \ \textit{strongly} \tag{7.15}$$

and hence if $\varphi \in \mathscr{D}(\Omega)$,

$$\int_\Omega v_n y_n \varphi \, dx = \int_\Omega v_n y \varphi \, dx + \int_\Omega v_n (y_n - y) \varphi \, dx \to \int_\Omega v y \varphi \, dx$$

(whence (7.14)), since

$$\left| \int_\Omega v_n (y_n - y) \varphi \, dx \right| \leqslant \text{constant } \|y_n - y\|_{L^2(\Omega)} \to 0.$$

Hence we have $y = y(v)$ and therefore

$$J(v) \leqslant \liminf J(v_n) = \inf_{v \in \mathscr{U}_{ad}} J(v)$$

and therefore v is an optimal control. □

Remark 7.4. Naturally, if we assume that $\mathscr{U}_{ad}$ is a bounded set in $L^\infty(\Omega)$, then it is unnecessary to have the term $\nu \|v\|_{L^\infty(\Omega)}$ in the cost function $J(v)$.

Remark 7.5. Systems governed by equations of the type (7.3) are often referred to as of "bilinear type" (since $v, y \to v y$ is bilinear!). Existence theorems analogous to that of Theorem 7.1 can be proved for the "non bilinear" case. An example of this will be considered in section 7.4.

Remark 7.6. (Unsolved problem) Let ξ_0, ξ_1 be given elements in $L^\infty(\Omega)$ with

$$0 \leqslant \beta \leqslant \xi_0(x) \leqslant \xi_1(x) \quad \text{almost everywhere in } \Omega, \tag{7.16}$$

and let

$$\mathscr{U}_{ad} = \{v \mid \xi_0(x) \leqslant v(x) \leqslant \xi_1(x) \text{ almost everywhere in } \Omega\}. \tag{7.17}$$

For $v \in \mathscr{U}_{ad}$, set

$$A(v)\cdot\psi = -\sum_{i,j=1}^{n} \frac{\partial}{\partial x_i}\left(a_{ij}(x)v(x)\frac{\partial}{\partial x_j}\psi\right) + a_0(x)\psi \tag{7.18}$$

where the functions a_0, a_{ij} satisfy (2.3).

Let us define the state $y(v)$ of the system as the solution of

$$A(v)\,y(v) = f, \quad f \text{ given in } L^2(\Omega), \quad y(v)|_\Gamma = 0. \tag{7.19}$$

This problem admits a unique solution. In fact, apply Theorem 1.1 with $V = H_0^1(\Omega)$ and $a(\varphi,\psi)$ replaced by

$$a_v(\varphi,\psi) = \sum_{i,j=1}^{n} \int_\Omega a_{ij}(x)v(x)\frac{\partial\varphi}{\partial x_j}\frac{\partial\psi}{\partial x_i}\,dx + \int_\Omega a_0\varphi\psi\,dx. \tag{7.20}$$

From (7.17) and (2.3) we have,

$$a_v(\varphi,\psi) \geqslant \alpha\beta\int_\Omega (\operatorname{grad}\psi)^2\,dx + \alpha\int_\Omega \psi^2\,dx \tag{7.21}$$

whence the result.

Let us now consider the cost function given for example by

$$J(v) = \int_\Omega (y(v) - z_d)^2\,dx. \tag{7.22}$$

It is not known whether there exists a $u \in \mathscr{U}_{ad}$ such that $J(u) \leqslant J(v)$ $\forall\, v \in \mathscr{U}_{ad}$. The difficulty is the following: let v_n be a minimizing sequence and let $y_n = y(v_n)$; by virtue of (7.21),

$$\|y_n\|_{H^1(\Omega)} \leqslant \text{constant}$$

and hence (as in the proof of Theorem 7.1) we may show that $y_n \to y$ weakly in $H^1(\Omega)$ and $v_n \to v$ in $L^\infty(\Omega)$ in the weak topology of the dual of $L^1(\Omega)$.

But this is not sufficient to pass to the limit in the terms

$$\frac{\partial}{\partial x_i}\left(a_{ij}v_n\frac{\partial y_n}{\partial x_j}\right).$$

There appears to be three possibilities for attacking this problem:

(i) write the equation $A(v)\,y(v) = f$ as a system of first order equations and embed the problem in a more general problem; this has been done by L. Cesari [1], in the spirit of the work of L. C. Young [1], McShane [1], Gamkrelidze [1], Warga [1];[41]

(ii) regularize the control: if, for example we assume $\mathscr{U}_{\text{ad}}$ is a convex set of functions satisfying (7.17) and also a closed, bounded subset of $H^1(\Omega)$, there is no difficulty in proving existence, since we may show that $v_n \to v$ strongly in $L^2(\Omega)$;

(iii) perturb the operator $A(v)$: this will be our objective in section 7.3.

7.3. Singular Perturbation of the System

Let

$$\left.\begin{aligned} &b(\varphi,\psi) = \text{continuous, bilinear form on } H_0^2(\Omega),\\ &b(\psi,\psi) \geqslant \beta \|\psi\|^2_{H_0^2(\Omega)}, \quad \beta > 0 \ \forall \psi \in H_0^2(\Omega), \end{aligned}\right\} \tag{7.23}$$

and let $B \in \mathscr{L}(H_0^2(\Omega); H^{-2}(\Omega))$ be the corresponding operator. For example, if

$$b(\varphi,\psi) = \int_\Omega \Delta\varphi \, \Delta\psi \, dx$$

it may be shown that (7.23) holds and hence $B = A^2$. We now assume that the state of the system is given by the solution $y_\varepsilon(v)$ in $H_0^2(\Omega)$ of

$$\left.\begin{aligned} &\varepsilon B y_\varepsilon(v) + A(v) y_\varepsilon(v) = f,\\ &y_\varepsilon(v) \in H_0^2(\Omega) \ \left(\text{that is, } y_\varepsilon(v)|_\Gamma = 0, \ \frac{\partial}{\partial n} y_\varepsilon(v)|_\Gamma = 0\right), \end{aligned}\right\} \tag{7.24}$$

where $\varepsilon > 0$ is "small".

We say that the operator $\varepsilon B + A(v)$ is obtained from $A(v)$ by "singular perturbation".

Let us note that problem (7.24) admits a unique solution since we may apply Theorem 1.1 with $V = H_0^2(\Omega)$ and with $a(\varphi,\psi)$ replaced by

$$\varepsilon b(\varphi,\psi) + a_v(\varphi,\psi) = a_{\varepsilon,v}(\varphi,\psi). \tag{7.25}$$

In fact,

$$a_{\varepsilon,v}(\psi,\psi) \geqslant \varepsilon \|\psi\|^2_{H^2(\Omega)} + c\|\psi\|^2_{H^1(\Omega)}, \quad c > 0. \tag{7.26}$$

The introduction of (7.24) is justified (to a certain extent) by the following results:

Theorem 7.2. *For v fixed in $\mathscr{U}_{\text{ad}}$, we have*

$$y_\varepsilon(v) \to y(v) \quad \text{in } H_0^1(\Omega) \quad \text{as } \varepsilon \to 0. \tag{7.27}$$

Theorem 7.3. *For $\varepsilon > 0$ fixed, there exists optimal control u_ε in $\mathscr{U}_{\text{ad}}$; that is u_ε is such that*

$$\int_\Omega (y_\varepsilon(u_\varepsilon) - z_d)^2 dx \leqslant \int_\Omega (y_\varepsilon(v) - z_d)^2 dx \ \ \forall v \in \mathscr{U}_{\text{ad}}. \tag{7.28}$$

---

[41] For other approaches to "relaxed" optimal control, cf. I. Ekeland [1] and A. D. Ioffe and V. M. Tikomirov [1].

Remark 7.7. Let us set

$$J_\varepsilon(v)=\int_\Omega (y_\varepsilon(v)-z_d)^2\,dx, \quad j_\varepsilon=\inf_{v\in\mathcal{U}_{ad}} J_\varepsilon(v), \quad j=\inf_{v\in\mathcal{U}_{ad}} J(v).$$

From Theorem 7.2, it follows that $\limsup_{\varepsilon\to 0} j_\varepsilon \leqslant j$; *it is however not clear whether* $j_\varepsilon\to j$. *Therefore the introduction of* (7.24) *in place of the original system is partially heuristic.* □

Proof of Theorem 7.2. We are here concerned with a standard result in singular perturbations which is a special case of results in D. Huet [1]. From (7.26) we obtain (we write y_ε, y in place of $y_\varepsilon(v)$, $y(v)$):

$$\sqrt{\varepsilon}\,\|y_\varepsilon\|_{H^2(\Omega)}\leqslant \text{constant}, \qquad \|y_\varepsilon\|_{H^1(\Omega)}\leqslant \text{constant}. \tag{7.29}$$

Therefore after extracting a "subsequence", we may assume that $y_\varepsilon\to z$ weakly in $H_0^1(\Omega)$. But this together with (7.29) is sufficient to pass to the limit in

$$\varepsilon b(y_\varepsilon,\psi)+a_v(y_\varepsilon,\psi)=(f,\psi),$$

whence

$$a_v(z,\psi)=(f,\psi).$$

Therefore $z=y$ and $y_\varepsilon\to y$ weakly in $H_0^1(\Omega)$. But

$$\varepsilon b(y_\varepsilon,y_\varepsilon)+a_v(y_\varepsilon-y,y_\varepsilon-y)=-a_v(y_\varepsilon,y)+a_v(y,y)\to 0,$$

whence (7.27). □

Proof of Theorem 7.3. We suppress the index "ε".

Let v_n be a minimizing sequence and let $y_n=y(v_n)$. Since ε is fixed, we may deduce from (7.26) that

$$\|y_n\|_{H^2(\Omega)}\leqslant \text{constant}.$$

Therefore, after extracting a subsequence, we may assume that $v_n\to v$ in $L^\infty(\Omega)$ in the weak topology of the dual of $L^1(\Omega)$ and that $y_n\to y$ weakly in $H^2(\Omega)$. Hence $\dfrac{\partial y_n}{\partial x_i}\to\dfrac{\partial y}{\partial x_i}$ weakly in $H^1(\Omega)$ and hence strongly in $L^2(\Omega)$.

Therefore, in particular

$$\frac{\partial}{\partial x_i}\left(a_{ij}v_n\frac{\partial y_n}{\partial x_j}\right)\to\frac{\partial}{\partial x_i}\left(a_{ij}v\frac{\partial y}{\partial x_j}\right) \quad \text{in } \mathscr{D}'(\Omega).$$

The proof is now completed as in Theorem 7.1. □

Remark 7.8. The preceding remarks are applicable to systems gouverned by elliptic operators of order $2m$, *where the control enters via derivatives of order* $\leqslant 2m-1$. □

7.4. Boundary Control

Let Ω be an open set with a regular boundary Ω. We take

$$\left.\begin{aligned}&\mathscr{U}_{\text{ad}}=\{v|v\in L^{\infty}(\Gamma), 0<\beta\leqslant\xi_0(x)\leqslant v(x)\leqslant\xi_1(x)\\&\text{almost everywhere on } \Gamma,\\&\xi_0,\xi_1\in L^{\infty}(\Gamma)\}.\end{aligned}\right\} \tag{7.30}$$

We assume that the state $y(v)$ is defined by the solution of

$$A\,y(v)=f, \quad f \text{ given in } L^2(\Omega),^{42} \tag{7.31}$$

together with the boundary condition

$$\frac{\partial y}{\partial \nu_A}(v)+v|y(v)|^{\rho-2}y(v)=0 \quad \text{on } \Gamma, \tag{7.32}$$

where $\rho\geqslant 2$ is fixed.

It is clear that the problem (7.31), (7.32) must be posed more precisely. Let us introduce the space,

$$V=\{\psi|\psi\in H^1(\Omega), \psi|_{\Gamma}\in L^{\rho}(\Gamma)\}^{43} \tag{7.33}$$

endowed with the norm

$$\|\psi\|_{H^1(\Omega)}+c\,\|\psi\|_{L^{\infty}(\Gamma)}, \qquad c>0. \tag{7.34}$$

(We may take $c=0$ if $q\geqslant\rho$, the notation being that of footnote 43). For $\varphi,\psi\in V$ (V being in general a Banach Space), we set

$$a_v(\varphi,\psi) = \sum_{i,j=1}^{n}\int_{\Omega} a_{ij}\frac{\partial\varphi}{\partial x_j}\frac{\partial\psi}{\partial x_i}dx + \int_{\Omega} a_0\varphi\psi\,dx + \int_{\Gamma} v|\varphi|^{\rho-2}\varphi\psi\,d\Gamma. \tag{7.35}$$

Problem (7.31), (7.32) may then be reformulated as:

Find $y(v)\in V$, a solution of

$$a_v(y(v),\psi) = \int_{\Omega} f\psi\,dx \;\; \forall\psi\in V. \tag{7.36}$$

Remark 7.9. Since $\varphi\to a_v(\varphi,\psi)$ is non-linear, problem (7.36) is clearly non-linear.

Theorem 7.4. *For v fixed in $\mathscr{U}_{\text{ad}}$, where $\mathscr{U}_{\text{ad}}$ is given by* (7.30), *if* (2.3) *holds, problem* (7.36) *admits a unique solution.*

Proof. We shall prove the theorem by specializing a general result in the theory of monotone operators.

---

[42] A as usual is given by (2.4) and hypothesis (2.3) is satisfied.

[43] In general if $g\in H^{\frac{1}{2}}(\Gamma)$, then the "fractional" Sobolev embedding theorem on a variety of $(n-1)$ dimensions gives: $g\in L^q(\Gamma)$ with $\frac{1}{q}=\frac{1}{2}-\frac{1}{2(n-1)}$ (if $n\geqslant 3$).

Let us set,

$$a(\varphi,\psi) = \sum_{i,j=1}^{n} \int_\Omega a_{ij} \frac{\partial \varphi}{\partial x_j} \frac{\partial \psi}{\partial x_i} dx + \int_\Omega a_0 \varphi \psi \, dx.$$

We have, $\forall \varphi_1, \varphi_2 \in V$:

$$a_v(\varphi_1, \varphi_1 - \varphi_2) - a_v(\varphi_2, \varphi_1 - \varphi_2) = a(\varphi_1 - \varphi_2, \varphi_1 - \varphi_2) + \int_\Omega v(\beta(\varphi_1) - \beta(\varphi_2))(\varphi_1 - \varphi_2) d\Gamma, \tag{7.37}$$

where $\beta(\varphi) = |\varphi|^{\rho-2} \varphi$.
Hence,

$$(\beta(\varphi_1) - \beta(\varphi_2))(\varphi_1 - \varphi_2) \geqslant 0,$$

and therefore

$$a_v(\varphi_1, \varphi_1 - \varphi_2) - a_v(\varphi_2, \varphi_1 - \varphi_2) \geqslant 0. \tag{7.38}$$

This shows that the operator $V \to V'$ associated with $a(\varphi,\psi)$ is monotone. We even have (strict monotonicity):

$$a_v(\varphi_1, \varphi_1 - \varphi_2) - a_v(\varphi_2, \varphi_1 - \varphi_2) \geqslant \alpha \|\varphi_1 - \varphi_2\|^2_{H^1(\Omega)}. \tag{7.39}$$

The monotonicity property given by (7.38) when combined with coercivity shows

$$\frac{a(\varphi,\varphi)}{\|\varphi\|_V} \to +\infty \quad \text{if } \|\varphi\|_V \to +\infty, \tag{7.40}$$

which in its turn implies the existence of $y(v)$ satisfying (7.36) (cf. for more general results Minty [2], Browder [2], Leray-Lions [1] and Brezis [1]).
Finally uniqueness is an immediate consequence of (7.39). □

Let us consider for example

$$J(v) = \int_\Omega [y(v) - z_d]^2 dx, \quad z_d \text{ given in } L^2(\Omega). \tag{7.41}$$

We have,

Theorem 7.5. *Assume that* (2.3) *holds. Let* $\mathcal{U}_{\text{ad}}$ *be given by* (7.30) *and let* $y(v)$ *be the solution of* (7.36). *If the cost function is given by* (7.41), *there exists at least one optimal control* $u \in \mathcal{U}_{\text{ad}}$.

Proof. Let v_n be a minimizing sequence and let $y_n = y(v_n)$. Since

$$a_v(\varphi,\varphi) \geqslant c\|\varphi\|_V^2, \quad c>0 \text{ independent of } v,$$

we have,

$$\|y_n\|_V \leqslant \text{constant}.$$

Since $\mathcal{U}_{\text{ad}}$ is bounded and closed in the weak topology of the dual of $L^1(\Omega)$, we may assume that $y_n \to y$ weakly in V and $v_n \to v$ in $L^\infty(\Omega)$, in the weak topology of the dual of $L^1(\Omega)$ and further $v \in \mathcal{U}_{\text{ad}}$. Hence,

$$y_n \to y \quad \text{weakly in } L^1(\Omega), \tag{7.42}$$

and

$$y_n|_\Gamma \to y|_\Gamma \quad \text{weakly in } L^\rho(\Gamma),\ \beta(y_n)\to\beta(y) \quad \text{weakly in } L^{\rho'}(\Gamma),$$
$$\frac{1}{\rho}+\frac{1}{\rho'}=1. \tag{7.43}$$

But from (7.42) and the trace theorem (Chapter 1, Section 3),

$$y_n|_\Gamma \to y|_\Gamma \quad \text{weakly in } H^{\frac{1}{2}}(\Gamma). \tag{7.44}$$

However the injection map of $H^{\frac{1}{2}}(\Gamma)$ into $L^2(\Gamma)$ is compact. Hence (7.44) implies,

$$y_n|_\Gamma \to y|_\Gamma \quad \text{strongly in } L^2(\Gamma), \tag{7.45}$$

and therefore we can assume that

$$y_n|_\Gamma \to y|_\Gamma \quad \text{almost everywhere on } \Gamma. \tag{7.46}$$

But (7.46) combined with (7.43) implies $\beta(y_n)\to\beta(y)$ strongly in $L^{\rho-\varepsilon}(\Gamma)$ $(\varepsilon>0)$ and hence

$$\int_\Gamma v_n\beta(y_n)\psi\,d\Gamma \to \int_\Gamma v\beta(y)\psi\,d\Gamma \quad \forall\psi\in V.$$

Hence,

$$a_v(y,\psi)=\int_\Omega f\psi\,dx \quad \forall\psi\in V,$$

which shows that $y=y(v)$.

Therefore

$$J(v)\leqslant \liminf J(v_n) = \inf_{v\in\mathscr{U}_{\text{ad}}} J(v)$$

and v is an optimal control.

7.5. Control of Systems Governed by Unilateral Problems

We have seen that problems of optimal control of systems governed by linear elliptic equations lead to unilateral problems. We may also formulate problems of optimal control of systems which themselves are governed by unilateral problems. In this direction (largely unexplored), we shall present an existence theorem for a particular example.

Let A be a second order elliptic operator. Let us define,

$$K=\{v|v\in H^1(\Omega),\ v\geqslant 0 \text{ almost everywhere on } \Gamma\}. \tag{7.47}$$

Let

$$\mathscr{U}=L^2(\Gamma), \qquad \mathscr{U}_{\text{ad}}\subset\mathscr{U}, \tag{7.48}$$

and assume that the state $y(v)$ is the solution in K of

$$a(y(v),k-y(v))\geqslant\int_\Gamma v(k-y(v))\,d\Gamma, \quad \forall k\in K. \tag{7.49}$$

In other words,

$$\left.\begin{aligned} A\,y(v)&=0 \quad \text{in } \Omega, \\ y(v)&\geqslant 0 \quad \text{almost everywhere on } \Gamma, \\ \frac{\partial y}{\partial \nu_A}(v)&\geqslant v \quad \text{almost everywhere on } \Gamma, \\ \left(\frac{\partial y}{\partial \nu_A}(v)-v\right) y(v)&=0 \quad \text{almost everywhere on } \Gamma. \end{aligned}\right\} \tag{7.50}$$

Let us consider, for example, the cost function

$$\left.\begin{aligned} &J(v)=\int_\Omega (y(v)-z_d)^2\,dx+(N\,v,v)_{\mathscr{U}}, \\ &N\in\mathscr{L}(\mathscr{U};\mathscr{U}), \qquad N\geqslant \nu\times(\text{Identity}), \qquad \nu>0. \end{aligned}\right\} \tag{7.51}$$

We then have,

Theorem 7.6. *If the state $y(v)$ and the cost function $J(v)$ are defined by* (7.49) *and* (7.51) *respectively, then there exists at least one optimal control.*

Proof. Let v_n be a minimizing sequence and let $y(v_n)=y_n$. Since $N\geqslant\nu\times$ (Identity), $\nu>0$, v_n belongs to a bounded set in $L^2(\Gamma)$ and hence y_n belongs to a bounded set in $H^1(\Omega)$. We remark that the mapping

$$\psi\to\psi|_\Gamma$$

of $H^1(\Omega)\to L^2(\Gamma)$ is compact. Hence we may extract a subsequence, again denoted by $\{v_n, y_n\}$ from the sequence $\{v_n, y_n\}$ such that

$$\begin{aligned} y_n&\to y \quad &&\text{weakly in } H^1(\Omega), \quad y\in K, \\ y_n|_\Gamma&\to y|_\Gamma \quad &&\text{strongly in } L^2(\Gamma), \\ v_n&\to v \quad &&\text{weakly in } H^1(\Omega), \\ v_n|_\Gamma&\to v|_\Gamma \quad &&\text{strongly in } L^2(\Gamma). \end{aligned}$$

Hence,

$$\int_\Gamma v_n\,y(v_n)\,d\Gamma\to\int_\Gamma v\,y\,d\Gamma,$$

and since

$$\liminf a(y_n, y_n)\geqslant a(y,y),$$

we may deduce from (7.49) for $v=v_n$ that:

$$a(y,k-y)\geqslant\int_\Gamma v(k-y)\,d\Gamma \quad \forall\,k\in K,\ y\in K,$$

and hence $y=y(v)$. Therefore $\liminf J(v_n)\geqslant J(v)$ and hence v is an optimal control. ▯

8. First Order Necessary Conditions

8.1. Statement of the Theorem

Let us assume

$$\left.\begin{array}{l}\mathscr{U}_{\mathrm{ad}}=\{v \mid x\to v(x) \text{ is a measurable function of } \Omega\to\mathbb{R}^M \\ \text{such that } v(x)\in K \text{ almost everywhere, where } K \text{ is a closed} \\ \text{convex subset of } \mathbb{R}^M\}.\end{array}\right\} \tag{8.1}$$

Let A be given by (2.4) and let (2.3) hold. Let

$$\begin{array}{l} x,\lambda\to b(x,\lambda) \quad \text{be a continuous function from } \bar{\Omega}\times K\to\mathbb{R} \\ \text{which is bounded with} \quad 0<\beta_0\leqslant b(x,\lambda)\leqslant\beta_1 \quad \forall x\in\bar{\Omega},\ \forall\lambda\in K. \end{array} \tag{8.2}$$

For $v\in\mathscr{U}_{\mathrm{ad}}$ let $b(x,v)$ be the function $x\to b(x,v(x))$ and let $A(v)$ be the operator defined by

$$A(v)\psi=A\psi+b(x,v)\psi \quad \forall\psi\in H^1(\Omega). \tag{8.3}$$

Further, let

$$\begin{array}{l} x,\lambda\to f(x,\lambda) \quad \text{be a continuous function from } \bar{\Omega}\times K\to\mathbb{R} \\ \text{which is bounded.} \end{array} \tag{8.4}$$

For given $v\in\mathscr{U}_{\mathrm{ad}}$, we assume that the state $y(v)$ of the system to be controlled is given by

$$A(v)\,y(v)=f(x,v), \tag{8.5}$$

$$y(v)|_{\Gamma}=0 \quad (\text{that is, } y(v)\in H_0^1(\Omega)). \tag{8.6}$$

In view of (8.2), the problem (8.5), (8.6) admits a unique solution.

Let us assume that the cost function is given by

$$J(v)=\int_{\Omega} h\,y(v)\,dx, \quad h \text{ given in } L^2(\Omega). \tag{8.7}$$

The operator $A^*(v)$ is given by

$$A^*(v)\psi=A^*\psi+b(x,v)\psi. \tag{8.8}$$

The adjoint state $p(v)$ is defined by

$$A^*(v)\,p(v)=h, \qquad p(v)\in H_0^1(\Omega). \tag{8.9}$$

We shall prove,

Theorem 8.1. *Let us assume that* (8.1), (8.2), (8.4) *hold and let* $A(v)$ *be given by* (8.3) *with* (2.3) *being satisfied. If the cost function is given by* (8.7), *assuming that the optimal control u exists, u satisfies*

$$\begin{aligned} &p(x;u(x))[f(x;u(x))-b(x,u(x))y(x;u(x))] \\ &=\inf_{k\in K} p(x;u(x))[f(x,k)-b(x,k)y(x;u(x))]. \end{aligned} \tag{8.10}$$

Remark 8.1. If $h\geqslant 0$ in Ω (that is, h is a positive measure in Ω), then from the maximum principle for elliptic equations[44],

$$p(x;u(x))>0,$$

and hence (8.10) reduces to

$$f(x,u(x))-b(x,u(x))y(x;u(x))=\inf_{k\in K}[f(x,k)-b(x,k)y(x;u(x))]. \tag{8.11}$$

8.2. Proof of the Theorem

8.2.1. "Algebraic" Transformation

Let u be an optimal control. In this proof, set

$$(f,g)=\int_\Omega f(x)g(x)dx.$$

We have,

$$J(u)\leqslant J(v)\quad \forall v\in\mathscr{U}_{\mathrm{ad}}.$$

Let,

$$(h,y(v)-y(u))\geqslant 0\quad \forall v\in\mathscr{U}_{\mathrm{ad}}. \tag{8.12}$$

But from (8.9) with $(v=u)$

$$\begin{aligned} (h,y(v)-y(u)) &=(p(u),A(u)y(v)-A(u)y(u)) \\ &=(p(u),A(v)y(v)-A(u)y(u))+(p(u),(A(u)-A(v))y(v)). \\ &=(p(u),f(v)-f(u))+(p(u),(A(u)-A(v))y(u)) \\ &\quad +(p(u),(A(u)-A(v))(y(v)-y(u))), \end{aligned}$$

and hence (8.12) is equivalent to

$$\left.\begin{aligned} &(p(u),f(v)-f(u))-(p(u),(A(v)-A(u))y(u)) \\ &\quad -(p(u),(A(v)-A(u))(y(v)-y(u)))\geqslant 0\quad \forall v\in\mathscr{U}_{\mathrm{ad}}. \end{aligned}\right\} \tag{8.13}$$

8.2.2. General Remarks on the Utilization of (8.13)

We have made no differentiability hypotheses on $f(x,\lambda)$, $b(x,\lambda)$ and hence we cannot replace v by $u+\theta(v-u)$, $\theta\in\,]0,1[$ in (8.13), divide by θ and let θ tend to zero.

---

[44] Not Pontryagin's Maximum Principle!

The basic idea, which is well known in the Calculus of Variations, is to choose

$$\begin{aligned} &v=(1-\chi_j)u+\chi_j k, \quad k\in K, \quad \text{where}, \\ &\chi_j = \text{characteristic function of a cube centered at } x_0\in K. \end{aligned} \tag{8.14}$$

Dividing by $\int_\Omega \chi_j dx$, we deduce from (8.13),

$$b_j - c_j - d_j \geqslant 0,$$

where

$$\left.\begin{aligned} b_j &= \frac{1}{\int_\Omega \chi_j dx}(p(u), f(v_j)-f(u)), \\ c_j &= \frac{1}{\int_\Omega \chi_j dx}(p(u), (A(v_j)-A(u))\,y(u)), \\ d_j &= \frac{1}{\int_\Omega \chi_j dx}(p(u), (A(v_j)-A(u))(y(v_j)-y(u))). \end{aligned}\right\} \tag{8.15}$$

Let E be the set of Lebesgue points in Ω of the functions $x\to p(x,u(x))$, $f(x,u(x))$, $b(x,u(x))$, $y(x,u(x))$ and let $x_0\in E$. Then as $j\to\infty$,

$$b_j \to p(x_0,u(x_0))\,[f(x_0,k)-f(x_0,u(x_0))],$$

and

$$c_j \to p(x_0,u(x_0))\,[b(x_0,k)-b(x_0,u(x_0))]\,y(x_0,u(x_0)).$$

The theorem will be proved (since E is of measure zero) if we can show that $d_j\to 0$, d_j being given as in (8.15).

8.2.3. Proof that $d_j \to 0$

Using a result of Stampacchia [2] (Theorem 4.2, p. 415), we have,

$$\|y(v)\|_{L^\infty(\Omega)} \leqslant c = \text{constant independent of } v\in\mathscr{U}_{\text{ad}}. \tag{8.16}$$

We have the same result for $p(v)$. Hence setting $\mu_j = \int_\Omega \chi_j dx$, σ_j = support of χ_j,

$$|d_j| \leqslant \frac{c}{\mu_j}\int_{\sigma_j} |b(v_j)-b(u)|\,|y(v_j)-y(u)|\,dx,$$

(c denoting various constants), and therefore,

$$|d_j| \leqslant \frac{c}{(\mu_j)^{\frac{1}{2}}}\left(\int_{\sigma_j} |y(v_j)-y(u)|^2\,dx\right)^{\frac{1}{2}}. \tag{8.17}$$

Set

$$\psi_j = y(v_j) - y(u). \tag{8.18}$$

We have,

$$A(u)\psi_j = f(v_j) - f(u) - (b(v_j) - b(u))\, y(v_j) = g_j.$$

Hence,

$$a(\psi_j, \psi_j) + \int_\Omega b(x,u)\psi_j^2\, dx = \int_\Omega g_j \psi\, dx,$$

whence,

$$\|\psi_j\|_{H^1(\Omega)} \leqslant c(\|f(v_j) - f(u)\|_{L^2(\Omega)} + \|(b(v_j) - b(u))\, y(v_j)\|_{L^2(\Omega)}). \tag{8.19}$$

In view of (8.16), the last term in (8.19) $\leqslant c\|b(v_j) - b(u)\|_{L^2(\Omega)}$, and hence from (8.2), (8.4) we deduce

$$\|\psi_j\|_{H^1(\Omega)} \leqslant c(\mu_j)^{\frac{1}{2}}. \tag{8.20}$$

From Sobolev's theorem [1], there exists a $\gamma > 2$ such that

$$\|\psi_j\|_{L^\gamma(\Omega)} \leqslant c(\mu_j)^{\frac{1}{2}}. \tag{8.21}$$

But Holder's Inequality applied to (8.12) gives us,

$$|d_j| \leqslant \frac{c}{(\mu_j)^{\frac{1}{2}}} \|\psi_j\|_{L^\gamma(\Omega)} \left(\int_{\sigma_j} dx \right)^{\frac{1}{2}s}, \quad \frac{1}{s} = 1 - \frac{2}{\gamma},$$

and hence

$$|d_j| \leqslant c(\mu_j)^{\frac{1}{2}s}.$$

Hence the result. □

Remark 8.2. In obtaining (8.16), the fact that A is of second order is used in an essential way.

Notes

The method of proof of Theorem 2.1 is well known in the Calculus of Variations—cf. in particular M. Hestenes [1]. We have attempted to demonstrate by means of examples the diverse situations that may arise. For a study of the general case where the system is governed by an elliptic differential operator of arbitrary order and general boundary conditions, cf. Lions [1], note II.

Formulation (1.31) may be considered as an analogue of the "Maximum Principle of Pontryagin". When the system is governed by ordinary differential equations, these results are documented in Pontryagin-Boltyanski-Gamkrelidze-Mischenko [1].

In the case of systems governed by ordinary differential equations, the "Bang-Bang" property has given rise to extensive developments. The "elliptic" case apparently has not been considered earlier. In the case considered in the present text, the result is a consequence of a classical but delicate theorem—uniqueness of the Cauchy problem (cf. proof of Lemma 4.1). The same situation will arise for evolution equations. For considerations on the uniqueness of the Cauchy problem for elliptic equations, the reader may consult Aronszajn [1], Cordes [1], Heinz [1], Landis [1], Muller [1], Pederson [1].

According to Theorem 5.1, for g given in $H^{-1}(\Gamma_2)$, there exists a u such that $\frac{\partial y}{\partial \nu_A}(u)$ is arbitrarily near (in the sense of $H^{-1}(\Gamma_2)$) of g. For a study on numerical approximation for problems of this type, cf. Lattes-Lions [1]. The proof of the fact that $\frac{\partial}{\partial \nu_A} y(u) \in H^{-1}(\Gamma)$ if $u \in L^2(\Gamma)$ (cf. section 5.2) has not been given in the text. More general results of this type may be found in Lions-Magenes [1], Chapter 2. Another method, which is less general, but uses more elementary arguments is given in *J.* Necas [1] (utilizing an identity due to Rellich).

A problem of the type considered in Remark 7.6 has been studied (in relationship to a physical problem) by Lurie. This author considers a system governed by two second order elliptic operators and he assumes that an optimal control exists (proof of this fact seems lacking). He then derives necessary conditions similar to those derived in Section 8.

Results more general than those in section 7.4 (in the parabolic case—the arguments may be adapted to the elliptic case) may be found in Lions [2].

Theorem 8.1 is a result of the type of the "Maximum Principle of Pontryagin". As we have indicated in the text it is certainly very difficult to obtain results of this type in cases which are more general—operators of order >2, system of operators. Sometimes proofs are given for these more general cases which make use of certain regularity hypotheses. But no proofs of these regularity hypotheses are known. These results should be viewed with suspicion. Axiomatizing the proof of Theorem 8.1 is obviously trivial.

Problems of the Calculus of Variations in which the "control" is exercised via the coefficients have been considered by C. Pucci [1], [2].

CHAPTER III

Control of Systems Governed by Parabolic Partial Differential Equations

1. Equations of Evolution

1.1. Data

As in Chapter 2, section 1, we are given Hilbert Spaces V and H. We use the same notation as in the previous chapter — V' is the dual of V, H' is identified with H and hence $V \subset H \subset V'$.

The variable t denotes 'time'. We assume $t \in]0, T[, T < \infty$; we shall sometimes consider the case where $T \to +\infty$.

A family of bi-linear forms on V is given:

$$\varphi, \psi \to a(t; \varphi, \psi) \quad \text{for each } t \in]0, T[.$$

Relative to this family we assume,

$$\begin{aligned}&\forall \varphi, \psi \in V \quad \text{the function } t \to a(t; \varphi, \psi) \quad \text{is measurable on} \\ &]0, T[\quad \text{and } |a(t; \varphi, \psi)| \leqslant c\|\varphi\| \cdot \|\psi\|,\end{aligned} \tag{1.1}$$

and

$$\begin{aligned}&\text{there exists a } \lambda \text{ such that}^1 \\ &a(t; \varphi, \psi) + \lambda|\varphi|^2 \geqslant \alpha\|\varphi\|^2, \quad \alpha > 0, \quad \forall \varphi \in V, \quad t \in]0, T[.\end{aligned} \tag{1.2}$$

For each t, we may write (as in section 1.7, Chapter 2),

$$a(t; \varphi, \psi) = (A(t)\varphi, \psi), \quad A(t)\varphi \in V', \tag{1.3}$$

the bracket denoting the scalar product between V and V'.

We now introduce the space $L^2(0, T; V)$ (as in Chapter 2, section 2.5 with $S =]0, T[, E = V$) such that

$$\left(\int_0^T \|f(t)\|^2 dt\right)^{\frac{1}{2}} < \infty.$$

---

[1] $|\varphi| = \|\varphi\|_H$, $\|\varphi\| = \|\varphi\|_V$.

In the same manner we define $L^2(0,T;V')$ and we note,

$$A(t)\in\mathscr{L}(L^2(0,T;V);L^2(0,T;V')), \tag{1.4}$$

that is, if $f\in L^2(0,T;V)$, $A(t)f$ is the function

$$t\to A(t)f(t)\in V'.$$

It may be verified that this function is measurable and satisfies

$$\|A(t)f(t)\|_{V'}\leqslant c\|f(t)\|_V \quad \text{(from (1.1))},$$

whence (1.4).

If $f\in L^2(0,T;V)$, we may define its derivative $\frac{df}{dt}$ in the following way: we first define

$$\mathscr{D}'(]0,T[;V)=\mathscr{L}(\mathscr{D}(]0,T[);V), \tag{1.5}$$

the space of distributions on $]0,T[$ with values in V (cf. L. Schwartz [3]). Then if $f\in\mathscr{D}'(]0,T[;V)$, for every $\varphi\in\mathscr{D}(]0,T[)$, $f(\varphi)\in V$ and $\varphi\to f(\varphi)$ is a continuous map of $\mathscr{D}(]0,T[)\to V$.

We shall write (writing distributions as functions),

$$f(\varphi)=\int_0^T f(t)\varphi(t)\,dt, \tag{1.6}$$

where the integral is a generalized integral with values in V.

The derivative $\frac{df}{dt}$ may be defined by

$$\varphi\to\frac{df}{dt}(\varphi)=-f\left(\frac{d\varphi}{dt}\right).$$

This formula defines a continuous linear map from $\mathscr{D}(]0,T[)\to V$. Hence

$$\frac{df}{dt}\in\mathscr{D}'(]0,T[;V).$$

We say that $f_n\to f$ in $\mathscr{D}'(]0,T[;V)$ if $f_n(\varphi)\to f(\varphi)$ $\forall\varphi\in\mathscr{D}(]0,T[)$. Hence

$$\frac{df_n}{dt}\to\frac{df}{dt} \quad \text{in } \mathscr{D}'(]0,T[;V).$$

If $f\in L^2(0,T;V)$, we define $f(\varphi)$ by (1.6) (usual Lebesgue integral with values in V) and $\varphi\to\tilde{f}(\varphi)$ is a continuous map of $\mathscr{D}(]0,T[)\to V$. In this manner we define $\tilde{f}\in\mathscr{D}'(]0,T[;V)$ and a linear map $f\to\tilde{f}$ of

$$L^2(0,T;V)\to\mathscr{D}'(]0,T[;V).$$

This linear map is a continuous injection. We may thus identity $\tilde{f}$ with f and we have

$$L^2(0,T;V) \subset \mathscr{D}'(]0,T[;V). \tag{1.7}$$

Then for $f \in L^2(0,T;V)$ we may define

$$\frac{df}{dt} \in \mathscr{D}'(]0,T[;V).$$

We then introduce,

$$W(0,T) = \left\{ f \mid f \in L^2(0,T;V), \frac{df}{dt} \in L^2(0,T;V') \right\}. \tag{1.8}$$

Endowed with the norm

$$\|f\|_{W(0,T)} = \left(\int_0^T \|f(t)\|^2 \, dt + \int_0^T \left\| \frac{df}{dt} \right\|_{V'}^2 dt \right)^{\frac{1}{2}} \tag{1.9}$$

$W(0,T)$ may be verified to be a Hilbert Space.

The following result may be proved (cf. Lions-Magenes, Chapter 1).

Theorem 1.1. *All functions $f \in W(0,T)$ are, with eventual modification on a set of measure zero, continuous from $[0,T] \to H$.*

Abbreviating, we shall write

$$W(0,T) \subset \mathscr{C}^0([0,T];H) = \text{space of continuous functions from } [0,T] \to H. \tag{1.10}$$

1.2. Evolution Problems

We now consider an evolution problem: find $y \in W(0,T)$ such that

$$A(t)y + \frac{dy}{dt} = f, \quad f \text{ given in } L^2(0,T;V), \tag{1.11}$$

and

$$y(0) = y_0, \quad y_0 \text{ given in } H.^2 \tag{1.12}$$

Examples of such problems are given in section 1.5. We shall first prove

Theorem 1.2. *Assuming that* (1.1), (1.2) *hold, problem* (1.11), (1.12) *admits a unique solution in* $W(0,T)$.

---

[2] From Theorem 1.1, this has a precise meaning.

Furthermore the solution *depends continuously on the data:* the *bilinear* map,

$$f, y_0 \to y$$

is *continuous* from $L^2(0,T;V') \times H \to W(0,T)$.

Let us remark that

$$\text{if } T<\infty, \text{ we may always reduce the problem to a case where (1.2) holds with } \lambda=0. \tag{1.13}$$

In fact, if we set

$$y=\exp(kt)z,$$

problem (1.11), (1.12) is equivalent to

$$(A(t)+kI)z + \frac{dz}{dt} = \exp(-kt)f, \qquad z(0)=y_0,$$

and hence we have replaced $A(t)$ by $A(t)+kI$. Choosing $k=\lambda$, (1.13) is obtained.

We shall now prove Theorem 1.2 assuming (1.13) holds.

1.3. Proof of Uniqueness

Let y satisfy (1.11), (1.12) with $f=0$, $y_0=0$. In (1.11), let us take the scalar product with $y(t)$ under the duality between V and V'. We obtain

$$a(t;y(t),y(t)) + \left(\frac{dy(t)}{dt}, y(t)\right)=0.$$

But we may verify (in fact while proving Theorem 1.1) that,

$$\int_0^T \left(\frac{d}{dt}y(t),y(t)\right)dt = \tfrac{1}{2}|y(T)|^2 \quad (\text{since } y(0)=0),$$

and hence

$$\int_0^T a(t;y(t),y(t))\,dt + \tfrac{1}{2}|y(T)|^2=0.$$

Therefore using (1.13)

$$\alpha \int_0^T \|y(t)\|^2\,dt + \tfrac{1}{2}|y(T)|^2 \leqslant 0,$$

and hence $y=0$.

1.4. Proof of Existence

To simplify the exposition somewhat, let us assume that V is *separable*. Therefore there exists a countable set which is dense in V. We may then find (in an infinity of ways) a "basis" $w_1, w_2, \ldots, w_m, \ldots$ in V in the following sense:

$$\begin{aligned}&\forall m, w_1, \ldots, w_m \text{ are linearly independent and the linear}\\ &\text{combinations } \sum_{\text{finite}} \xi_j w_j,\ \xi_j \in \mathbb{R}, \text{ are dense in } V.\end{aligned} \tag{1.14}$$

Let us define an "approximate solution" of (1.11), (1.12) by

$$y_m(t) = \sum_{i=1}^{m} g_{im}(t) w_i, \tag{1.15}$$

where the g_{im} are chosen such that

$$\left(\frac{d}{dt} y_m(t), w_j\right) + a(t; y_m(t), w_j) = (y(t), w_j), \quad 1 \leqslant j \leqslant m \tag{1.16}$$

and

$$y_m(0) = y_{0m} = \sum_{i=1}^{m} \xi_{im} w_i, \quad \sum_{i=1}^{m} \xi_{im} w_i \to y_0 \text{ in } H \text{ as } m \to \infty. \tag{1.17}$$

System (1.16), (1.17) is a system of m linear differential equations in $g_{im}(t)$ of the form

$$\mathscr{W}_m \frac{d\boldsymbol{g}_m}{dt} + \mathscr{A}_m(t) \boldsymbol{g}_m = \boldsymbol{f}_m, \quad \boldsymbol{g}_m(0) = \{\xi_{im}\}$$

where

$$\mathscr{W}_m = \|(w_i, w_j)\|, \quad \mathscr{A}_m(t) = \|a(t; w_i, w_j)\|,$$
$$\boldsymbol{g}_m(t) = \{g_{im}(t)\}, \quad \boldsymbol{f}_m(t) = \{(f(t), w_j)\}.$$

Since $\det \mathscr{W}_m \neq 0$, *problem* (1.16), (1.17) *admits a unique solution.* We shall show that as $m \to \infty$, $y_m \to y$, y being a solution of (1.11), (1.12)[3]. Multiplying (1.16) by $g_{jm}(t)$ and summing over j, we obtain

$$\left(\frac{d}{dt} y_m(t), y_m(t)\right) + a(t; y_m(t), y_m(t)) = (f(t), y_m(t)),$$

that is,

$$\frac{1}{2} \frac{d}{dt} |y_m(t)|^2 + a(t; y_m(t), y_m(t)) = (f(t), y_m(t)),$$

[3] Let us remark that once a "basis" $w_1 \ldots w_m \ldots$ is chosen this proof is constructive.

whence using (1.13),

$$|y_m(T)|^2+2\alpha\int_0^T \|y_m(t)\|^2\,dt \leqslant |y_{0m}|^2+2\int_0^T |(f(t),y_m(t))|\,dt$$

$$\leqslant |y_{0m}|^2+2\int_0^T \|f(t)\|_{V'}\,\|y_m(t)\|\,dt$$

$$\leqslant |y_{0m}|^2+\int_0^T \|y_m(t)\|^2\,dt+\frac{1}{\alpha}\int_0^T \|f(t)\|_{V'}^2\,dt.$$

Finally (since $|y_{0m}|\geqslant c|y_0|$, cf. (1.17)),

$$\int_0^T \|y_m(t)\|^2\,dt \leqslant c\left(|y_0|^2+\int_0^T \|f(t)\|_{V'}^2\,dt\right). \tag{1.18}$$

Therefore y_m ranges in a bounded set in $L^2(0,T;V)$ and we may extract a subsequence y_μ such that

$$y_\mu \to z \quad \text{weakly in } L^2(0,T;V). \tag{1.19}$$

Let j be fixed but arbitrary and let $\mu>j$. Then (1.16) is valid with $m=\mu$. Multiply both sides of (1.16) by $\varphi(t)$ where

$$\varphi(t)\in C^1[0,T], \qquad \varphi(T)=0, \tag{1.20}$$

and integrate over $(0,T)$. Setting $\varphi_j(t)=\varphi(t)w_j$, we have,

$$\int_0^T [-(y_\mu(t),\varphi_j'(t))+a(t;y_\mu(t),\varphi_j(t))]\,dt=\int_0^T (f(t),\varphi_j(t))\,dt+(y_{0\mu},\varphi_j(0)). \tag{1.21}$$

By virtue of (1.19), we may pass to the limit in (1.21). We then have,

$$\int_0^T [-(z,\varphi_j')+a(t;z,\varphi_j)]\,dt=\int_0^T (f,\varphi_j)\,dt+(y_0,\varphi_j(0)). \tag{1.22}$$

But the above is true for any φ satisfying (1.20). Therefore we may take $\varphi\in\mathscr{D}(]0,T[)$ and hence (1.22) gives

$$\frac{d}{dt}(z(t),w_j)+a(t;z(t),w_j)=(f(t),w_j) \tag{1.23}$$

where the derivative is taken in $\mathscr{D}'(]0,T[)$.

But in (1.23) j is arbitrary and since finite linear combinations of w_j are dense in V, we deduce

$$\frac{dz}{dt} + A(t)z = f. \tag{1.24}$$

Therefore,

$$\frac{dz}{dt} = f - A(t)z \in L^2(0,T;V') \quad \text{and hence} \quad z \in W(0,T).$$

This enables us to integrate by parts in t. Then taking into account (1.24) we obtain

$$(z(0), w_j)\varphi(0) = (y_0, w_j)\varphi(0) \quad \forall j, \ \forall \varphi.$$

Hence $(z(0), w_j) = (y_0, w_j)$ $\forall j$ and thus $z(0) = y_0$. Hence z is a solution and using (1.3) $z = y$ is a solution. We may then replace (1.19) by

$$y_m \to y \quad \text{weakly in } L^2(0,T;V)^{4} \tag{1.25}$$

The estimate (1.18) gives us

$$\int_0^T \|y(t)\|^2 dt \leqslant c\left(|y_0|^2 + \int_0^T \|f(t)\|_{V'}^2 dt\right). \tag{1.26}$$

Further, since $\frac{dy}{dt} = f - A(t)y$, we have with (1.26)

$$\left\|\frac{dy}{dt}\right\|^2_{L^2(0,T;V')} \leqslant c'\left(|y_0|^2 + \int_0^T \|f(t)\|_{V'}^2 dt\right).$$

1.5. Some Examples

We shall use the following notation:

$$\left.\begin{array}{l} Q = Q_T = \Omega \times]0,T[, \quad \Omega \text{ an open subset of } \mathbb{R}^n; \\ \Sigma = \Sigma_T = \Gamma \times]0,T[, \\ \Gamma = \text{boundary of } \Omega, \ \Sigma = \text{lateral boundary of } Q. \end{array}\right\} \tag{1.27}$$

[4] In fact, we may go further: $y_m \to y$ strongly in $L^2(0,T;V)$. To see this,

$$\int_0^T a(t; y_m - y, y_m - y)dt + \tfrac{1}{2}|y_m(T) - y(T)|^2$$

$$= \int_0^T (f, y_m)dt - \int_0^T a(t; y_m, y_m)dt - \tfrac{1}{2}(y_m(T), y(T))$$

$$- \int_0^T a(t; y, y_m - y)dt - \tfrac{1}{2}(y(T), y_m(T) - y(T)) \to 0.$$

Example 1.1. Let a_{ij} be given functions in $\Omega \times]0,T[=Q$ with

$$\left.\begin{aligned} &a_{ij} \in L^\infty(Q), \\ &\sum_{i,j=1}^{n} a_{ij}(x,t)\xi_i\xi_j \geqslant \alpha(\xi_1^2+\cdots+\xi_n^2), \qquad \alpha>0, \qquad \xi_i \in \mathbb{R}, \\ &\text{almost everywhere in } \Omega. \end{aligned}\right\} \tag{1.28}$$

For $\varphi, \psi \in H^1(\Omega)$, we set

$$a(t;\varphi,\psi) = \int_\Omega a_{ij}(x,t)\frac{\partial \varphi}{\partial x_i}\,\frac{\partial \psi}{\partial x_j}\,dx. \tag{1.29}$$

Let us take (cf. Chapter 1, section 3),

$$V = H_0^1(\Omega).$$

Under these conditions Theorem 1.2 is applicable giving us: there exists a unique $y \in W(0,T)$ which satisfies,

$$A(t)y + \frac{\partial y}{dt} = f \ \text{ in } Q, \text{ where } A(t)y = -\sum_{i,j=1}^{n} \frac{\partial}{\partial x_i}\left(a_{ij}(x,t)\frac{\partial y}{\partial x_j}\right), \tag{1.30}$$

with

$$y(x,0) = y_0(x) \quad \text{in } \Omega. \tag{1.31}$$

The condition "$y \in W(0,T)$" signifies,

$$y \in L^2(0,T;V), \qquad \frac{dy}{dt} \in L^2(0,T;V').$$

But if $y \in L^2(0,T;V)$ and satisfies (1.30), $\frac{dy}{dt}$ is in $L^2(0,T;V')$. Hence in the statement of the theorem we may only retain the condition "$y \in L^2(0,T;V)$"—or, again

$$y, \frac{\partial y}{\partial x_i} \in L^2(Q), \qquad y=0 \quad \text{on } \Sigma.^5 \tag{1.32}$$

Summarizing,

Under the hypothesis (1.28), *if* f *is given in* $L^2(0,T;V')$ *and* y_0 *is given in* $L^2(\Omega)$, *there exists a unique* y *satisfying* (1.32), (1.30), (1.31).

Example 1.2. We use the same notation as in the previous example but with $V = H^1(\Omega)$. In order not to commit any errors in the interpretation of V', it is preferable to replace (1.11) by the equivalent form,

$$\left(\frac{d}{dt}y(t),\psi\right) + a(t;y(t),\psi) = (f(t),\psi) \ \ \forall \psi \in V. \tag{1.33}$$

---

[5] We have this condition since $V = H_0^1(\Omega)$ (cf. Chapter 1, section 3.3).

Let us take

$$f = f(x,t) \quad \text{given in } L^2(Q) \tag{1.34}$$

and let us define,

$$(f(t), \psi) = \int_\Omega f(x,t)\psi(x)\,dx, \qquad \psi \in V.$$

In this way we obtain an element of $L^2(0,T;V')$ and (1.33) may be formally (as in Chapter 1, section 3.4, Example 3.3) interpreted as,

$$\frac{\partial y}{\partial t} + A(t)y = f \quad \text{in } Q, \tag{1.35}$$

$$\frac{\partial y}{\partial \nu_A} = 0 \quad \text{on } \Sigma, \tag{1.36}$$

$$y(x,0) = y_0(x) \quad \text{on } \Omega. \tag{1.37}$$

Remark 1.1. Problems (1.30), (1.31), (1.32) and (1.35), (1.36), (1.37) are *mixed problems* (in the sense of J. Hadamard). The operator $\dfrac{\partial}{\partial t} + A$ is a *second order parabolic operator*. In order to define a well-posed boundary value problem, the following data must be given:

(i) the initial condition (1.31) or (1.37);

(ii) a boundary condition on the lateral boundary Σ; in (1.32) this is a Dirichlet condition and in (1.37) a Neumann condition; clearly there are many other possibilities and we shall see some of these in the examples which follow.

Remark 1.2. Up to now, the boundary conditions were "homogeneous": $y|_\Sigma = 0$ or $\left.\dfrac{\partial y}{\partial \nu_A}\right|_\Sigma = 0$. Later on we shall examine the non-homogeneous case.

Remark 1.3. The general theorem, Theorem 1.2, is applicable to parabolic operators of order >2, to systems of equations etc. We shall study various examples later. But it is clear that all examples considered in Chapter 2 which were "stationary" give rise to corresponding examples of "evolution".

1.6. Semi-groups

Let us reconsider the case of Theorem 1.2 with

$$A(t) = A, \quad \text{independent of } t. \tag{1.38}$$

Let us first take $f=0$ in (1.11). Problem (1.11), (1.12) then reduces to

$$\frac{dy}{dt} + A y = 0, \qquad y(0) = y_0, \tag{1.39}$$

which admits a unique solution in $W(0,T)$, for T arbitrary but finite.

From Theorem 1.1 and 1.2, we may define a linear map

$$y_0 \to y(t) \tag{1.40}$$

of $H \to H$ which is continuous. Hence

$$y(t) = G(t)\, y_0, \qquad G(t) \in \mathscr{L}(H; H) \tag{1.41}$$

and $G(t)$ possesses the following properties:

$$\left.\begin{array}{l} \text{(i)} \quad y_0 \in H, \text{ the function } t \to G(t)\, y_0 \text{ is a continuous} \\ \text{function of } t \geqslant 0 \to H; \\ \text{(ii)} \quad G(0) = I; \\ \text{(iii)} \quad G(t)\, G(s) = G(s)\, G(t) = G(t+s) \quad t, s \geqslant 0. \end{array}\right\} \tag{1.42}$$

The family of operators $G(t)$ constitutes a semigroup in H.

Remark 1.4. Clearly the fact that H is a Hilbert Space plays no role whatsoever in the *definition* of a semigroup by properties (1.42). Hence if H is a Banach space[6], all families of operators $G(t) \in \mathscr{L}(H; H)$ satisfying (1.42) is termed a *semigroup*.

Remark 1.5. Formally,

$$A = -\frac{d}{dt} G(t)|_{t=0}; \tag{1.43}$$

we say that $(-A)$ is the *infinitesimal generator* of a semigroup. The operator $(-A)$ is an *unbounded operator* in H. Its domain $D(A)$ is the set of points $h \in H$ such that

$$\frac{1}{\sigma}(G(\sigma)h - h) \quad \text{converges in } H \text{ as } \sigma \to 0.$$

In the present case, we may verify that

$$D(A) = \{h \mid h \in V, A h \in H\},$$

cf. Hille-Phillips [1], T. Kato [1], K. Yosida [1].

---

[6] For extensions to more general cases, cf. K. Yosida [1].

Remark 1.6. The general solution of (1.11), (1.12) may be represented in the form,

$$y(t)=\int_0^t G(t-\sigma)f(\sigma)d\sigma+G(t)y_0. \tag{1.44}$$

Example 1.3. Let the operator A be symmetric: $a(\varphi,\psi)=a(\psi,\varphi)$. Then A considered as an unbounded operator in H with domain $D(A)=\{h|h\in V, Ah\in H\}$ is self-adjoint. If the injection map of V into H is compact, the operator A has a basis of eigenfunctions,

$$\left.\begin{aligned}&Aw_j=\lambda_j w_j, \quad w_j\in V, \quad \lambda_j\leqslant\lambda_{j+1}, \quad \lambda_j\to+\infty \quad \text{as } j\to\infty,\\ &(w_j,w_k)=\delta_j^k,\end{aligned}\right\} \tag{1.45}$$

the w_j's forming a complete orthonormal system in H.

We may then easily verify that

$$G(t)y_0=\sum_{j=1}^{\infty} e^{-\lambda_j t}(y_0,w_j). \tag{1.46}$$

Remark 1.7. Theorem 1.2 implies that if A is defined by $a(\varphi,\psi)$ with

$$a(\psi,\psi)+\lambda|\psi|^2\geqslant\alpha\|\psi\|^2 \quad \forall\psi\in V, \tag{1.47}$$

then $(-A)$ is the infinitesimal generator of a semigroup. The converse is however incorrect.

There exist operators A with $-A$ being the infinitesimal generator of a semigroup in a Hilbert space H but without (1.47) being true—cf. for a study of this question T. Kato [1]. A necessary and sufficient condition for $-A$ to be an infinitesimal generator of a semigroup (Hille-Yosida theorem) is that there exists a λ such that for $\xi>\lambda$ we have,

$$\|(A+\xi I)^{-k}\|\leqslant\frac{M}{(\xi-\lambda)^k} \quad \forall k.$$

Remark 1.8. In the case where $A(t)$ depends on t, a formula similar to (1.44) can be obtained in the following way. Consider the equation,

$$\left.\begin{aligned}&\frac{dz}{dt}+A(t)z=0 \quad \text{in }]\sigma,t[,\\ &z(\sigma)=\xi, \quad \xi\in H\end{aligned}\right\} \tag{1.48}$$

which admits a unique solution $z\in L^2(\sigma,t;V)$ (and hence $\frac{dz}{dt}\in L^2(\sigma,t;V')$). Then,

$$\left.\begin{aligned}&z(t)=\Phi(t,\sigma)\xi,\\ &\Phi(t,\sigma)\in\mathscr{L}(H;H)\end{aligned}\right\} \tag{1.49}$$

and the solution y of (1.11), (1.12) may be represented as

$$y(t)=\int_0^t \Phi(t,\sigma)f(\sigma)d\sigma+\Phi(t,0)y_0. \tag{1.50}$$

2. Problems of Control

2.1. Notation. Immediate Properties

Let us denote by $\mathscr{U}$ the Hilbert space of controls (compare with Chapter 2, section 2.2). We are given an operator,

$$B\in\mathscr{L}(\mathscr{U};L^2(0,T;V')). \tag{2.1}$$

Let f and y_0 with $f\in L^2(0,T;V')$, $y_0\in H$ be given. We assume that (1.1) and (1.2) hold and we denote by $y(v)$ a solution of

$$\frac{dy(v)}{dt}+A(t)y(v)=f+Bv, \tag{2.2}$$

$$y(v)|_{t=0}=y_0, \tag{2.3}$$

$$y(v)\in L^2(0,T;V). \tag{2.4}$$

Notation. $y(v)$ is a function $t\to y(v)(t)$ which we will denote as $y(t;v)$. Hence (2.3) may also be written as

$$y(0;v)=y_0. \tag{2.3 bis}$$

In applications, $y(v)$ also depends on x; this function of x and t will be denoted by $y(x,t;v)$. Therefore (2.3 bis) may also be written as

$$y(x,0;v)=y_0(x),\qquad x\in\Omega. \tag{2.3 ter}$$

The function $y(v)$ is the state of the system. The observation is given by (again compare with Chapter 2, section 1.1)

$$\begin{aligned} &z(v)=Cy(v),\\ &C\in\mathscr{L}(W(0,T);\mathscr{H})^{7}; \end{aligned} \tag{2.5}$$

N is given (as in (1.12), Chapter 2),

$$N\in\mathscr{L}(\mathscr{U};\mathscr{U}),\qquad (Nu,u)_{\mathscr{U}}\geqslant\nu\|u\|^2_{\mathscr{U}},\qquad \nu>0. \tag{2.6}$$

The cost function is given by,

$$J(v)=\|Cy(v)-z_d\|^2_{\mathscr{H}}+(Nv,v)_{\mathscr{U}}. \tag{2.7}$$

---

[7] $W(0,T)$ is defined in (1.8); $y(v)\in W(0,T)$.

Remark 2.1. The control does not appear in the initial conditions. For this case, cf. section 1.1.

We are also given (as in Chapter 2, section 1)

$$\mathscr{U}_{\mathrm{ad}} = \text{closed, convex subset of } \mathscr{U} \text{ (set of admissible controls).} \tag{2.8}$$

We seek,

$$\operatorname*{Inf}_{v \in \mathscr{U}_{\mathrm{ad}}} J(v). \tag{2.9}$$

It is clear that considerations similar to Chapter 2, section 1.2 are valid, the only pertinent property coming into play being the continuity of the affine map $v \to y(v)$ of $\mathscr{U} \to W(0,T)$ (which replaces V). Hence

$$\text{under hypothesis (2.6), there exists a unique optimal control;} \tag{2.10}$$

$$\left.\begin{array}{l}\text{if } N=0 \text{ and if } \mathscr{U}_{\mathrm{ad}} \text{ is bounded, there exists a non-empty}\\ \text{closed, convex set consisting of optimal controls}^{8}.\end{array}\right. \tag{2.11}$$

The remaining problem is to study more deeply the properties of the optimal control (it is here that the situation differs from Chapter 2).

2.2. Set of Inequalities Characterizing the Optimal Control

As in Chapter 2, section 1.3, the control $u \in \mathscr{U}_{\mathrm{ad}}$ is optimal if and only if

$$J'(u)\cdot(v-u) \geqslant 0 \quad \forall\, v \in \mathscr{U}_{\mathrm{ad}},$$

that is,

$$(Cy(u)-z_d, C(y(v)-y(u)))_{\mathscr{H}} + (Nu, v-u)_{\mathscr{U}} \geqslant 0 \quad \forall\, v \in \mathscr{U}_{\mathrm{ad}}.$$

Denoting (as in (1.25) and (1.28), Chapter 2)

$$\left.\begin{array}{l}\Lambda = \text{canonical isomorphism of } \mathscr{H} \text{ onto } \mathscr{H}',\\ \Lambda_{\mathscr{U}} = \text{canonical isomorphism of } \mathscr{U} \text{ onto } \mathscr{U}',\end{array}\right\} \tag{2.12}$$

the above formula reduces to,

$$\left.\begin{array}{l}(C^* \Lambda(Cy(u)-z_d), y(v)-y(u)) + (Nu, v-u)_{\mathscr{U}} \geqslant 0 \quad \forall\, v \in \mathscr{U}_{\mathrm{ad}},\\ \text{where the first bracket denotes the duality between}\\ W(0,T)' \text{ and } W(0,T).\end{array}\right\} \tag{2.13}$$

[8] Cf. section 7 for examples with $N=0$ and $\mathscr{U}_{\mathrm{ad}}$ not necessarily bounded.

It remains to interpret (2.13) by introducing the adjoint state. We shall develop this by considering the following cases[9]:

(i) $C \in \mathscr{L}(L^2(0,T;V);\mathscr{H})$,[10]

(ii) $Cy(v) = Dy(T;v), D \in \mathscr{L}(H;H)$.[11]

2.3. Case (i). Set of Inequalities

In this case, $C^* \in \mathscr{L}(\mathscr{H}';L^2(0,T;V'))$ and (2.13) may be written as,

$$\int_0^T (C^* \Lambda(Cy(u)-z_d), y(v)-y(u))\,dt + (Nu, v-u)_{\mathscr{U}} \geqslant 0 \quad \forall v \in \mathscr{U}_{\mathrm{ad}}. \tag{2.14}$$

We introduce the adjoint state $p(v)$ by

$$-\frac{d}{dt}p(v) + A^*(t)p(v) = C^* \Lambda(Cy(v)-z_d) \quad \text{in }]0,T[, \tag{2.15}$$

$$p(T;v) = 0, \tag{2.16}$$

$$p(v) \in L^2(0,T;V). \tag{2.17}$$

Problem (2.15), (2.16), (2.17) admits a unique solution, a fact which follows by applying Theorem 1.2 with the flow of time reversed (change t to $T-t$).

Multiplying both sides of (2.15) (with $v=u$) by $y(v)-y(u)$ and noting that

$$\int_0^T \left(-\frac{d}{dt}p(u), y(v)-y(u)\right)dt = \int_0^T \left(p(u), \frac{d}{dt}y(v) - \frac{d}{dt}y(u)\right)dt$$

and

$$\int_0^T (A^*(t)p(u), y(v)-y(u))\,dt = \int_0^T (p(u), A(t)y(v) - A(t)y(u))\,dt$$

we obtain,

---

[9] The general case requires consideration of the space $W(0,T)'$ in the adjoint problem. This is possible but requires considerations which are somewhat complicated. Hence it is better to divide the problem into cases (i) and (ii). Naturally C may be a linear combination of case (i) and (ii).

[10] Hence (2.5) clearly holds.

[11] We say that we observe the final state.

$$\int_0^T (C^* \Lambda(Cy(u)-z_d), y(v)-y(u))\,dt = \int_0^T \left(p(u), \left(\frac{d}{dt} + A(t)\right)(y(v)-y(u))\right) dt$$

$$= \int_0^T (p(u), Bv - Bu)\,dt = (B^* p(u), v-u)$$

(the bracket denoting the scalar product between $\mathcal{U}$ and $\mathcal{U}'$),

$$= (\Lambda_{\mathcal{U}}^{-1} B^* p(u), v-u)_{\mathcal{U}}.$$

Hence (2.14) may be written as,

$$(\Lambda_{\mathcal{U}}^{-1} B^* p(u) + N u, v-u)_{\mathcal{U}} \geq 0 \quad \forall v \in \mathcal{U}_{\text{ad}}.\text{[12]} \tag{2.18}$$

Therefore we have proved

Theorem 2.1. *We assume that* (1.1), (1.2) *as well as* (2.6) *hold. Let us suppose that* $C \in \mathcal{L}(L^2(0,T;V);\mathcal{H})$. *The optimal control u is characterized by the following system of partial differential equations and inequalities:*

$$\left.\begin{aligned} &\frac{d}{dt} y(u) + A(t) y(u) = f + Bu, \\ &y(0;u) = y_0, \end{aligned}\right\} \tag{2.19}$$

$$\left.\begin{aligned} &-\frac{d}{dt} p(u) + A^*(t) p(u) = C^* \Lambda(C y(u) - z_d), \\ &p(T;u) = 0, \end{aligned}\right\} \tag{2.20}$$

$$\left.\begin{aligned} &(\Lambda_{\mathcal{U}}^{-1} B^* p(u) + N u, v-u) \geq 0 \quad \forall v \in \mathcal{U}_{\text{ad}}, \\ &u \in \mathcal{U}_{\text{ad}}, \end{aligned}\right\} \tag{2.21}$$

together with

$$\left.\begin{aligned} &y(u) \in L^2(0,T;V), \\ &p(u) \in L^2(0,T;V). \end{aligned}\right\} \tag{2.22}$$

Remark 2.2. The conditions $y(0;u) = y_0$ and $p(T;u) = 0$ are the boundary conditions for the coupled system in $\{y,p\}$; when considering particular examples, cf. section 3 following—we have to add boundary conditions on Σ.

Remark 2.3. The case of "no constraints" — $\mathcal{U}_{\text{ad}} = \mathcal{U}$.

---

[12] The previous calculation shows $\frac{1}{2} J'(u) = B^* p(u) + \Lambda_{\mathcal{U}} N u$.

If $\mathcal{U}_{ad}=\mathcal{U}$, (2.21) reduces to

$$\Lambda_{\mathcal{U}}^{-1} B^* p(u)+N u=0; \tag{2.23}$$

u may now be eliminated giving us

$$\left.\begin{aligned} \frac{dy}{dt} + A(t)y+BN^{-1}\Lambda_{\mathcal{U}}^{-1}B^*p=f, \\ \frac{dp}{dt} + A^*(t)p = C^*\Lambda(Cy-z_d), \end{aligned}\right\} \tag{2.24}$$

$$y(0)=y_0, \quad p(T)=0. \text{[13]} \tag{2.25}$$

Further,

$$u=-N^{-1}\Lambda_{\mathcal{U}}^{-1}B^*p. \tag{2.26}$$

The system (2.24), (2.25) will be studied in detail in section 4.

2.4. Case (ii). Set of Inequalities

If C is given as in (ii), section 2.2, the cost function is

$$J(v)=|Dy(T;v)-z_d|^2+(Nv,v)_{\mathcal{U}} \tag{2.27}$$

and the optimality condition (2.13) is equivalent to

$$\left.\begin{aligned} &(Dy(T;u)-z_d, Dy(T;v)-Dy(T;u))+(Nu,v-u)_{\mathcal{U}}\geqslant 0,\\ &\forall v\in\mathcal{U}_{ad} \text{ (where the bracket round the first term}\\ &\text{denotes the scalar product in } H). \end{aligned}\right\} \tag{2.28}$$

We introduce the adjoint state $p(v)$ by,

$$-\frac{d}{dt}p(v)+A^*(t)p(v)=0 \quad \text{in }]0,T[, \tag{2.29}$$

$$p(T;v)=D^*(Dy(T;v)-z_d), \tag{2.30}$$

$$p(v)\in L^2(0,T;V). \tag{2.31}$$

According to Theorem 1.2, the above problem admits a unique solution (after changing t into $T-t$).

Let us set $v=u$ in the above equations and scalar multiply both sides of (2.29) by $y(v)-y(u)$ and integrate from 0 to T. This gives us,

$$(Dy(T;u)-z_d, Dy(T;v)-Dy(T;u)) = \int_0^T (p(u), B(v-u))\,dt$$

---

[13] Here $y(0)$ is $y(t)|_{t=0}$. Same for $p(T)$.

(since,

$$\int_0^T \left(-\frac{d}{dt}p(u), y(v)-y(u)\right)dt = \int_0^T \left(p(u), \frac{d}{dt}(y(v)-y(u))\right)dt$$
$$-\left(D^*(Dy(T;u)-z_d), y(T;v)-y(T;u)\right)\big).$$

Hence (2.28) is equivalent to

$$(\Lambda_{\mathscr{U}}^{-1} B^* p(u) + Nu, v-u)_{\mathscr{U}} \geqslant 0 \quad \forall v \in \mathscr{U}_{\text{ad}}. \tag{2.32}$$

Therefore we have,

Theorem 2.2. *Assume that* (1.1), (1.2), (2.6) *hold and* $J(v)$ *is given by* (2.27) *where* $D \in \mathscr{L}(H; H)$. *Then the optimal control is determined by*

$$\left.\begin{aligned} \frac{d}{dt}y(u) + A(t)y(u) = f + Bu, \\ y(0;u) = y_0, \end{aligned}\right\} \tag{2.33}$$

$$\left.\begin{aligned} -\frac{dp}{dt}(u) + A^*(t)p(u) = 0, \\ p(T;u) = D^*(Dy(T;u)-z_d), \end{aligned}\right\} \tag{2.34}$$

$$(\Lambda_{\mathscr{U}}^{-1} B^* p(u) + Nu, v-u)_{\mathscr{U}} \geqslant 0 \quad \forall v \in \mathscr{U}_{\text{ad}}, \quad u \in \mathscr{U}_{\text{ad}}, \tag{2.35}$$

with

$$\left.\begin{aligned} y(u) \in L^2(0,T;V), \\ p(u) \in L^2(0,T;V). \end{aligned}\right\} \tag{2.36}$$

Remark 2.4. (Analogous to Remark 2.3.) In the case where there are no constraints on u, that is $\mathscr{U}_{\text{ad}} = \mathscr{U}$, by eliminating u, condition (2.35) reduces to (2.23). We thus obtain,

$$\left.\begin{aligned} \frac{dy}{dt} + A(t)y + BN^{-1}\Lambda_{\mathscr{U}}^{-1}B^*p = f, \\ -\frac{dp}{dy} + A^*(t)p = 0, \end{aligned}\right\} \tag{2.37}$$

$$y(0) = y_0, \quad p(T) = D^*(Dy(T)-z_d). \tag{2.38}$$

Further,

$$u = -N^{-1}\Lambda_{\mathscr{U}}^{-1}B^*p.$$

2.5. Orientation

Examples of application of Theorem 2.1 and 2.2 will be presented in section 3.

We shall then return to the case where there are no constraints on the control.

It is clear that Theorem 2.1 of Chapter 2 may be used in the present context. This fact is however important for applications.

Remark 2.5. In Theorems 2.1 and 2.2, for the case *(singular)* where

$$\Lambda_{\mathscr{U}}^{-1} B^* p(u) + N u = 0 \tag{2.39}$$

condition (2.21) or (2.25) is automatically satisfied. But (2.39) corresponds to the case where there are *no constraints*. This implies that the constraints are not binding, that is the u minimizing $J(v)$ when v ranges over the whole space $\mathscr{U}$ is contained in $\mathscr{U}_{\text{ad}}$.

Remark 2.6. Assume that

$$\mathscr{U}_{\text{ad}} = \{v \mid \|v\|_{\mathscr{U}} \leqslant R\}. \tag{2.40}$$

Then if we set,

$$\xi = \Lambda_{\mathscr{U}}^{-1} B^* p(u) + N u, \tag{2.41}$$

and if we assume that $\xi \neq 0$ (cf. Remark 2.5), condition (2.21) (or (2.35)) is equivalent to

$$u = -R \frac{\xi}{\|\xi\|_{\mathscr{U}}}. \tag{2.42}$$

Remark 2.7. Let us now suppose that

$$\mathscr{U} = L^2(0, T; E), \qquad E = \text{Hilbert space} \tag{2.43}$$

and that

$$\mathscr{U}_{\text{ad}} = \{v \mid \|v(t)\|_E \leqslant R \text{ almost everywhere}\}. \tag{2.44}$$

Then with the same notation as in (2.41), condition (2.21) (or (2.35)) is equivalent to (cf. Theorem 2.1, Chapter 2),

$$(\xi(t), u(t))_E = \inf_{\|k\|_E \leqslant R} (\xi(t), k).$$

Hence

$$u(t) = -R \frac{\xi(t)}{\|\xi(t)\|} \quad \text{provided } \xi(t) \neq 0. \tag{2.45}$$

A theory analogous to that developed in this section can be obtained if the cost function is given by

$$J(v) = (C_1 y(v), y(v))_{\mathscr{H}} + (N v, v)_{\mathscr{U}}, \tag{2.46}$$

instead of (2.7). In (2.46) $C_1 \in \mathscr{L}(\mathscr{H};\mathscr{H})$ but is not necessarily $\geqslant 0$. However the theory is valid provided N is taken to be "sufficiently large" so that $J(v)$ defined by (2.46) has the coercivity property:

$$J(v) \to +\infty \quad \text{if } \|v\|_{\mathscr{U}} \to +\infty.$$

3. Examples

3.1. Mixed Dirichlet Problem for a Second Order Parabolic Equation

We take $V = H_0^1(\Omega)$ and $A(t)$ as in Example 1.1, section 1.5. Further, $\mathscr{U} = L^2(Q)$ ($Q = \Omega \times]0, T[$) (hence $\mathscr{U}' = \mathscr{U}$) and $B =$ identity map.

The state of the system is given by

$$\left.\begin{aligned} \frac{\partial}{\partial t} y(u) + A(t) y(u) &= f + u \quad \text{in } Q, \\ y(u)|_\Sigma &= 0, \\ y(x,0;u) &= y_0(x) \quad x \in \Omega; \end{aligned}\right\} \tag{3.1}$$

the control is therefore distributed over Q. Let us consider several different cases.

3.1.1. $C =$ Injection Map of $L^2(0,T;V) \to L^2(Q)$

Hence $\mathscr{H} = L^2(Q) = \mathscr{H}'$, $\Lambda =$ identity map and the adjoint state is defined by,

$$\left.\begin{aligned} -\frac{\partial p}{dt}(u) + A^*(t) p(u) &= y(u) - z_d \quad (z_d \in L^2(Q)), \\ p(u)|_\Sigma &= 0, \\ p(x,T;u) &= 0. \end{aligned}\right\} \tag{3.2}$$

The optimal control is characterized by (3.1), (3.2) and

$$\int_Q (p(u) + Nu)(v-u)\,dx\,dt \geqslant 0 \ \forall v \in \mathscr{U}_{\mathrm{ad}}, \quad u \in \mathscr{U}_{\mathrm{ad}}. \tag{3.3}$$

Example 3.1. $\mathscr{U}_{\mathrm{ad}} = \mathscr{U}$.

Hence $u = -N^{-1} p$.

Regularity properties of the optimal control can be deduced (cf. also Remark 2.3, Chapter 2): if the boundary Γ of Ω and the coefficients

$a_{ij}(x,t)$ of $A(t)$ are sufficiently regular, then for the solution $p(u)$ of (3.2) we may verify that

$$p(u) \in H^{2,1}(Q) \tag{3.4}$$

where

$$H^{2,1}(Q) = \left\{ g \mid g, \frac{\partial g}{\partial x_i}, \frac{\partial^2 g}{\partial x_i \partial x_j}, \frac{\partial g}{\partial t} \in L^2(Q) \right\}. \text{ [14]} \tag{3.5}$$

If, for example $N = \nu \times$ Identity, we obtain,

$$\text{the optimal } u \in H^{2,1}(Q). \tag{3.6}$$

Example 3.2. $\mathscr{U}_{\text{ad}} = \{v \mid v \geqslant 0$ almost everywhere in $Q\}$.

Then (3.3) reduces to

$$\left.\begin{aligned} p(u) + Nu \geqslant 0 \quad &\text{in } Q, \\ u \geqslant 0 \quad &\text{in } Q, \\ (p(u) + Nu)u = 0 \quad &\text{in } Q. \end{aligned}\right\} \tag{3.7}$$

We may eliminate u; the optimal control u is given by the simultaneous solution of the system of equations,

$$\left.\begin{aligned} \frac{\partial y}{\partial t} + Ay - f \geqslant 0 \quad &\text{in } Q (A = A(t)), \\ -\frac{\partial p}{\partial t} + A^* p = y - z_d \quad &\text{in } Q, \\ p + N\left(\frac{\partial y}{\partial t} + Ay - f\right) \geqslant 0 \quad &\text{in } Q, \\ \left[p + N\left(\frac{\partial y}{\partial t} + Ay - f\right)\right]\left[\frac{\partial y}{\partial t} + Ay - f\right] = 0 \quad &\text{in } Q, \\ y|_\Sigma = 0, \quad p|_\Sigma = 0&, \\ y(x,0) = y_0(x), \quad p(x,T) = 0&. \end{aligned}\right\} \tag{3.8}$$

This is a problem of "unilateral" type.

Clearly, from the last condition of (3.7), either $u = 0$ and $p + Nu = p > 0$, or $p + Nu = 0$ and $u > 0$, or $u = 0$ and $p + Nu = 0$. If the last

---

[14] $H^{2,1}(Q)$ is a Hilbert space when endowed with the norm

$$\left(\int_Q \left[|g|^2 + \sum_i \left(\frac{\partial g}{\partial x_i}\right)^2 + \sum_{i,j} \left(\frac{\partial^2 g}{\partial x_i \partial x_j}\right)^2 + \left(\frac{\partial g}{\partial t}\right)^2 \right] dx\, dt \right)^{\frac{1}{2}}.$$

condition obtains, $y = z_d$ and $\frac{\partial y}{\partial t} + A(t)y = f$. If we then assume that

$$\frac{\partial z_d}{\partial t} + A(t)z_d \neq f \quad \text{almost everywhere in } Q,$$

only the first two cases are possible (as in Example 2.2, Chapter 2). If $N = \nu \times$ Identity, we find that

$$u = -\frac{1}{\nu}\inf(0, p) \quad \text{almost everywhere in } Q. \tag{3.9}$$

Hence the optimal control is determined by the simultaneous solution of

$$\left.\begin{aligned} \frac{\partial y}{\partial t} + Ay + \frac{1}{\nu}\inf(0,p) &= f \quad \text{in } Q, \\ -\frac{\partial p}{\partial t} + A^* p &= y - z_d \quad \text{in } Q, \\ y|_\Sigma = 0, \quad & p|_\Sigma = 0, \\ y(x,0) = y_0(x), \quad & p(x,T) = 0. \end{aligned}\right\} \tag{3.10}$$

Condition (3.9) then determines the optimal control. Let us remark that if the coefficients a_{ij} and the boundary Γ of Ω are sufficiently regular, (3.4) holds and hence we obtain from (3.9)

$$y \in H^1(Q). \tag{3.11}$$

Example 3.3. Let us now take

$$\mathcal{U}_{\text{ad}} = \{v \mid \xi_0(x,t) \leqslant v(x,t) \leqslant \xi_1(x,t) \text{ almost everywhere in } Q,\ \xi_i \in L^\infty(Q)\}. \tag{3.12}$$

Hence from Theorem 2.1, Chapter 2, (3.3) is equivalent to

$$(p(x,t;u) + Nu(x,t))(\xi - u(x,t)) \geqslant 0 \quad \forall \xi \in [\xi_0(x,t), \xi_1(x,t)],$$

which we may interpret as in Chapter 2, Example 2.3.

3.1.2. C = Identity Map of $L^2(0,T;V)$ into itself

Then $\mathcal{H} = L^2(0,T;V)$ and $\mathcal{H}' = L^2(0,T;V')$. For the case $V = H_0^1(\Omega)$, we have

$$\Lambda = (-\Delta + I), \quad \text{that is}$$

$$\Lambda f(x,t) = -\Delta_x f(x,t) + f(x,t).$$

Hence the adjoint state is defined by,

$$\left.\begin{aligned} -\frac{\partial p}{\partial t}(u)+A^*(t)p(u)=(-\Delta_x+I)(y(u)-z_d), \\ (z_d \text{ given in } L^2(0,T;V)), \qquad p(u)|_\Sigma=0, \\ p(x,T;u)=0. \end{aligned}\right\} \tag{3.13}$$

Examples analogous to the preceeding ones may be constructed for particular choices of $\mathscr{U}_{\mathrm{ad}}$.

3.1.3. Observation of the Final State

Within the framework of section 2.4, let $D\in\mathscr{L}(H;H)$, $H=L^2(\Omega)$, and

$$J(v)=\int_\Omega (Dy(x,T;v)-z_d(x))^2\,dx+(Nv,v)_{\mathscr{U}}. \tag{3.14}$$

The adjoint state is then given by (cf. (2.29), (2.30), (2.31)),

$$\left.\begin{aligned} -\frac{\partial p}{\partial t}(u)+A^*(t)p(u)=0 \quad \text{in } Q, \\ p(u)|_\Sigma=0, \\ p(x,T;u)=D^*(Dy(x,T;u)-z_d(x)) \quad \text{in } \Omega. \end{aligned}\right\} \tag{3.15}$$

The optimal control is given by (3.1), (3.15) and (3.3).

Example 3.4. $\mathscr{U}_{\mathrm{ad}}=\mathscr{U}$.

Therefore $u=-N^{-1}p(u)$. This is obtained by solving (3.1) and (3.15). But in this case the solution of (3.15) in general is not an element of $H^{2,1}(Q)$ but $p(u)\in L^2(0,T;V)$ and hence if $N=\nu\times$ Identity, we

$$u\in L^2(0,T;H_0^1(\Omega)).$$

Again examples similar to those of 3.2 and 3.3 can be constructed.

Remark 3.1. In the present framework the case when "$N=0$" and "$\mathscr{U}_{\mathrm{ad}}=$ a bounded set" will be studied in section 7.

3.2. Mixed Neumann Problem for a Parabolic Equation of Second Order

In this case we take

$$V=H^1(\Omega),$$
$$\mathscr{U}=L^2(\Sigma),$$

and $A(t)$ is given as in the preceeding example.

The state $y(u)$ is given by the solution of

$$\left.\begin{aligned}&\frac{d}{dt}(y(u),\psi)+a(t;y(u),\psi)=(f(t),\psi)_{L^2(\Omega)}+(u(t),\psi|_\Gamma)_{L^2(\Gamma)},\\&\forall\psi\in H^1(\Omega)\quad (f \text{ given in } L^2(Q))\end{aligned}\right\}\tag{3.16}$$

with the initial condition,

$$y(0;u)=y_0 \quad \text{in } \Omega,\ \ y_0 \text{ given in } L^2(\Omega).\tag{3.17}$$

Equation (3.16) may be interpreted as (compare section 2.4, Chapter 2)

$$\frac{\partial}{\partial t}y(u)+A(t)\,y(u)=f \quad \text{in } Q,\tag{3.18}$$

$$\frac{\partial}{\partial \nu_A}y(u)=u \quad \text{on } \Sigma,\tag{3.19}$$

and (3.17) is equivalent to

$$y(x,0;u)=y_0(x),\qquad x\in\Omega.\tag{3.20}$$

The state is therefore given by the solution of a mixed problem (in the sense of Hadamard) with a Neumann boundary condition, control being exercised through the boundary.

For the observation, we consider the following two cases:

(i) $C=$ injection map of $L^2(0,T;V)\to L^2(Q)$ (as in 3.1.1),
(ii) observation of final state (as in 3.1.3)[15].

3.2.1. Case (i)

The cost function then is

$$T(v)=\int_Q (y(v)-z_d)^2\,dx\,dt+(N v,v)_{L^2(\Sigma)},\qquad z_d\in L^2(Q).\tag{3.21}$$

The adjoint state $p(u)$ is given by the solution of

$$\left.\begin{aligned}&-\frac{\partial}{\partial t}p(u)+A^*p(u)=y(u)-z_d \quad \text{in } Q,\qquad (A^*=A^*(t)),\\&\frac{\partial p}{\partial \nu_{A^*}}(u)=0 \quad \text{on } \Sigma,\\&p(x,T;u)=0,\qquad x\in\Omega,\end{aligned}\right\}\tag{3.22}$$

---

[15] Clearly examples analogous to 3.1.2 may be constructed, the details of which are omitted.

and the optimality condition is

$$\int_{\Sigma}(p+Nu)(v-u)\,d\Sigma \geqslant 0 \quad \forall v \in \mathscr{U}_{ad}. \tag{3.23}$$

Example 3.5. The Case Where There Are No Constraints ($\mathscr{U}_{ad}=\mathscr{U}$).

Then (3.23) reduces to

$$p+Nu=0 \quad \text{on } \Sigma. \tag{3.24}$$

The optimal control is obtained by the simultaneous solution of the following system of partial differential equations:

$$\left.\begin{aligned}
&\frac{\partial y}{\partial t}+Ay=f \quad \text{in } Q,\\
&-\frac{\partial p}{\partial t}+A^*p=y-z_d \quad \text{in } Q,\\
&\left.\frac{\partial y}{\partial \nu_A}\right|_{\Sigma}+N^{-1}p|_{\Sigma}=0 \quad \text{on } \Sigma,\\
&\frac{\partial p}{\partial \nu_{A^*}}=0 \quad \text{on } \Sigma,\\
&y(x,0)=y_0(x), \quad x\in\Omega,\\
&p(x,T)=0, \quad x\in\Omega;
\end{aligned}\right\} \tag{3.25}$$

further,

$$u=-N^{-1}(p|_{\Sigma}). \tag{3.26}$$

For this example the regularity property of the optimal control is also worth noting. If the coefficients of A and the boundary Γ are sufficiently regular, then p, the solution of

$$-\frac{\partial p}{\partial t}+A^*p=y-z_d, \quad \left.\frac{\partial p}{\partial \nu_{A^*}}\right|_{\Sigma}=0, \quad p(x,T)=0$$

is in $H^{2,1}(Q)$ (cf. definition of $H^{2,1}(Q)$ in (3.5)) and hence

$$p|_{\Sigma}\in H^{\frac{1}{2}}(\Sigma).\ ^{16} \tag{3.27}$$

If, for example, $N=\nu \times$ Identity, we obtain

$$u\in H^{\frac{1}{2}}(\Sigma). \tag{3.28}$$

Example 3.6. $\mathscr{U}_{ad}=\{v \mid v\in L^2(\Sigma),\ v\geqslant 0$ almost everywhere on $\Sigma\}$.

[16] More precisely (cf. Lions-Magenes [1], Chapter 4): $p|_{\Sigma}\in H^{\frac{3}{2},\frac{3}{4}}(\Sigma)$.

The optimal control is obtained by the solution of the problem (of unilateral type),

$$\left.\begin{aligned}
&\frac{\partial y}{\partial t}+Ay=f, \qquad -\frac{\partial p}{\partial t}+A^*p=y-z_d \quad \text{in } Q,\\
&\frac{\partial y}{\partial \nu_A}\geqslant 0 \quad \text{on } \Sigma, \qquad \frac{\partial p}{\partial \nu_{A^*}}=0 \quad \text{on } \Sigma,\\
&p+N\frac{\partial y}{\partial \nu_A}\geqslant 0 \quad \text{on } \Sigma, \quad \frac{\partial y}{\partial \nu_A}\left[p+N\frac{\partial y}{\partial \nu_A}\right]=0 \quad \text{on } \Sigma,\\
&y(x,0)=y_0(x), \quad p(x,T)=0, \quad x\in\Omega;
\end{aligned}\right\} \tag{3.29}$$

hence

$$u=\left.\frac{\partial y}{\partial \nu_A}\right|_{\Sigma}. \tag{3.30}$$

3.2.2. Case (ii)

The cost function now is

$$J(v)=\int_\Omega (y(x,T;v)-z_d)^2\,dx+(Nv,v)_{L^2(\Sigma)}, \qquad z_d\in L^2(\Omega). \tag{3.31}$$

The adjoint state is defined by,

$$\left.\begin{aligned}
&-\frac{\partial p}{\partial t}(u)+A^*(t)p(u)=0 \quad \text{in } Q,\\
&\frac{\partial p}{\partial \nu_{A^*}}(u)=0 \quad \text{on } \Sigma,\\
&p(x,T;u)=y(x,T;u)-z_d(x) \quad \text{on } \Omega,
\end{aligned}\right\} \tag{3.32}$$

and the optimality condition is

$$\int_\Sigma (p+Nu)(v-u)\,d\Sigma\geqslant 0. \tag{3.33}$$

Example 3.7. Case Where There Are No Constraints ($\mathcal{U}_{\mathrm{ad}}=\mathcal{U}$).

In this case,

$$u=-N^{-1}(p|_\Sigma); \tag{3.34}$$

the optimal control is obtained by the simultaneous solution of the system of partial differential equations,

$$\left.\begin{array}{ll}\dfrac{\partial y}{\partial t}+A(t)y=f, & -\dfrac{\partial p}{\partial t}+A^*p=0 \quad \text{in } Q,\\[2ex] \dfrac{\partial y}{\partial \nu_A}+N^{-1}(p|_\Sigma)=0, & \dfrac{\partial p}{\partial \nu_{A^*}}=0 \quad \text{on } \Sigma,\\[2ex] y(x,0)=y_0(x), & p(x,T)=y(x,T)-z_d(x) \quad \text{on } \Omega,\end{array}\right\} \tag{3.35}$$

u being given by (3.34).

Example 3.8. $\mathscr{U}_{\mathrm{ad}}=\{v|v\in L^2(\Sigma),\ v\geqslant 0$ almost everywhere on $\Sigma\}$.

Then (3.33) is equivalent to,

$$u\geqslant 0, \quad p(u)+Nu\geqslant 0, \quad u(p(u)+Nu)=0 \quad \text{on } \Sigma. \tag{3.36}$$

But we have the following

Lemma 3.1. *Let us suppose that the coefficients of $A(t)$ are analytic in $\bar{Q}$ and the boundary Γ (and hence Σ) is analytic. If $p(u)$, the solution of (3.32) is zero on $\Sigma_0\subset\Sigma$, measure $(\Sigma_0)>0$, then $p(u)\equiv 0$.*

Proof. Under the given hypotheses, we know (cf. for example H. Tanabe [1]) that $p(u)$ is analytic in Q and on Σ. But since $p(u)$ is analytic on Σ and zero on a set of non-zero measure, $p(u)|_\Sigma=0$. Hence the data corresponding to the Cauchy problem is zero on Σ, which implies according to the Cauchy-Kowaleska theorem (since $p(u)$ is analytic) that $p(u)\equiv 0$.

Let us now return to (3.36). Suppose that on a set $\Sigma_0\subset\Sigma$ of positive measure we have

$$u=0 \quad \text{and} \quad p(u)+Nu=0;$$

hence $u=0$ and $p(u)=0$ and from Lemma 3.1 this is impossible unless $p(u)\equiv 0$.

To simplify matters, let us assume that

$$N=\nu \times \text{Identity}. \tag{3.37}$$

If $p(u)=0$, we must have $\nu u^2=0$ on Σ and hence $u=0$ and

$$p(x,T;0)=y(x,T;0)-z_d(x)=0.$$

Hence,

$$\left.\begin{array}{l}\text{if } y(x,T;0)=z_d(x), \quad \text{then } u=0 \quad \text{is the optimal control;}\\ \text{otherwise } p(u)|_\Sigma\neq 0 \quad \text{almost everywhere.} \quad \square\end{array}\right. \tag{3.38}$$

If we therefore consider the case where $y(T;0)\neq z_d$, only two possibilities remain:

$$u>0, \quad p(u)+Nu=0; \quad u=0, \quad p(u)>0.$$

Hence

$$u=-\frac{1}{\nu}\inf(0,p|_\Sigma), \tag{3.39}$$

and the optimal control is given by

Theorem 3.1. *We assume that the coefficients of A are analytic in $\bar{Q}$ and* (1.28) *holds. We also assume that Γ is analytic and that the state of the system is given by* (3.18), (3.19), (3.20). *The cost function is assumed to be given by* (3.31) *and assume that $y(T;0)\neq z_d$.*[17] *Finally let $\mathcal{U}_{\text{ad}}$ be given by,*

$$\mathcal{U}_{\text{ad}}=\{v\,|\,v\geqslant 0 \text{ on } \Sigma\}\subset\mathcal{U}=L^2(\Sigma).$$

Then the optimal control is given by the solution of the following non-linear system of equations:

$$\left.\begin{aligned}
&\frac{\partial y}{\partial t}+Ay=f, \qquad -\frac{\partial p}{\partial t}+A^*p=0 \quad \text{in } Q \qquad (A=A(t)),\\
&\frac{\partial y}{\partial \nu_A}+\frac{1}{\nu}\inf(0,p)=0 \quad \text{on } \Sigma,\\
&\frac{\partial p}{\partial \nu_{A^*}}=0 \quad \text{on } \Sigma,\\
&y(x,0)=y_0(x), \qquad p(x,T)=y(x,T)-z_d(x),
\end{aligned}\right\} \tag{3.40}$$

and by (3.39).

Remark 3.2. A *direct* study of a non-linear problem of the type (3.40) appears to be an open problem. More general examples may be found in Lions [5] (cf. also section 3.4 following).

Remark 3.3. The case where $N=0$ and "$\mathcal{U}_{\text{ad}}$ bounded" will be studied in section 7.

Remark 3.4. In (3.6) we assumed $u\in L^2(\Sigma)$. This implies that the map

$$\psi\to(u(t),\psi|_\Gamma)_{L^2(\Gamma)} \quad \text{defines an element in } L^2(0,T;V').$$

More generally, we may assume that

$$u\in L^2(0,T;H^{-\frac{1}{2}}(\Gamma))$$

[17] This is equivalent to assuming $\inf\limits_{v\in\mathcal{U}_{\text{ad}}} J(v)>0$.

(compare with section 2.4, Chapter 2). We may go even further; we shall be seeing examples of this. For the general theory of non-homogeneous problems we refer the reader to Lions-Magenes [1], Chapter 4.

Remark 3.5. Using analogous methods we shall treat the case where the state of the system is given by the solution of a mixed problem of transmission (for the stationary case, cf. Chapter 1, section 3.4, Example 3.4).

We shall now briefly consider two examples of "evolution equations" which correspond to examples in Chapter 2, section 2.6 and 2.7.

3.3. System of Equations and Equations of Higher Order

3.3.1. System of Equations

We consider an example of an "evolution equation" which is analogous to that considered in section 2.6, Chapter 2, but with Neumann boundary conditions and boundary control.

Let $y(u)=\{y_1(u), y_2(u)\}$ be the state of the system which is given by,

$$\left.\begin{aligned} &\frac{\partial y_1}{\partial t}(u)-\Delta y_1(u)+y_1(u)-y_2(u)=f_1 \quad \text{in } Q, \quad f_1\in L^2(Q),\\ &\frac{\partial y_2}{\partial t}(u)-\Delta y_2(u)+y_2(u)+y_1(u)=f_2 \quad \text{in } Q, \quad f_2\in L^2(Q),\\ &\left.\frac{\partial y_1}{\partial n}\right|_{\Sigma}=u_1, \quad \left.\frac{\partial y_2}{\partial n}\right|_{\Sigma}=u_2, \end{aligned}\right\} \tag{3.41}$$

$$y_1(x,0;u)=y_{0,1}(x)\in L^2(\Omega), \qquad y_2(x,0;u)=y_{0,2}(x)\in L^2(\Omega), \qquad x\in\Omega.$$

The control u is taken in $L^2(\Sigma)\times L^2(\Sigma)$:

$$u=\{u_1,u_2\}\in\mathcal{U}=L^2(\Sigma)\times L^2(\Sigma). \tag{3.42}$$

Problem (3.41) admits a unique solution. To see this, we apply Theorem 1.2, with,

$$V=H^1(\Omega)\times H^1(\Omega), \qquad \varphi=\{\varphi_1,\varphi_2\}\in V,$$

$$\begin{aligned} a(t;\varphi,\psi)=a(\varphi,\psi) &= \int_\Omega (\operatorname{grad}\varphi_1 \operatorname{grad}\psi_1+\operatorname{grad}\varphi_2 \operatorname{grad}\psi_2)\,dx\\ &\quad+\int_\Omega [\varphi_1\psi_1+\varphi_2\psi_2-\varphi_2\psi_1+\varphi_1\psi_2]\,dx, \end{aligned}$$

$$(f(t),\psi)=\int_\Omega (f_1(x,t)\psi_1(x)+f_2(x,t)\psi_2(x))\,dx+\int_\Gamma (u_1(t)\psi_1+u_2(t)\psi_2)\,d\Gamma.$$

Let us consider the case where we have partial observation of the final state,

$$z(v)=y_1(x,T;v) \tag{3.43}$$

and the cost function is given by,

$$J(v)=\int_\Omega [y_1(x,T;v)-z_d(x)]^2\,dx+(Nv,v)_{\mathscr{U}}, \qquad z_d\in L^2(\Omega). \tag{3.44}$$

The adjoint state is given by (cf. Theorem 2.2, the operator D mapping $H=L^2(\Omega)\times L^2(\Omega)\to H$ being given by $D\{h_1,h_2\}=\{h_1,0\}$):

$$\left.\begin{aligned}
&-\frac{\partial p_1}{\partial t}(u)-\Delta p_1(u)+p_1(u)+p_2(u)=0 \quad \text{in } Q,\\
&-\frac{\partial p_2}{\partial t}(u)-\Delta p_2(u)+p_2(u)-p_1(u)=0 \quad \text{in } Q,\\
&\qquad \frac{\partial p_1}{\partial n}(u)=0, \quad \frac{\partial p_2}{\partial n}(u)=0 \quad \text{on } \Sigma,\\
&\qquad\qquad p_1(x,T;u)=y_1(x,T;u)-z_d(x) \quad \text{in } \Omega,\\
&\qquad\qquad p_2(x,T;u)=0 \quad \text{in } \Omega.
\end{aligned}\right\} \tag{3.45}$$

The optimality condition is

$$\left.\begin{aligned}
&\int_\Sigma [p_1(v_1-u_1)+p_2(v_2-u_2)]\,d\Sigma+(Nu,v-u)_{\mathscr{U}}\geqslant 0,\\
&\forall v\in\mathscr{U}_{\mathrm{ad}} \qquad u\in\mathscr{U}_{\mathrm{ad}}.
\end{aligned}\right\} \tag{3.46}$$

Example 3.9. In the case of no constraints on the control ($\mathscr{U}_{\mathrm{ad}}=\mathscr{U}$) and if $N=\{N_1,N_2\}$ is a diagonal matrix of operators, we obtain

$$u_1=-N_1^{-1}(p_1(u)|_\Sigma), \qquad u_2=-N_2^{-1}(p_2(u)|_\Sigma). \tag{3.47}$$

The optimal control is determined by simultaneously solving (3.41), (3.45) (where we eliminate u_1 with the aid of (3.47)) and then utilizing (3.47).

Remark 3.6. If we take,

$$\mathscr{U}_{\mathrm{ad}}=\{v\,|\,v_i\geqslant 0 \text{ on } \Sigma,\ i=1,2\},$$

and $N=\nu\times$ Identity, (3.46) gives us,

$$\left.\begin{aligned}
&u_1\geqslant 0, \quad p_1(u)+\nu_1u_1\geqslant 0, \quad u_1(p_1(u)+\nu_1u_1)=0 \quad \text{on } \Sigma,\\
&u_2\geqslant 0, \quad p_2(u)+\nu_2u_2\geqslant 0, \quad u_2(p_2(u)+\nu_2u_2)=0 \quad \text{on } \Sigma.
\end{aligned}\right\} \tag{3.48}$$

It is not known whether we have uniqueness properties analogous to that in Lemma 3.1. It is however clear, by using a similar proof, that if

$p_1(u)$ and $p_2(u)$ are zero on two different sets of positive measure, then $p_i(u)=0$ on Σ, $i=1,2$ and $p_i\equiv 0$, $i=1,2$. But could it happen, for example, $p_1(u)|_\Sigma=0$, no conditions on $p_2(u)|_\Sigma$ and $p_i\not\equiv 0$?

3.3.2. Higher Order Equations

We consider the state $y(u)$ to be given by

$$\frac{\partial}{\partial t}y(u)+\Delta(a_1(x,t)\Delta y(u))=f+u \quad \text{in } Q, \qquad u,f\in L^2(Q), \tag{3.49}$$

$$y(u)|_\Sigma=0, \qquad \frac{\partial}{\partial n}y(u)|_\Sigma=0, \tag{3.50}$$

$$y(x,0;u)=y_0(x) \quad \text{in } \Omega, \qquad y_0\in L^2(\Omega).\ ^{18} \tag{3.51}$$

We assume that

$$a_1\in L^\infty(Q), \qquad a_1\geqslant\alpha>0 \quad \text{almost everywhere.} \tag{3.52}$$

The control is distributed:

$$\mathscr{U}=L^2(Q). \tag{3.53}$$

Let us suppose that

$$J(v)=\int_Q (y(v)-z_d)^2\,dx\,dt+(Nv,v)_{\mathscr{U}}. \tag{3.54}$$

The adjoint state is given by

$$-\frac{\partial}{\partial t}p(u)+\Delta(a_1(x,t)\Delta p(u))=y(u)-z_d \quad \text{in } Q, \tag{3.55}$$

$$p(u)|_\Sigma=0, \qquad \frac{\partial}{\partial n}p(u)|_\Sigma=0, \tag{3.56}$$

$$p(x,T;u)=0, \qquad x\in\Omega. \tag{3.57}$$

The optimality condition is,

$$\left.\begin{aligned}&\int_Q (p(x,t;u)+Nu(x,t))(v(x,t)-u(x,t))\,dx\,dt\geqslant 0 \qquad \forall v\in\mathscr{U}_{ad},\\ &u\in\mathscr{U}_{ad}.\end{aligned}\right\} \tag{3.58}$$

[18] Problem (3.49) – (3.50) admits a unique solution; apply Theorem 1.2 with

$$V=H_0^2(\Omega), \qquad a(t;\varphi,\psi)=\int_\Omega a_1(x,t)\Delta\varphi\,\Delta\psi\,dx.$$

3.4. Additional Results [19]

The problem of *direct*[20] solution of the non-linear system (3.10) may be posed.

Let us eliminate y and apply the operator $\left(\frac{\partial}{\partial t}+A\right)$ to the right hand side of equation (3.10). We obtain:

$$\left(\frac{\partial}{\partial t}+A\right)\left(-\frac{\partial}{\partial t}+A^*\right)p+\frac{1}{\nu}\inf(0,p)=f-\left(\frac{\partial}{\partial t}+A\right)z_d, \tag{3.59}$$

with the boundary condition,

$$p|_\Sigma=0, \quad \left(-\frac{\partial}{\partial t}+A^*\right)p|_\Sigma=-z_d|_\Sigma. \tag{3.60}$$

The second condition in (3.60) does not have any meaning unless z_d belongs to a space which is *strictly smaller* than $L^2(Q)$. Thus, the direct solution which we shall obtain requires *additional regularity hypotheses on the data.*

To be precise, let us suppose that

$$\left.\begin{aligned} &z_d\in L^2(0,T;H^1(\Omega)),\\ &\left(\frac{\partial}{\partial t}+A\right)z_d\in L^2(Q).\end{aligned}\right\} \tag{3.61}$$

By virtue of (3.61), $z_d|_\Sigma\in L^2(\Sigma)$ (in particular) and the second condition in (3.60) makes sense.

Finally, we have

$$\left.\begin{aligned} &\left(-\frac{\partial}{\partial t}+A^*\right)p(0)=y_0-z_d(0),\\ &p(T)=0.\end{aligned}\right\} \tag{3.62}$$

Our objective is to solve directly the non-linear problem (3.59), (3.60), (3.62).

Let us set

$$\beta(p)=\frac{1}{\nu}\inf(0,p) \tag{3.63}$$

and note that

$$p\beta(p)\geqslant 0. \tag{3.64}$$

---

[19] The reading of this section is not essential for the sequel.

[20] That is without reference to the control problem to which it is equivalent.

Let us define (cf. (3.5))

$$\mathscr{V}=\{\psi \mid \psi\in H^{2,1}(Q),\ \psi|_{\Sigma}=0,\ \psi(x,T)=0,\ x\in\Omega\}. \tag{3.65}$$

The space $\mathscr{V}$ is a Hilbert space when normed by the norm of $H^{2,1}(Q)$. For $\varphi,\psi\in\mathscr{V}$, let us set

$$\pi(\varphi,\psi)=\int_Q\left(-\frac{\partial}{\partial t}+A^*(t)\right)\varphi\cdot\left(-\frac{\partial}{\partial t}+A^*\right)\psi\,dx\,dt+\int_Q\beta(\varphi)\psi\,dx\,dt, \tag{3.66}$$

and note that the problem to be solved is equivalent to

$$\pi(p,\psi)=\int_Q\left(f-\left(\frac{\partial}{\partial t}+A\right)z_d\right)\psi\,dx\,dt+-\int_\Omega(y_0-z_d(x,0))\,\psi(x,0)\,dx$$
$$-\int_\Sigma z_d(x,t)\frac{\partial\psi}{\partial\nu_{A^*(t)}}\,d\Sigma\quad\forall\psi\in\mathscr{V}. \tag{3.67}$$

Now using results of Leray-Lions [1], for example, and utilizing the following:

(i) $$\pi(\psi,\psi)\geqslant\int_Q\left|\left(-\frac{\partial}{\partial t}+A^*\right)\psi\right|^2dx\,dt\geqslant c\,\|\psi\|^2_{H^{2,1}(Q)}$$

(the first inequality is obtained from (3.64) and the second from Lions-Magenes [1], Chapter 4, for example; more general results are given in Agranovich-Visik);

(ii) $$\pi(\varphi_1,\varphi_1-\varphi_2)-\pi(\varphi_2,\varphi_1-\varphi_2)\geqslant c\,\|\varphi_1-\varphi_2\|^2_{H^{2,1}(Q)}$$

this being due to the monotonicity of the function $p\to\beta(p)$;

(iii) the right hand side of (3.67) is a continuous linear form on $\mathscr{V}$,

we conclude that the problem admits a unique solution.

Remark 3.8. The *monotonicity* of the operator $p\to\beta(p)$ is obviously not miraculous; it is in fact related to the optimality condition (3.3).

3.5. Orientation

We shall now undertake a detailed study of the case where there are no constraints on the control and see how known results for systems described by ordinary differential equations can be extended to the present case. This will be our task in Sections 4 and 5.

We shall then return to the case where there are constraints on the control and then to other cases.

4. Decoupling and Integro-Differential Equation of Riccati Type (I)

4.1. Notation and Assumptions

We shall work within the framework of Section 2.3 where there are no constraints on the control (Remark 2.3) and we shall undertake a detailed study of the system of equations (2.24), (2.25) under the following hypotheses:

$$\mathscr{U} = L^2(0,T;E), \qquad E = \text{separable Hilbert Space}^{21}, \tag{4.1}$$

$$\mathscr{H} = L^2(0,T;F), \qquad F = \text{separable Hilbert Space}; \tag{4.2}$$

let

$$\left.\begin{array}{l} B(t)\in\mathscr{L}(E;V'), \quad C(t)\in\mathscr{L}(V;F), \quad t\in]0,T[, \\ t\to(B(t)e,\psi) \text{ and } t\to(C(t)\varphi,f') \text{ be measurable} \\ \forall e\in E, \quad \psi\in V, \quad \varphi\in V, \quad f'\in F', \text{ and} \\ \|B(t)\|_{\mathscr{L}(E;V')}\leqslant c, \quad \|C(t)\|_{\mathscr{L}(V;F)}\leqslant c. \end{array}\right\} \tag{4.3}$$

The operators B (resp. C) are defined by,

$$\left.\begin{array}{ll} Bu = \text{“}t\to B(t)u(t)\text{”}, & u\in\mathscr{U}, \\ Cf = \text{“}t\to C(t)f(t)\text{”}, & f\in L^2(0,T;V); \end{array}\right\} \tag{4.4}$$

the above formulae give us $B\in\mathscr{L}(\mathscr{U};L^2(0,T;V'))$ and $C\in\mathscr{L}(L^2(0,T;V),\mathscr{H})$.

We shall write $B(t)u$ instead of Bu, $C(t)f$ instead of Cf and we further assume that

$$\left.\begin{array}{l} Nu = N(t)u \quad (\text{function } t\to N(t)u(t)), \\ \text{where,} \\ N(t)\in\mathscr{L}(E;E), \quad (N(t)e,e_1) \text{ measurable}, \quad \|N(t)\|_{\mathscr{L}(E;E)}\leqslant c \quad\text{and} \\ (N(t)e,e)_E\geqslant\nu\|e\|_E^2 \quad \forall e\in E. \end{array}\right\} \tag{4.5}$$

Let

$$\left.\begin{array}{l} \Lambda_E \text{ (resp. } \Lambda_F\text{) be the canonical isomorphism of } E \text{ (resp. } F\text{)} \\ \text{onto its dual space.} \end{array}\right\} \tag{4.6}$$

Then $\Lambda_{\mathscr{U}}u(t)=\Lambda_E u(t)$ almost everywhere, $\Lambda f(t)=\Lambda_F f(t)$ almost everywhere.

To simplify the notation, we set

$$\left.\begin{array}{l} D_1(t)=B(t)N(t)^{-1}\Lambda_E^{-1}B(t)^*, \\ D_2(t)=C(t)^*\Lambda_F C(t). \end{array}\right\} \tag{4.7}$$

---

[21] To simplify some questions of measurability—but this is not at all essential.

We have,

$D_1(t) \in \mathscr{L}(V; V')$, $\quad D_2(t) \in \mathscr{L}(V; V')$,

$t \to (D_i(t)\varphi, \psi)$ measurable $\forall \varphi, \quad \psi \in V, \quad i=1,2, \quad \|D_i(t)\|_{\mathscr{L}(V;V')} \leqslant c$,

and

$$D_1(t)^* = D_1(t), \quad D_2(t)^* = D_2(t) \tag{4.8}$$

(note that $(N(t)^{-1} \Lambda_E^{-1})^* = N(t)^{-1} \Lambda_E^{-1}$).

With the above notation equations (2.24), (2.25) may be written as:

$$\left.\begin{aligned} &\frac{\partial y}{\partial t} + A(t)y + D_1(t)p = f, \quad t \in]0, T[, \\ -&\frac{\partial p}{\partial t} + A^*(t)p - D_2(t)y = g, \quad t \in]0, T[, \end{aligned}\right\} \tag{4.9}$$

where

$$g(t) = -C^*(t)\Lambda_F z_d(t), \tag{4.10}$$

with

$$y(0) = y_0, \quad p(T) = 0. \tag{4.11}$$

4.2. Operator $P(t)$, Function $r(t)$

The following observation is essential for the "decoupling" of the system of equations (4.9)–(4.11):

Lemma 4.1. *Under the hypotheses of Section 4.1. the system*

$$\left.\begin{aligned} &\frac{d\varphi}{dt} + A(t)\varphi + D_1(t)\psi = f \quad \text{in }]s, T[, \quad 0 < s < T, \\ -&\frac{d\psi}{dt} + A(t)^*\psi - D_2(t)\varphi = g \quad \text{in }]s, T[, \end{aligned}\right\} \tag{4.12}$$

$$\left.\begin{aligned} &\varphi(s) = h, \quad h \text{ given in } H, \\ &\psi(T) = 0, \end{aligned}\right\} \tag{4.13}$$

admits a unique solution.

Proof. From Section 2, we have, that the system of equations (4.12), (4.13) is precisely the system which determines the optimal control of a system whose state is given by

$$\left.\begin{aligned} &\frac{d\varphi}{dt} + A(t)\varphi = f + B(t)v \quad \text{in }]s, T[, \\ &\varphi(s) = h, \end{aligned}\right\} \tag{4.14}$$

where $\varphi=\varphi(v)$, and where the cost function is given by,

$$J_s^h(v)=\int_s^T \|C(t)\varphi(t;v)-z_d(t)\|_F^2 dt+\int_s^T (N(t)v(t),v(t))_E dt. \tag{4.15}$$

For this system $v\in\mathscr{U}(s,T)=L^2(s,T;E)$.

Indeed, the adjoint state $\psi(v)$ corresponding to $\varphi(v)$ is given by,

$$\left.\begin{aligned}-\frac{d\psi}{dt}+A^*(t)\psi&=C^*(t)\Lambda_E C(t)(\varphi(v)-z_d),\\ \psi(T)&=0\end{aligned}\right\} \tag{4.16}$$

and u is an optimal control if and only if

$$\Lambda_E^{-1}B^*(t)\psi+N(t)u=0. \tag{4.17}$$

Hence eliminating u we obtain (4.12), (4.13). □

We also have,

Lemma 4.2. *The mapping* $h\to\{\varphi,\psi\}$ = *solution of* (4.12), (4.13) *is a continuous mapping of* $H\to W(s,T)\times W(s,T)$, *where*

$$W(s,T)=\left[\varphi\,|\,\varphi\in L^2(s,T;V),\ \frac{\partial\varphi}{dt}\in L^2(s,T;V')\right]. \tag{4.18}$$

This mapping if affine.

Proof. The linear part of the mapping corresponds to the case where $f=0$ and $z_d=0$ (and hence $g=0$). Let us denote by $\varphi_n(v)$ the state of the system given by (4.14) when $\varphi(s)=h_n$. For a fixed v, if $h_n\to h$,

$$\varphi_n(v)\to\varphi(v)\quad\text{in } W(s,T). \tag{4.19}$$

Let u_n (resp. u) be the optimal control for $J_s^{h_n}(v)$ (resp. $J_s^h(v)$). Hence $J_s^{h_n}(u_n)=\inf_{v\in\mathscr{U}(s,T)} J_s^{h_n}(v)\leqslant J_s^{h_n}(u)$ and $J_s^{h_n}(u)\to J_s^h(u)$ from (4.19) (with $v=u$).

Hence,

$$\limsup J_s^{h_n}(u_n)\leqslant J_s^h(u)=\inf_{v\in\mathscr{U}(s,T)} J_s^h(v). \tag{4.20}$$

But

$$J_s^{h_n}(u_n)\geqslant \nu\int_s^T \|u_n\|_E^2 dt$$

which when combined with (4.20) shows that

$$u_n \text{ belongs to a bounded subset of } \mathscr{U}(s,T) \text{ as } h_n\to h. \tag{4.21}$$

We may then extract a subsequence u_μ such that

$$u_\mu \to w \quad \text{weakly in } \mathscr{U}(s,T). \tag{4.22}$$

Therefore $\varphi_\mu(u_\mu) \to \varphi(w)$ weakly in $W(s,T)$ and hence

$$\liminf J_s^{h_\mu}(u_\mu) \geqslant J_s^h(w),$$

which when combined with (4.20) shows $J_s^h(w) \leqslant J_s^h(u)$ and hence necessarily $w = u$. Therefore,

$$\left.\begin{aligned} u_n \to u \quad &\text{weakly in } \mathscr{U}(s,T), \\ J_s^{h_n}(u_n) \to J_s^h(u), & \end{aligned}\right\} \tag{4.23}$$

$$\left.\begin{aligned} \varphi_n(u_n) \to \varphi(u) \quad &\text{weakly in } W(s,T) \quad \text{and} \\ \psi_n(u_n) \to \psi(u) \quad &\text{weakly in } W(s,T). \end{aligned}\right\} \tag{4.24}$$

This proves the continuity of the linear part of the map $h \to \{\varphi, \psi\}$, mapping $H \to W(s,T) \times W(s,T)$ where the continuity is with respect to the *strong* topology of H and the *weak* topology of $W(s,T) \times W(s,T)$. This proves the lemma. □

Corollary 4.1. *For $h \in H$, let $\{\varphi, \psi\}$ be the solution of* (4.12), (4.13). *The mapping*

$$h \to \psi(s) \tag{4.25}$$

is a continuous affine mapping of $H \to H$.

Proof. The proof follows from the fact that the mapping defined by (4.25) is the composition of the mapping $h \to \{\varphi, \psi\}$ and the mapping $\{\varphi, \psi\} \to \psi(T)$. But from Theorem 1.1, the mapping $\{\varphi, \psi\} \to \psi(T)$ is a continuous mapping of $W(s,T) \times W(s,T) \to H$. □

Corollary 4.2. *The mapping* (4.25) *may be written in a unique manner as*

$$\left.\begin{aligned} \psi(s) = P(s)h + r(s), \quad &\textit{where} \\ P(s) \in \mathscr{L}(H;H), \quad &r(s) \in H. \end{aligned}\right\} \tag{4.26}$$

We are now in a position to prove the "fundamental identity":

Lemma 4.3. *Let $\{y, p\}$ be a solution of* (4.9)−(4.11). *We have*

$$p(t) = P(t)y(t) + r(t) \quad \forall t \in]0, T[, \tag{4.27}$$

where $P(t)$ and $r(t)$ are given in the following manner:

1. *we solve*

$$\left.\begin{aligned} &\frac{d\beta}{dt} + A(t)\beta + D_1(t)\gamma = 0 \quad \text{in }]s,T[, \\ -&\frac{d\gamma}{dt} + A^*(t)\gamma - D_2(t)\beta = 0 \quad \text{in }]s,T[, \\ &\beta(s) = h, \quad \gamma(T) = 0; \end{aligned}\right\} \tag{4.28}$$

then

$$P(s)h = \gamma(s); \tag{4.29}$$

2. *we solve,*

$$\left.\begin{aligned} &\frac{d\eta}{dt} + A(t)\eta + D_1(t)\xi = f \quad \text{in }]s,T[, \\ -&\frac{d\xi}{dt} + A^*(t)\xi - D_2(t)\eta = g \quad \text{in }]s,T[, \\ &\eta(s) = 0, \quad \xi(T) = 0; \end{aligned}\right\} \tag{4.30}$$

then

$$r(s) = \xi(s). \tag{4.31}$$

Proof. The operator $P(s)$ and the function $r(s)$ introduced in (4.26) are given by the rules (1) and (2) of the statement of the theorem. It suffices to decompose the mapping $h \to \psi(s)$ into the linear and constant part.

It now remains to prove the identity (4.27). Consider (4.12), (4.13) with $h = y(s)$. Then, if we define,

$\tilde{\varphi}$ (resp. $\tilde{\psi}$) = restriction of y (resp. p) on $]s,T[$,
$\tilde{\varphi}$ and $\tilde{\psi}$ satisfiy (4.12), (4.13) and hence $\varphi = \tilde{\varphi}$, $\psi = \tilde{\psi}$.

Therefore

$$\psi(s) = \tilde{\psi}(s) = p(s) \quad \text{and from (4.26),} \quad p(s) = P(s)y(s) + r(s).$$

Since s is fixed but arbitrary, we obtain the desired result. □

Let us now indicate some of the properties of $P(t)$:

Lemma 4.4. *We have,*

$$P(t)^* = P(t), \tag{4.32}$$

and

$$t \to (P(t)h, \bar{h}) \quad \textit{is continuous on} \quad [0,T], \qquad \forall h, \bar{h} \in H. \tag{4.33}$$

Proof. 1. Let $\bar{h} \in H$; applying (4.28), (4.29) and calling $\bar{\beta}, \bar{\gamma}$ the solution of (4.28) corresponding to $\bar{h}$, we obtain $P(s)\bar{h} = \bar{\gamma}(s)$.

Forming the scalar product of the second equation in (4.28) with $\bar{\beta}$ and integrating between s and T, we obtain,

$$0=\int_s^T\left(-\frac{d\gamma}{dt}+A^*(t)\gamma-D_2(t)\beta,\bar{\beta}(t)\right)dt$$

$$=(\gamma(s),\bar{\beta}(s))-\int_s^T(D_2(t)\beta,\bar{\beta})\,dt+\int_s^T\left(\gamma,\frac{d\bar{\beta}}{dt}+A(t)\bar{\beta}\right)dt$$

$$=(\gamma(s),\bar{\beta}(s))-\int_s^T(D_2(t)\beta,\bar{\beta})\,dt-\int_s^T(\gamma,D_1(t)\gamma)\,dt$$

whence, (using (4.8)),

$$(P(s)h,\bar{h})=\int_s^T(D_2(t)\beta,\bar{\beta})\,dt+\int_s^T(D_1(t)\gamma,\bar{\gamma})\,dt. \tag{4.34}$$

From (4.7) we see that this formula is symmetric. This proves (4.32).

2. To prove (4.33), we shall transform the problem to one which takes place in a fixed interval by effecting the change of variable

$$t=s+\tau(T-s),\qquad \tau\in[0,1].$$

Let us set

$$\beta(s+\tau(T-s))=\beta_s(\tau), \tag{4.35}$$

the same for γ_s, f_s, etc., and

$$A(s+\tau(T-s))=A_s(\tau),$$

and the same for $D_{1s}(\tau), D_{2s}(\tau)$.

The system of equations (4.28) becomes:

$$\left.\begin{aligned}\frac{d\beta_s}{d\tau}+(T-s)A_s(\tau)\beta_s+(T-s)D_{1s}(\tau)\gamma_s=0,&\quad 0<\tau<1,\\ -\frac{d\gamma_s}{d\tau}+(T-s)A_s^*(\tau)\gamma_s-(T-s)D_{2s}(\tau)\beta_s=0,&\\ \beta_s(0)=h,\quad \gamma_s(1)=0.&\end{aligned}\right\} \tag{4.36}$$

System (4.36) may be considered as determining the optimal control for the following problem:

—the *state* of the new system is given by,

$$\left.\begin{aligned}\frac{d\lambda}{d\tau}+(T-s)A_s(\tau)\lambda=(T-s)B_s(\tau)w, \qquad w\in L^2(0,1;E),\\ \lambda(0)=h,\end{aligned}\right\}\tag{4.37}$$

and the cost function is given by,

$$\mathcal{J}_s(w)=(T-s)\int_0^1 \|C_s(\tau)\lambda(\tau)\|_F^2\,d\tau+(T-s)\int_0^1 (N_s(\tau)\,w,w)_E\,d\tau. \tag{4.38}$$

The adjoint state then is defined by,

$$\left.\begin{aligned}-\frac{d\mu}{d\tau}+(T-s)A_s^*(\tau)\,\mu=C_s^*(\tau)\,\Lambda_F\,C_s(\tau)\lambda,\\ \mu(1)=0,\end{aligned}\right\}\tag{4.39}$$

and w is an optimal control if and only if

$$\Lambda_E^{-1}\,B_s^*(\tau)\,\mu+N_s(\tau)\,w=0, \tag{4.40}$$

whence we obtain (4.36) by eliminating w.

Let us remark that (this is precisely the reason why we carried the factor $(T-s)$ in (4.38))

$$\operatorname*{Inf}_{w\in L^2(0,1;E)} \mathcal{J}_s(w)=\operatorname*{Inf}_{v\in\mathcal{U}(s,T)} J_s^h(v) \qquad (f=0,\ z_d=0). \tag{4.41}$$

We thus have,

$$P(s)h=\gamma_s(0). \tag{4.42}$$

Consider a sequence $s_n\to s\neq T$. Let us denote by $\lambda_s(w)$ the state given by (4.37) and let us first prove that,

$$\lambda_{s_n}(w)\to\lambda_s(w) \quad \text{weakly in } W(0,1).\,^{22} \tag{4.43}$$

We may easily verify that

$$\lambda_{s_n} \quad \text{ranges in a bounded set in } L^2(0,1;V),$$

and then utilising (4.37) that

$$\lambda_{s_n}\in\text{bounded set in } W(0,1) \quad \text{as } s_n\to s. \tag{4.44}$$

We may then extract a subsequence s_μ such that

$$\lambda_{s_\mu}=\lambda_{s_\mu}(w)\to\tilde{\lambda} \quad \text{weakly in } W(0,1). \tag{4.45}$$

---

[22] Let us recall that $W(0,1)=\left\{\lambda\,|\,\lambda\in L^2(0,1;V),\ \dfrac{d\lambda}{dt}\in L^2(0,1;V')\right\}$.

If ψ is fixed in V, we have,

$$((T-s_\mu)A_{s_\mu}(\tau)\lambda_{s_\mu}(\tau),\psi)=(\lambda_{s_\mu}(\tau),(T-s_\mu)A^*_{s_\mu}(\tau)\psi)$$

and

$$(T-s_\mu)A^*_{s_\mu}(\tau)\psi\to(T-s)A^*_s(\tau)\psi \quad \textit{strongly} \text{ in } L^2(0,1;V),$$

hence "$\tau\to(\lambda_{s_\mu}(\tau),(T-s_\mu)A^*_{s_\mu}(\tau))$" converges for example in $\mathscr{D}'(]0,1[)$, towards "$\tau\to(\tilde{\lambda}(\tau),(T-s)A^*_s(\tau)\psi)$". Therefore from (4.37) (setting $s=s_\mu$) we deduce

$$\frac{d\tilde{\lambda}}{dt}+(T-s)A_s(\tau)\tilde{\lambda}=(T-s)B_s(\tau)w.$$

Since $\tilde{\lambda}(0)=h$, we deduce that $\tilde{\lambda}=\lambda_s(w)$ whence (4.43).

Let us now show that

$$\lambda_{s_n}(w)\to\lambda_s(w)=\lambda_s \quad \textit{strongly} \text{ in } W(0,1). \tag{4.46}$$

For this, we consider,

$$X_n=\int_0^1(T-s_n)(A_{s_n}(\lambda_{s_n}-\lambda_s),\lambda_{s_n}-\lambda_s)d\tau+\tfrac{1}{2}|\lambda_{s_n}(1)-\lambda_s(1)|^2.$$

Using (4.37) (with s, s_n, w fixed),

$$X_n=\int_0^1(T-s_n)(B_{s_n}w,\lambda_{s_n})d\tau-\left[\int_0^1(T-s_n)(A_{s_n}\lambda_{s_n},\lambda_s)d\tau+\tfrac{1}{2}(\lambda_{s_n}(1),\lambda_s(1))\right]$$
$$-\int_0^1(T-s_n)(A_{s_n}\lambda_s,\lambda_{s_n}-\lambda_s)d\tau-\tfrac{1}{2}(\lambda_s(1),\lambda_{s_n}(1)-\lambda_s(1))$$

and by virtue of (4.43), we deduce that

$$X_n\to\int_0^1(T-s)(B_sw,\lambda_s)d\tau-\int_0^1(T-s)(A_s\lambda_s,\lambda_s)d\tau-\tfrac{1}{2}|\lambda_s(1)|^2=0.$$

Since

$$X_n\geqslant\alpha\inf(T-s_n)\,\|\lambda_{s_n}-\lambda_s\|^2_{L^2(0,1;V)},$$

it follows that $\lambda_{s_n}\to\lambda$ *strongly* in $L^2(0,1;V)$. We finally deduce (4.46) by using the fact that

$$\frac{d}{d\tau}\lambda_{s_n}=-(T-s_n)A_{s_n}(\tau)\lambda_{s_n}+(T-s_n)B_{s_n}(\tau)w.$$

Let w_s be the optimal control corresponding to problem (4.37), (4.38) as w ranges over $L^2(0,1;E)$. We have,

$$J_{s_n}(w_{s_n})\leqslant J_{s_n}(w_s)\to J_s(w_s) \quad \text{from (4.46)}.$$

Hence

$$\limsup J_{s_n}(w_{s_n}) \leqslant J_s(w_s). \tag{4.47}$$

We now show, as in the proof of Lemma 4.2, that w_{s_n} ranges in a bounded set of $L^2(0,1;E)$; hence we may extract a subsequence $w_{s_\mu} \to \tilde{w}$ *weakly* in $L^2(0,1;E)$ and we note that $\lambda_{s_\mu}(w_{s_\mu}) \to \lambda_s(\tilde{w})$ *weakly* in $W(0,1)$. Now as in Lemma 4.2, we prove that $\tilde{w} = w_s$ and $\lambda_{s_n} \to \lambda_s$ weakly in $W(0,1)$ and

$$\mu_{s_n} \to \mu_s \quad \textit{weakly} \text{ in } W(0,1). \tag{4.48}$$

But $\lambda_{s_n} = \beta_{s_n}$, $\mu_{s_n} = \gamma_{s_n}$ and hence from (4.48),

$$\mu_{s_n}(0) = \gamma_{s_n}(0) = P(s_n)h \to \mu_s(0) = \gamma_s(0) = P(s)h \quad \textit{weakly} \text{ in } H. \quad \square$$

Lemma 4.5. *There exists a constant C_1 such that*

$$|P(s)h| \leqslant C_1 |h| \quad \forall h \in H, \qquad \forall s \in [0,T]. \tag{4.49}$$

Proof. The solution $\{\beta, \gamma\}$ of (4.28) corresponds to the solution of an optimal control problem whose state equation is given by (4.14) and cost function (4.15) with $f=0$ and $z_d=0$. Let $J_s^h(v)$ be the cost function corresponding to a control v. If u denotes the optimal control, we have

$$J_s^h(u) \leqslant J_s^h(0) \leqslant C_2 |h|^2. \tag{4.50}$$

Now the optimal control u is determined

$$B^*\gamma + \Lambda_E N u = 0 \quad \text{in }]s,T[. \tag{4.51}$$

From (4.28), we may deduce,

$$\int_s^T (C\beta, C\beta)\,dt = \int_s^T (D_2\beta, \beta)\,dt = \int_s^T \left(-\frac{d\gamma}{dt} + A^*\gamma, \beta\right) dt$$

$$= (\gamma(s), \beta(s)) + \int_s^T \left(\gamma, \frac{d\beta}{dt} + A\beta\right) dt$$

$$= (P(s)h, h) + \int_s^T (\gamma, -D_1\gamma)\,dt,$$

and hence from (4.51)

$$\int_s^T (Nu,u)_E\,dt = \int_s^T (B^*\gamma, N^{-1}\Lambda_E^{-1}B^*\gamma)\,dt = \int_s^T (\gamma, D_1\gamma)\,dt$$

whence

$$J_s^h(u)=(P(s)h,h). \tag{4.52}$$

Utilising (4.50), we obtain the desired result (4.49). ▯

Lemma 4.6. *We have formula* (4.52) *and hence*

$$(P(s)h,h)\geqslant 0 \quad \forall h\in H. \tag{4.53}$$

4.3. Formal Calculations

With the identity (4.27) as our starting point, we shall now carry out certain formal calculations which will be justified in a later section. Differentiating (4.27) formally, and setting, in general, $dX/dt=X'$, and substituting this in (4.9), we obtain,

$$-P'y-Py'-r'+A^*p-D_2y=g.$$

Utilising the first equation in (4.9), the above becomes

$$-P'y-P[f-A(t)y-D_1(t)p]-r'+A^*p-D_2y=g,$$

and further replacing p by (4.27), we obtain,

$$(-P'+PA)y-Pf+PD_1(Py+r)-r'+A^*(Py+r)-D_2y=g,$$

whence

$$(-P+PA+A^*P+PD_1P-D_2)y+(-r'+A^*r+PD_1r-Pf-g)=0. \tag{4.54}$$

Since y is arbitrary (a fact that needs to be justified!), we deduce that (4.49) is equivalent to

$$-\frac{dP}{dt}+PA+A^*P+PD_1P=D_2 \quad \text{in }]0,T[, \tag{4.55}$$

$$-\frac{dr}{dt}+A^*r+PD_1r=Pf+g \quad \text{in }]0,T[. \tag{4.56}$$

Further from (4.27) setting $t=T$

$$p(t)=P(T)y(T)+r(T)=0,$$

which is equivalent to

$$P(T)=0, \tag{4.57}$$

$$r(T)=0. \tag{4.58}$$

We have thus arrived at the following result in a formal manner: *$P(t)\in\mathscr{L}(H;H)$ is determined by the solution of the non-linear equation (of Riccati type)* (4.55) *with the boundary condition* (4.57)*; $r(t)$ is determined by the solution of the (abstract) parabolic equation* (4.56) *with the boundary condition* (4.58).

Remark 4.1. In the sequel we shall justify these formal calculations. Let us however emphasize that the *existence of $P(t)$ is demonstrated by virtue of corollary 4.2 and not from an eventual direct solution of* (4.55)–(4.57).

4.4. The Finite Dimensional Case; Approximation

Before justifying the formal calculations of Section 4.3, we shall consider a finite dimensional approximation to the problem. The original problem will be reduced to the *finite dimensional problem* by using the method of Faedo-Galerkin.

As in Section 1.4, let $w_1,\dots w_m,\dots$ be a 'basis' in V. Let $y_m(t)$ be the "*m^{th} order approximation*" to the state of the system given by

$$\left.\begin{aligned}&\frac{d}{dt}(y_m(t;v),w_j)+a(t;y_m(t;v),w_j)=(f(t),w_j)+(B(t)v(t),w_j),\\&\qquad\qquad 1\leqslant j\leqslant m,\\&y_m(0;v)=y_{0m}=\sum_{i=i}^{m}\xi_{im}w_i\to y_0\quad\text{in } H\quad\text{as } m\to\infty,\end{aligned}\right\}\tag{4.59}$$

(where $y_m(t;v)$ belongs to the space spanned by $w_1,\dots w_m$).

We consider the cost function

$$J_m(v)=\int_0^T\|C(t)y_m(t;v)-z_d(t)\|_F^2\,dt+\int_0^T(N(t)v,v)_E\,dt\tag{4.60}$$

and it is desired to find

$$\operatorname{Inf} J_m(v),\qquad v\in L^2(0,T;E)=\mathscr{U}.$$

Let u_m be the "*m^{th} order approximate control*". We introduce "*the mth order adjoint state*" by

$$\left.\begin{aligned}&-\left(\frac{dp_m}{dt}(t;v),w_j\right)+a^*(t;p_m(t;v),w_j)=(C\,y_m(t;v)-z_d,C(t)w_j)_E,\\&\qquad 1\leqslant j\leqslant m,\quad p_m(T;v)=0,\end{aligned}\right\}\tag{4.61}$$

where $p_m(t;v)$ belongs to the space spanned by $w_1,\dots w_m$.

The optimal control u_m is given by the simultaneous solution of (4.59), (4.61) together with $v=u_m$ satisfying

$$B^* p_m + \Lambda_E N u_m = 0. \tag{4.62}$$

Eliminating u_m, we obtain (using the notation of Section 4.1):

$$\left.\begin{aligned} &\left(\frac{d}{dt} y_m, w_j\right) + a(t; y_m, w_j) + (D_1(t) p_m, w_j) = (f(t), w_j),\\ &-\left(\frac{d}{dt} p_m, w_j\right) + a^*(t; p_m, w_j) - (D_2(t) y_m, w_j) = (g(t) w_j),\\ &\qquad y_m(0) = y_{0m} \to y_0,\\ &\qquad p_m(T) = 0. \end{aligned}\right\} \tag{4.63}$$

We now prove

Theorem 4.1. *Let the hypotheses and notation of Section 4.1 hold. Let $J_m(v)$ be the cost function given by* (4.60) *for the "m*[th] *order approximate system"* (4.59). *The optimal control u_m is given by the solution of* (4.63) *and* (4.62). *We have:*

$$\begin{aligned} &\operatorname{Inf} J_m(v) = J_m(u_m) \to \operatorname{Inf} J(v) = J(u),\\ &u_m \to u \quad \text{strongly in } L^2(0,T;E)\ (=\mathscr{U}),\\ &y_m \to y, \quad p_m \to p \quad \text{stongly in } L^2(0,T;V). \end{aligned}$$

Proof. We know (see footnote [4], page 106) that

$$y_m(t;v) \to y(t;v) \quad \text{strongly in } L^2(0,T;V).$$

Then $J_m(u_m) \leqslant J_m(u) \to J_m(u)$, and hence $\limsup J_m(u_m) \leqslant J(u)$. Since

$$J_m(u_m) \geqslant \nu \int_0^T \|u_m\|_E^2 dt,$$

we may extract a subsequence $u_\mu \to \tilde{u}$ weakly in $\mathscr{U}$. Hence $y_\mu(u_\mu) \to y(\tilde{u})$ weakly in $L^2(0,T;V)$ and $\liminf J_\mu(u_\mu) \geqslant J(\tilde{u})$. Hence $\tilde{u}=u$ necessarily, and $J_m(u_m) \to J(u)$ and $u_m \to u$ weakly in $\mathscr{U}$. Therefore $y_m = y_m(u_m) \to y(u)$ weakly in $L^2(0,T;V)$ and $p_m(u_m) \to p(u)$ weakly in $L^2(0,T;V)$.

But since $J_m(u_m) \to J(u)$ we necessarily have (using a contradiction argument)

$$\int_0^T (N u_m, u_m) dt \to \int_0^T (N u, u) dt,$$

which when combined with the fact that $u_m \to u$ weakly in $\mathscr{U}$ implies $u_m \to u$ strongly in $\mathscr{U}$. Hence the theorem. □

We may now "decouple" the system of equation (4.63) in the same manner as we have done for the infinite dimensional case in the previous section. Let

$$V_m = \text{space spanned by } w_1, \dots, w_m. \tag{4.64}$$

We consider the analogue of (4.12), (4.13) in V_m:

$$\left.\begin{aligned} &\left(\frac{d}{dt}\varphi_m, w_j\right) + a(t;\varphi_m, w_j) + (D_1(t)\psi_m, w_j) = (f(t), w_j),\\ &-\left(\frac{d}{dt}\psi_m, w_j\right) + a^*(t;\psi_m, w_j) - (D_2(t)\varphi_m, w_j) = (g(t), w_j),\\ &\varphi_m(s) = h, \quad h \text{ given in } V_m, \psi_m(T) = 0. \end{aligned}\right\} \tag{4.65}$$

Then

$$\left.\begin{aligned} &\psi_m(s) = P_m(s)h + r_m(s),\\ &P_m(s) \in \mathscr{L}(V_m; V_m), \quad r_m(s) \in V_m, \quad P_m(s)^* = P_m(s), \end{aligned}\right\} \tag{4.66}$$

and we have the identity,

$$p_m(t) = P_m(t)\, y_m(t) + r_m(t). \tag{4.67}$$

A formal calculation analogous to that of Section 4.3 leads to

$$\left.\begin{aligned} &-\left(\frac{dP_m}{dt} w_i, w_j\right) + a(t; w_i, P_m w_j) + a^*(t; P_m w_i, w_j)\\ &+ (D_1 P_m w_i, P_m w_j) = (D_2 w_i, w_j) \quad \forall\, i,j \quad 1 \leqslant i, \quad j \leqslant m,\\ &P_m(T) = 0, \end{aligned}\right\} \tag{4.68}$$

and

$$\left.\begin{aligned} &-\left(\frac{dr_m}{dt}, w_j\right) + a^*(t; r_m, w_j) + (P_m D_1 r_m, w_j) = (P_m f + g, w_j),\\ &1 \leqslant j \leqslant m, \quad r_m(T) = 0. \end{aligned}\right\} \tag{4.69}$$

Conversely, let P_m be a local solution of the system of equations (of Riccati type), that is, P_m is a solution in $]s, T[$, for $T - s$ sufficiently small. Let r_m be a solution of (4.69) in the same interval. Then, if we define

$$p_m(s) = P_m(s)\, y_m(s) + r_m(s),$$

we obtain results analogous (in m-dimensions) to that of (4.52) and (4.49) giving us

$$|P_m(s)h| \leqslant C_1 |h| \quad \forall h \in V_m, \tag{4.70}$$

where C_1 is a constant independent of s and m.

Therefore a solution of (4.68) globally exists in $]0, T[$ and r_m defined by (4.69) also exists in $]0, T[$. This justifies the formulae (4.68), (4.69).

Hence we have,

Theorem 4.2. *Let the hypotheses and notation of Theorem 4.1 hold. Then $P_m(t)$ is a function from $[0, T] \to \mathscr{L}(V_m; V_m)$ which is continuous and has its first derivative in L^∞; similarly $r_m(t)$ is a function from $[0, T] \to V_m$ which is continuous and has its first derivative in L^∞; P_m is a solution of the system of Riccati equations* (4.68) *and r_m is a solution of* (4.69). *Furthermore* (4.70) *holds.*

We shall now let m tend to infinity.

4.5. Passage to the Limit

We apply the analogue of Theorem 4.1 to the system $\{\varphi_m, \psi_m\}$ with $\varphi_m(s) = h \in V_m$, $m \geqslant m_0$; then $\varphi_m \to \varphi$, $\psi_m \to \psi$ strongly in $L^2(s, T; V)$, where φ, ψ is the solution of (4.14), (4.16) and (4.17).

Using the second condition in (4.65) we may verify that

$$|\psi_m(s)| \leqslant \text{constant} \tag{4.71}$$

and hence we may assume that $\psi_\mu(s) \to \chi$ weakly in H. But

$$(\psi_\mu(s), w_j) = \int_s^T [a^*(t; \psi_\mu, w_j) - (D_2 \varphi_\mu, w_j) - (g, w_j)] \, dt$$

hence $(\chi, w_j) = (\psi(s), w_j) \, \forall j$. Thus

$$\chi = \psi(s) \tag{4.72}$$

and

$$\psi_m(s) = P_m(s)h + r_m(s) \to \psi(s) = P(s)h + r(s) \quad \text{weakly in } H. \tag{4.73}$$

Therefore setting $h = 0$ in (4.73) and then making h arbitrary, we have

$$r_m(s) \to r(s) \quad \text{weakly in } H, \quad \forall s, \tag{4.74}$$

$$P_m(s)h \to P(s)h \quad \text{weakly in } H, \quad \forall h \in V_m, \quad \forall s. \tag{4.75}$$

From (4.75) and from Lebesgue's theorem we deduce that

$$\left.\begin{array}{l}(P_m(s)h,\bar{h}) \to (P(s)h,\bar{h}) \quad \text{in } L^1(0,T),\\ h\in V_m\,, \quad \bar{h}\in H\end{array}\right\} \tag{4.76}$$

whence,

Theorem 4.3. *Under the hypotheses and notation of Theorem 4.2, as* $m\to\infty$, $P_m\to P$ *in the sense that* $(P_m h,\bar{h})\to(Ph,\bar{h})$ *in* $L^1(0,T)$ $\forall h\in V_{m_0}$ *(m_0 fixed but arbitrary),* $\bar{h}\in H$; $r_m\to r$ *weakly in* $L^2(0,T;H)$.

4.6. Integro-Differential Equation of Riccati Type

The remaining problem is to the pass to the limit in equations (4.68), (4.69) in a manner such that we may obtain (4.55)–(4.58). The difficulty lies in the fact that there is a non-linear term in (4.68) (and due to the term $P_m D_1 r_m$ in (4.69)).

We shall now obtain this result under an *additional hypothesis* which is rather *restrictive*[23]. In the sequel we shall consider an example where this hypothesis does not hold but even then by using direct arguments we are able to justify (4.55)–(4.58).

The additional hypothesis is the following:

$$B(t)\in\mathscr{L}(E;H) \quad (\text{instead of } \mathscr{L}(E;V')), \tag{4.77}$$

and we also assume that (this is not restrictive for the applications at least when Ω is *bounded*)

$$\text{the injection map of } V \text{ into } H \text{ is compact.} \tag{4.78}$$

We shall prove,

Lemma 4.7. *Under the hypotheses of Theorem 4.1 together with* (4.77), (4.78), *if* $f\in L^2(0,T;H)$, *we have*

$$r\in W(0,T) \quad (\text{i.e. } r\in L^2(0,T;V),\ \frac{dr}{dt}\in L^2(0,T;V')). \tag{4.79}$$

Proof. In (4.69), we have

$$(P_m D_1 r_m, w_j)=(D_1 r_m, P_m w_j), \qquad (P_m f, w_j)=(f, P_m w_j).$$

We may deduce that

$$-\left(\frac{dr_m}{dt}, r_m\right) + a^*(t; r_m, r_m) = -(D_1 r_m, P_m r_m)+(f, P_m r_m)+(g, r_m)$$

---

[23] This hypothesis is probably not required.

and by virtue of (7.40), integrating from s to T, we deduce,

$$\tfrac{1}{2}|r_m(s)|^2+\alpha\int_s^T \|r_m(\sigma)\|^2\,d\sigma-\lambda\int_s^T |r_m(\sigma)|^2\,d\sigma$$

$$\leqslant C\int_s^T (|r_m|^2+|f|\cdot|r_m|+\|g\|_{V'}\cdot\|r_m\|)\,d\sigma,$$

whence

$$|r_m(s)|^2+\int_s^T \|r_m(\sigma)\|^2\,d\sigma\leqslant C\int_s^T |r_m(\sigma)|^2\,d\sigma+C\int_s^T (|f(\sigma)|^2+\|g(\sigma)\|_{V'}^2)\,d\sigma. \tag{4.80}$$

From the Gronwall inequality, we deduce

$$\int_0^T \|r_m(\sigma)\|^2\,d\sigma \leqslant \text{constant}. \tag{4.81}$$

Hence from (4.69) we have

$$\left.\begin{aligned}&\left(\frac{dr_m}{dt}, w_j\right)=(\rho_m(t),w_j)_V, \qquad r_m(T)=0,\\ &\rho_m \quad \text{bounded in } L^2(0,T;V).\end{aligned}\right\} \tag{4.82}$$

Let us extend ρ_m by 0 outside $(0,T)$. Then, since $r_m(T)=0$, we may assume

$$\left(\frac{dr_m}{dt}, w_j\right)=(\rho_m(t),w_j)_V \quad \text{on } \mathbb{R}.$$

By considering the Fourier Transform in t ($\hat{\varphi}=$ Fourier transform of φ) of the above, we obtain

$$i\tau(\hat{r}_m(\tau),w_j)=(\hat{\rho}_m(\tau),w_j)_V, \tag{4.83}$$

with

$$\hat{r}_m(\tau)=\sum_{i=1}^m \hat{g}_{im}(\tau)w_i.$$

Multiplying (4.83) by $\overline{\hat{g}_{jm}(\tau)}$ and summing over j, we get

$$|\tau|\cdot|\hat{r}_m(\tau)|^2\leqslant\|\hat{\rho}_m(\tau)\|\cdot\|\hat{r}_m(\tau)\|$$

whence

$$\int_{-\infty}^{+\infty}|\tau|\,|\hat{r}_m(\tau)|^2\,d\tau\leqslant\text{constant}. \tag{4.84}$$

The estimate (4.84) signifies that the derivative in t of r_m of order $\frac{1}{2}$ is bounded in $L^2(0,T;H)$ as $m\to\infty$. Hence, from a compactness result

(Lions [3], Proposition 4.2, page 60), we may extract a sequence r_μ such that

$$r_\mu \to r \quad \text{weakly in } L^2(0,T;V),\ [24]$$

But then,

$$r_\mu \to r \quad \text{strongly in } L^2(0,T;H).\ [25]$$

$$\int_0^T (P_\mu D_1 r_\mu, w_j \varphi)\,dt = \int_0^T (D_1 r_\mu, P_\mu w_j \varphi)\,dt$$

$$\to \int_0^T (D_1 r, P w_j \varphi)\,dt \quad \forall \varphi \in C^\circ[0,T]$$

and hence we have, for example,

$$(P_\mu D_1 r_\mu, w_j) \to (P D_1 r, w_j) \quad \text{in } \mathscr{D}'(]0,T[).$$

Therefore we may pass to the limit in (4.69) to obtain

$$-\left(\frac{dr}{dt}, w_j\right) + a^*(t; r, w_j) + (P D_1 r, w_j) = (Pf + g, w_j) \quad \forall j,$$

which proves (4.56). We also have $\frac{dr}{dt} \in L^2(0,T;V')$ whence (4.79). □

We are now in a position to prove

Theorem 4.4. *Let the hypotheses of Theorem* 4.1 *together with* (4.77) *and* (4.78) *hold. Let* $f \in L^2(0,T;H)$. *Let* $\{y, p\}$, *the solution of* (4.9), (4.11), *determine the optimal control. Then*

$$p = Py + r \tag{4.85}$$

where P and r have the following properties:

$$\left.\begin{array}{l} P(t) \in \mathscr{L}(H;H), \quad P(t)^* = P(t); \\ \text{if } \eta \in W(0,T) \text{ with } \dfrac{d\eta}{dt} + A(t)\eta \in L^2(0,T;H), \text{ then} \\ P(t)\eta \in W(0,T); \end{array}\right\} \tag{4.86}$$

P satisfies (4.55) *in the sense*

$$-\left(\frac{d}{dt} P(t)\right)\eta + PA\eta + A^* P\eta + PD_1 P\eta = D_2 \eta, \tag{4.87}$$

$$\text{for all } \eta \text{ as in (4.86) with } A\eta \in L^2(0,T;H),$$

and we have $P(T) = 0$;
r is the solution in $W(0,T)$ *of* (4.56), (4.58);
P and r thus defined are unique.

---

[24] These arguments, together with (4.84) were introduced for non-linear problems in Lions [6].

[25] From Theorem 4.3, we already know that $r_m \to r$ weakly in $L^2(0,T;H)$.

Proof. Let η be as in (4.86). Let us define p as the solution of

$$-\frac{dp}{dt} + A^*(t)p = D_2\eta + g, \quad p(T)=0,$$

and f by

$$f = \frac{d\eta}{dt} + A(t)\eta + D_1 p; \quad f \in L^2(0,T;H).$$

Then, if we define $y_0 = \eta(0)$, $\{\eta, p\}$ is a solution of (4.9), (4.11) and therefore

$$p = P\eta + r$$

and since (Lemma 4.7) $r \in W(0,T)$, we have (4.86).

Now let Θ denote the space of functions η satisfying (4.86) together with $A\eta \in L^2(0,T;H)$ (hence $\eta' \in L^2(0,T;H)$). The mapping

$$\eta \to \frac{d}{dt}(P(t)\eta) - P(t)\left(\frac{d\eta}{dt}\right)$$

will be shown to be a continuous linear mapping of $\Theta \to L^2(0,T;V')$, and on the space $\Theta \cap \mathscr{D}(]0,T[;H)$ it coincides with $P'(t)\cdot\eta = (dP(t)/dt)\eta$, where $P'(t)$ denotes the distributional derivative of $t \to P(t): [0,T] \to \mathscr{L}(H;H)$. We then set

$$\frac{d}{dt}(P(t)\eta) - P(t)\left(\frac{d\eta}{dt}\right) = P'(t)\eta \quad \text{or} \quad \left(\frac{dP(t)}{dt}\right)\eta$$

(which is in fact not completely justified unless $\Theta_0 \cap \mathscr{D}(]0,T[;H)$ is dense in Θ_0 where $\Theta_0 = \{\eta \mid \eta \in \Theta, \eta(0) = \eta(T) = 0\}$).

The calculations performed in Section 4.3 (with $y = \eta$) are now valid and using the same notation we arrive at (4.87) from which (4.56) follows. Since we must have $P(T)y(T) + r(T) = 0$ for arbitrary $y(T)$, we obtain (4.57) and (4.58).

It remains to show that (4.86), (4.87), (4.57) and (4.56), (4.58) define P and r in a unique manner. Let us suppose P and r satisfy these equations. Let y be the solution of

$$\frac{dy}{dt} + Ay + D_1 Py = f - D_1 r,$$

$$y(0) = y_0$$

and set

$$p = Py + r.$$

Then we may verify that p satisfies

$$-\frac{dp}{dt}+A^*p-D_2y=g,$$

and hence $\{y,p\}$ is a solution of (4.9), (4.11), which proves the uniqueness of P and r. ▯

Remark 4.2. The non-linear equation (4.55) (or (4.87)) is an evolution equation of Riccati type. For concrete examples, see the following sections.

Remark 4.3. The operator $P(t)$ and the function $r(t)$ permit us to synthesize an *optimal feedback controller* for this problem: the optimal control is an *affine function* of the state given by

$$u(t)=N(t)^{-1}\Lambda_E^{-1}B^*(t)[P(t)y(t)-r(t)]. \tag{4.88}$$

We have here obtained an extension of results in the finite dimensional case. For finite dimensional results see Kalman [1], [2].

Remark 4.4. If the domain of $A(t)$ in H is independent of t, that is, $D(A(t))=D\ \forall t\in[0,T]$, then in (4.87) we may take $\eta\in D$ to be independent of t.

4.7. Connections with the Hamilton-Jacobi Theory

To simplify the exposition, we shall take "$f=0$" and "$z_d=0$" in the preceeding problem.

Let us consider a system whose state is given by,

$$\left.\begin{aligned}\frac{dy(v)}{dt}+A(t)y(v)=B(t)v \quad \text{in }]s,T[,\\ y(s;v)=h;\end{aligned}\right\} \tag{4.89}$$

the cost function is given by (cf. proof of Lemma 4.5)

$$J_s^h(v)=\int_s^T[\|Cy(t;v)\|_F^2+(Nv,v)_E]\,dt. \tag{4.90}$$

We set

$$\mathscr{W}(h,s)=\begin{cases}\frac{1}{2}\inf J_s^h(v),\\ v\in L^2(s,T;E)\end{cases} \tag{4.91}$$

and we propose to obtain an "equation" satisfied by $\mathscr{U}$, $\dfrac{\partial\mathscr{W}}{\partial s}$ and $\dfrac{\partial\mathscr{W}}{\partial h}$ (Fréchet derivatives).

We remark (cf. (4.52)) that

$$\mathscr{W}(h,s)=\tfrac{1}{2}(P(s)h,h), \tag{4.92}$$

hence

$$\frac{\partial \mathscr{W}}{\partial s}(h,s)=\frac{1}{2}\left(\frac{dP}{ds}(s)h,h\right),\quad \frac{\partial \mathscr{W}}{\partial h}(h,s)=P(s)h.$$

Utilising (4.87) (assumed valid!) we have (if $h\in V$)

$$-\frac{d}{ds}(P(s)h,h)+(A(s)h,P(s)h)+(P(s)h,A(s)h)$$
$$+(N(s)^{-1}\Lambda_E^{-1}B^*(s)P(s)h,B^*(s)P(s)h)=(D_2(s)h,h)=\|C(s)h\|_F^2.$$

If we now define (h,p being the "independent variables" in V)

$$H^0(h,p,s)=\tfrac{1}{2}\|C(s)h\|_F^2-(p,A(s)h)-(N(s)^{-1}\Lambda_E^{-1}B^*(s)p,B^*(s)p) \tag{4.93}$$

we have

$$\frac{\partial \mathscr{W}}{\partial s}(h,s)+H^0\left(h,\frac{\partial \mathscr{W}}{\partial h}(h,s)s\right)=0,\quad h\in V,\quad s\in]0,T[. \tag{4.94}$$

Now, introducing the Hamiltonian

$$H(h,p,s,e)=\tfrac{1}{2}\|C(s)h\|_F^2-(p,A(s)h)+\tfrac{1}{2}(N(s)e,e)+(p,B(s)e) \tag{4.95}$$

it is easily verified that

$$H^0(h,p,s)=\inf_{e\in E}H(h,p,s,e). \tag{4.96}$$

Equation (4.94) may thus be written in the equivalent form

$$\frac{\partial \mathscr{W}}{\partial s}(h,s)+\inf_{e\in E}H\left(h,\frac{\partial \mathscr{W}}{\partial h}(h,s),s,e\right)=0,\quad h\in V,\quad 0<s<T. \tag{4.94 bis}$$

Equation (4.94 bis) (or (4.94)) is the *Hamilton-Jacobi equation*. Clearly the boundary condition

$$\mathscr{W}(h,T)=0 \tag{4.95}$$

must be adjoined to the above equation.

Remark 4.5. Equation (4.94) is a partial differential equation of the Frechet-Volterra type. In the present case (no constraints) the operator $h\rightarrow\left(\frac{\partial \mathscr{W}}{\partial h},s\right)$ is linear, but this is clearly not true in general. Cf. Section 4.8 following.

4.8. The Case where Constraints Are Present

A natural question now arises: Can equation (4.94 bis) be derived when constraints on the control are present?

We shall show that by using the technique of dynamic programming (cf. Bellman [1]), we may *formally* derive such an equation.

Therefore, let

$$\begin{aligned} \mathscr{U}_{ad} &= \{v \mid v \in L^2(0,T;E), v(t) \in E_{ad}\}, \\ E_{ad} &= \text{closed, convex subset of } E. \end{aligned} \tag{4.96}$$

We now define (compare with 4.96)

$$H^0(h,p,s) = \inf_{e \in E_{ad}} H(h,p,s,e) \tag{4.97}$$

and we shall verify *(in a purely formal manner)* that

$$\frac{\partial \mathscr{W}}{\partial s}(h,s) + H^0\left(h, \frac{\partial \mathscr{W}}{\partial h}(h,s), s\right) = 0, \tag{4.98}$$

where now

$$\mathscr{W}(h,s) = \inf_{v \in \mathscr{U}_{ad}} J_s^h(v). \tag{4.99}$$

Let δ be a positive number with $\delta < T - s$, where δ will eventually be made to tend to zero. We have,

$$\begin{aligned} 2\mathscr{W}(h,s) = \inf_v \Bigg[& \int_s^{s+\delta} \|Cy(t;v)\|_F^2 dt + \int_s^{s+\delta} (Nv,v)_E dt \\ & + \int_{s+\delta}^T (\|Cy(t;v)\|_F^2 + (Nv,v)_E) dt \Bigg]. \end{aligned}$$

Let us set $v(s) = e (e \in E_{ad})$. Then from (4.89) we deduce (let us again emphasize the formal nature of the calculations)

$$y(s+\delta; v) = y(s) + \delta(B(s)e - A(s)h) + 0(\delta)$$

where $\delta^{-1}\|0(\delta)\| \to 0$ as $\delta \to 0$, the norm being taken on an appropriate space (here V'). Then,

$$\left.\begin{aligned} 2\mathscr{W}(h,s) = \inf \Bigg\{ & \delta[\|C(s)h\|_F^2 + (N(s)e,e)_E] \\ & + \int_{s+\delta}^T [\|Cy(t;v)\|_F^2 + (Nv,v)_E] dt + 0(\delta) \Bigg\} \end{aligned}\right\} \tag{4.100}$$

where v may be considered as the pair $\{e, v|_{s-\delta}^T\}$.

But in (4.100), $y(t;v)$ may be considered as the state of the system (4.89) in $]s+\delta, T[$ with the initial condition

$$y(s+\delta; v) = h + \delta(B(s)e - A(s)h) + 0(\delta)$$

such that

$$\left.\begin{aligned} &\operatorname*{Inf}_v \int_{s+\delta}^{T} [\|Cy(t;v)\|_F^2 + (Nv, v)_E]\,dt \\ &= 2\mathscr{W}(h+\delta(B(s)v - A(s)h), s+\delta) + 0(\delta). \end{aligned}\right\} \tag{4.101}$$

But from Bellman's "Principle of Optimality" [1]:

$$\begin{aligned} 2\mathscr{W}(h,s) = \inf_{e\in E_{\text{ad}}} \Big\{ &\delta[\|C(s)h\|_F^2 + (N(s)e,e)_E] \\ &+ \inf_{\substack{v\in L^2(s+\delta, T; E)\\ v(t)\in E_{\text{ad}}}} \int_{s+\delta}^{T} [\|Cy(v)\|_F^2 + (Nv,v)]\,dt + 0(\delta) \Big\} \end{aligned}$$

whence

$$\begin{aligned} 2\mathscr{W}(h,s) &= \inf_{e\in E_{\text{ad}}} \{\delta[\|C(s)h\|_F^2 + (N(s)e,e)_E] \\ &\quad + \mathscr{W}(h+\delta(B(s)e - A(s)h), s+\delta) + 0(\delta)\} \\ &= \inf_{e\in E_{\text{ad}}} \Big\{ \delta[\|C(s)h\|_F^2 + (N(s)e,e)_E] + 2\mathscr{W}(h,s) \\ &\quad + 2\delta\left(\frac{\partial\mathscr{W}}{\partial h}(h,s), B(s)e - A(s)h\right) + 2\delta\frac{\partial\mathscr{W}}{\partial s}(h,s) + 0(\delta) \Big\} \end{aligned}$$

Simplifying and dividing by 2δ,

$$\begin{aligned} \inf_{e\in E_{\text{ad}}} \Big\{ &\frac{\partial\mathscr{W}}{\partial s}(h,s) + \left(\frac{\partial\mathscr{W}}{\partial h}(h,s), B(s)e - A(s)h\right) \\ &+ \frac{1}{2}\|C(s)h\|_F^2 + \frac{1}{2}(N(s)e,e)_E \Big\} = 0 \end{aligned}$$

whence (4.98).

4.9. Various Remarks

4.9.1. Direct Study of the "Riccati Equation"

We shall indicate how some systems related to (4.9) may be solved "directly" (that is, without any reference to some control problem).

We consider two Hilbert spaces V_1, V_2, satisfying

$$V_i \subset H, \quad V_i \text{ dense in } H, \quad i = 1, 2$$

and two operators A_i, $i=1,2$, satisfying

$$\begin{aligned} &A_i \in \mathscr{L}(V_i; V_i'), \\ &(A_i \varphi, \varphi) \geqslant \alpha \|\varphi\|_{V_i}^2, \quad \alpha>0, \quad \forall \varphi \in V_i. \end{aligned} \tag{4.102}$$

We consider the following "two point boundary value problem":

$$\left.\begin{aligned} \frac{dy}{dt} + A_1 y - p = f, \\ -\frac{dp}{dt} + A_2 p + y = g, \end{aligned}\right\} \tag{4.103}$$

$$y(0)=0, \quad p(T)=0. \tag{4.104}$$

Remark 4.6. The system (4.103) is in one sense more general than (4.9) since one can have $V_1 \neq V_2$, $A_2 \neq A_1^*$, but less general than (4.9) in another sense, since $D_1(t) = -I$, $D_2(t) = I$. Let us add that the remarks which follow may be easily extended to cases where A_1 and A_2 depend on t.

The system (4.103), together with boundary conditions (4.104), can be solved by two methods:

1. We set

$$w = \{y, p\}, \quad \Lambda = \begin{pmatrix} \partial/\partial t & 0 \\ 0 & -\partial/\partial t \end{pmatrix}, \quad A = \begin{pmatrix} A_1 & -I \\ I & A_2 \end{pmatrix}, \quad F = \{f, g\},$$

and observe that (4.103) is equivalent to

$$\Lambda w + A w = F.$$

We also observe that, if $y(0)=0$, $p(T)=0$ then

$$\int_0^T (\Lambda w, w)\, dt = \tfrac{1}{2}(|y(T)|^2 + |p(0)|^2) \geqslant 0$$

and that

$$\begin{aligned} \int_0^T (A w, w)\, dt &= \int_0^T [(A_1 y, y) + (A_2 p, p)]\, dt \\ &\qquad \text{(due to (4.102))} \\ &\geqslant \alpha \int_0^T (\|y(t)\|_{V_1}^2 + \|p(t)\|_{V_2}^2)\, dt, \end{aligned}$$

hence the existence and uniqueness of a solution of (4.103), (4.104) follows (by applying Theorem 1.1 of Lions-Magenes [1], Vol. 1); more precise results are given in Lions [3 bis].

2. A second method, due to Cooper [1], is applicable under more restrictive hypotheses on A_i: we assume that (4.102) is true and moreover that both A_i's are symmetric:

$$(A_i\varphi,\psi)=(\varphi,A_i\psi)\quad\forall\,\varphi,\psi\in V_i. \tag{4.105}$$

However this method gives *stronger solutions*. This time we set

$$\mathscr{A}=\begin{pmatrix}A_1 & -I\\ -I & -A_2\end{pmatrix} \tag{4.106}$$

and we observe that (4.103) is equivalent to

$$\frac{dw}{dt}+\mathscr{A}w=G(=\{f,-g\}). \tag{4.107}$$

We take scalar products of (4.107) *with* $\mathscr{A}w$, and we observe that

$$\int_0^T\left(\frac{dw}{dt},w\right)dt=\tfrac{1}{2}[(A_1w_1(T),w_1(T))+(A_2w_2(0),w_2(0))]\geqslant 0.$$

Due to the fact that $\mathscr{A}$ is *onto*, one can deduce from this inequality, the existence and uniqueness of a solution of (4.103), (4.104). See more details in Cooper [1] and Lions [3 bis].

4.9.2. Another Approach to the Direct Study of the "Riccati Equation"

We wish to indicate here, without giving any technical details, some results of Da Prato [1], [2]. This author considers directly (that is again without reference to a control problem) "abstract" partial differential equations of type (4.87), and more generally

$$-\frac{dp(t)}{dt}+PA_1+A_2P+\varphi(p)=0, \tag{4.108}$$

where A_1 and A_2 need not necessarily be adjoint to each other and where φ is "any" non-linear analytic function. In Da Prato [1], the author considers the case when A_i does not depend on t and proves a *local* existence and uniqueness theorem (i.e., $P(t)$ exists in *some* interval $[T-\alpha,T]$); the local solution is actually global in case one can obtain an a priori estimate. Cases where A_i depend on t and examples are given in Da Prato [2]. For the proofs, Da Prato introduces a new class of semi groups in the space of continuous linear mappings.

4.9.3. Yet Another Approach to the Direct Study of the "Riccati Equation"

Yet another approach to the direct study of "Riccati equations" of the type (4.87) is due to Temam [1], [2]; the approach gives *global* results in t for general classes of such equations. Moverover this method (briefly indicated below) gives *a constructive approximation* procedure;

somewhat related constructions together with numerical computations may be found in Nedelec [1].

Let us consider an equation of the type (4.87); in order to be able to use a more standard notation, we reverse time; after introducing a new unknown operator Q, we consider the equation

$$\frac{dQ}{dt} + A^* Q + QA + Q \cdot Q = F, \tag{4.109}$$

$$Q(0) = Q_0 \quad \text{(given)}. \tag{4.110}$$

The method which follows applies equally well to the case where in (4.109) $Q \cdot Q$ is replaced by $Q \cdot Q \cdot \dots \cdot Q$ (with an arbitrary number of factors).

We use the space $\mathscr{H}$ of *Hilbert-Schmidt* operators from $H \to H$ [26] and we make the *additional assumption*:

$$F \text{ is given in } L^2(0,T;\mathscr{H}),\ F(t) \geqslant 0; \quad Q_0 \text{ is given in } \mathscr{H}, Q_0 \geqslant 0. \tag{4.111}$$

We define "approximate" values of Q at time nt in the following fashion—which we denote by Q^n.

We begin with

$$Q^0 = Q_0.$$

Assuming Q^n to be known, $Q^n \in \mathscr{H}$, $Q^n \geqslant 0$, we define Q^{n+1} in three *steps* (this is the adaptation of the "fractional steps methods" used in Numerical Analysis; cf. N. N. Yanenko [1], R. Temam [3]):

a) *definition of* $Q^{n+\frac{1}{3}}$:

$$\frac{Q^{n+\frac{1}{3}} - Q^n}{\Delta t} + Q^n \cdot Q^{n+\frac{1}{3}} = 0, \tag{4.112}$$

which uniquely defines $Q^{n+\frac{1}{3}} \in \mathscr{H}$, $Q^{n+\frac{1}{3}} \geqslant 0$;

b) *definition of* $Q^{n+\frac{2}{3}}$

$$\frac{Q^{n+\frac{2}{3}} - Q^{n+\frac{1}{3}}}{\Delta t} + Q^{n+\frac{2}{3}} \cdot A = \frac{1}{\Delta t} \int_{n\Delta t}^{(n+1)\Delta t} F(\sigma)\, d\sigma, \tag{4.113}$$

which uniquely defines $Q^{n+\frac{2}{3}}$ as a Hilbert-Schmidt operator from $V' \to H$;

c) *definition of* Q^{n+1}:

$$\frac{Q^{n+1} - Q^{n+\frac{2}{3}}}{\Delta t} + A^* Q^{n+1} = 0, \tag{4.114}$$

[26] Provided with its natural Hilbert space structure.

which uniquely defines $Q^{n+1} \in \mathscr{H}, Q^{n+1} \geqslant 0$ (and Q^{n+1} is Hilbert Schmidt from $H \to V$).

One can then obtain "energy estimates" (cf. Temam [2]); defining $Q_{\Delta t}$ as the step function equal to Q^n in $[n\Delta t, (n+1)\Delta t[$, one can prove that, in an appropriate topology, $Q_{\Delta t}$ converges, as $t \to 0$, to *the solution Q of* (4.109) (4.110) *having the properties*:

$$\left.\begin{aligned} &Q \in L^2(0,T;\mathscr{H}) = \text{Hilbert-Schmidt operators from } \\ &\qquad\qquad V' \to H \text{ and from } H \to V; \\ &Q \in L^\infty(0,T;\mathscr{H}), \\ &Q(t) \geqslant 0. \end{aligned}\right\} \tag{4.115}$$

Remark 4.7. Regularity properties of the solution (under regularity hypotheses on the data) are given in Temam [2].

Remark 4.8. In the examples (see Section 5 below) the method of Temam applies to operators A of *any even order*. For the special case where A is a *second order* elliptic operator, yet another "direct" approach to the Riccati equation is given in Kushner [1].

Remark 4.9. The "fractional step" method is a useful method of solving *some* classes of non-linear partial differential equations (Demidov-Marchouk [1], Temam [4]; see also Lions [13], Chapter 4, No. 2).

5. Decoupling and Integro-Differential Equation of Riccati Type (II)

5.1. Application of the Schwartz-Kernel Theorem

We consider the example in Section 3.1, case 3.1.1. Then $E = H = L^2(\Omega)$, $B(t)$ = identity map (of the type that Theorem 4.4 is applicable), $N(t) = N$, $C(t)$ = injection map of $V \to H$, $D_2(t)$ = identity map, $D_1(t) = N^{-1}$.

From the Kernel Theorem due to L. Schwartz [3], [4], $P(t)$ has the representation

$$P(t)\varphi = \int_\Omega P(x,\xi,t)\varphi(\xi)\,d\xi \quad \forall \varphi \in \mathscr{D}(\Omega) \tag{5.1}$$

where

$P(x,\xi,t)$ = kernel of $P(t)$ = distribution on $\Omega_x \times \Omega_\xi$, defined uniquely by $P(t)$.

We then have

Theorem 5.1. *Let the assumptions and data of Section 3.1, case 3.1.1 hold. Then the kernel $P(x,\xi,t)$ corresponding to the operator $P(t)$ characterised in Theorem 4.4 satisfies the equation*

$$\frac{\partial P}{\partial t}(x,\xi,t)+(A_x^*+A_\xi^*)P(x,\xi,t)+\int_\Omega P(x,\zeta,t)N^{-1}P(\zeta,\xi,t)\,d\zeta=\delta(x-\xi), \tag{5.2}$$

$$P(x,\xi,t)=P(\xi,x,t), \tag{5.3}$$

$$\left.\begin{aligned} &P(x,\xi,t)=0 \quad \text{if } x\in\Gamma, \quad \xi\in\Omega,\\ &(\text{and hence } P(x,\xi,t)=0 \quad \text{if } x\in\Omega, \quad \xi\in\Gamma, \quad \text{from (5.3)}) \end{aligned}\right\} \tag{5.4}$$

and

$$P(x,\xi,T)=0. \tag{5.5}$$

Remark 5.1. The kernel $P(x,\xi,t)$ has regularity properties corresponding to (4.86). In this particular case it has stronger properties. We may in fact verify that if the coefficients of $A\left(x,t,\dfrac{\partial}{\partial x}\right)$ are sufficiently regular, then

$$P(t)\in\mathscr{L}(L^2(\Omega);H_0^1(\Omega)). \tag{5.6}$$

We have thus established the existence of a global solution to the non-linear problem (5.2)–(5.5) *in the class of functions defined in a very special manner (by considering these functions as kernels of linear mappings). The choice of the space where one expects to find a solution plays an essential role for non-linear problems (far more important than in the case of linear problems where one has a greater degree of latitude).*

Proof of Theorem 5.1. Condition (5.3) follows from the fact that $P(t)^*=P(t)$ and (5.5) corresponds to $P(T)=0$. If $\varphi\in\mathscr{D}(\Omega)$ (for example), we must have $P(t)\varphi\in L^2(0,T;V)$, and hence $P(t)=0$ on $\Sigma=\Gamma\times]0,T[$ (since $V=H_0^1(\Omega)$ and hence we must have (5.4)).

We now note that (as usual if $\varphi\in\mathscr{D}(\Omega)$)

$$\begin{aligned} PA\varphi(x)&=\int_\Omega P(x,\xi,t)A_\xi\varphi(\xi)\,d\xi\\ &=\int_\Omega A_\xi^*P(x,\xi,t)\varphi(\xi)\,d\xi \end{aligned}$$

and hence the kernel of PA is A_ξ^*P. Hence we deduce (5.2) from (4.87) (by noting that $\delta(x-\xi)$ is the kernel of the identity operator). □

Remark 5.2. The system of equations (5.2)–(5.5) is a non-linear partial differential equation problem of parabolic type, the non-linearity being of the "Riccati type".

Remark 5.3. The results of Section 4.4 and 4.5 show that P may be approximated by the method of Galerkin.

5.2. Example of a Mixed Neumann Problem with Boundary Control

Let us now consider a system governed by equation (3.16) and let the cost function be given by (3.21).

Therefore the system of equations determining $\{y,p\}$ is (3.25). We may apply the results of Section 4 excepting those (fundamental) of section 4.6, since in this case $B(t)\in\mathscr{L}(L^2(\Gamma);V')$[27] and is defined by

$$(Bu,\psi)=(B(t)u,\psi)=\int_\Gamma u\psi\,d\Gamma$$

and (4.77) does not hold. In spite of this, we shall see that the final results are still true. This will be shown by using the general arguments and taking into account the structure of the particular problem.

For this problem, let us write down explicitly the system of equations (4.28), (4.30) determining $P(s)$ and $r(s)$.

We immediately have

$$\left.\begin{aligned}
&\frac{\partial\beta}{\partial t}+A\beta=0 \quad \text{in } \Omega\times]s,T[,\\
&-\frac{\partial\gamma}{\partial t}+A^*\gamma-\beta=0 \quad \text{in } \Omega\times]s,T[,\\
&\frac{\partial\beta}{\partial\nu_A}+N^{-1}\gamma=0 \quad \text{on } \Gamma\times]s,T[,\\
&\frac{\partial\gamma}{\partial\nu_{A^*}}=0 \quad \text{on } \Gamma\times]s,T[,\\
&\beta(x,s)=h(x),\quad x\in\Omega,\quad (h \text{ given in } L^2(\Omega)),\\
&\gamma(x,T)=0,\quad x\in\Omega.
\end{aligned}\right\}\tag{5.7}$$

[27] In fact we have a slightly stronger result: $B\in\mathscr{L}(L^2(\Gamma);(H^{\frac{1}{2}+\varepsilon}(\Omega))')$ $\forall\varepsilon>0$ —but even this is not sufficient to apply the results of Section 4.6.

Then

$$P(s)h=\gamma(0,s), \tag{5.8}$$

and also

$$\left.\begin{aligned}
&\frac{\partial\eta}{\partial t}+A\eta=f \quad \text{in } \Omega\times]s,T[,\\
&-\frac{\partial\xi}{\partial t}+A^*\xi-\eta=-z_d \quad \text{in } \Omega\times]s,T[,\\
&\frac{\partial\eta}{\partial\nu_A}+N^{-1}\xi=0 \quad \text{on } \Gamma\times]s,T[,\\
&\frac{\partial\xi}{\partial\nu_{A^*}}=0 \quad \text{on } \Gamma\times]s,T[,\\
&\eta(x,s)=0, \quad \xi(x,T)=0, \quad x\in\Omega
\end{aligned}\right\} \tag{5.9}$$

and hence

$$r(x,s)=\xi(x,s), \qquad x\in\Omega. \tag{5.10}$$

We shall now prove the following result:

Theorem 5.2. *We assume that the coefficients of* $A\left(x,t,\frac{\partial}{\partial x}\right)$ *are regular*[28] *in* $\bar{\Omega}\times[0,T]$ *and the boundary of* Γ *is regular. The operator* $P(t)$ *satisfies*

$$\left.\begin{aligned}
&-(P'\varphi,\psi)+a(t;\varphi,P\psi)+a^*(t;P\varphi,\psi)+(N^{-1}P\varphi,P\psi)_{L^2(\Gamma)}=(\varphi,\psi),\\
&\forall\,\varphi,\psi\in H^1(\Omega) \quad ('=d/dt),
\end{aligned}\right\} \tag{5.11}$$

with

$$\left.\begin{aligned}
&P(T)=0,\\
&P^*(t)=P(t)
\end{aligned}\right\} \tag{5.12}$$

and the function $r(t)$ *satisfies*

$$-(r',\psi)+a^*(t;r(t),\psi)+(N^{-1}r,P\psi)=(Pf-z_d,\psi)\;\forall\,\psi\in H^1(\Omega) \tag{5.13}$$

with

$$r(T)=0. \tag{5.14}$$

To simplify matters let $N=\nu\times$ *identity map. Then the kernel* $P(x,\xi,t)$[29] *of the operator* $P(t)$ *satisfies*

[28] For example twice continuously differentiable.

[29] See Section 5.1 for the use of the Kernel Theorem.

$$-\frac{\partial P}{\partial t}+(A_x^*+A_\xi^*)P+\nu^{-1}\int_\Gamma P(x,\zeta,t)P(\zeta,\xi,t)\,d\Gamma_\zeta=\delta(x-\xi)$$

$$\text{in } Q_x\times Q_\xi\times\,]0,T[, \tag{5.15}$$

with

$$P(x,\xi,t)=P(\xi,x,t), \tag{5.16}$$

$$\left.\begin{aligned}&\frac{\partial P}{\partial \nu_{A^*,x}}(x,\xi,t)=0, \quad x\in\Gamma, \quad \xi\in\Omega \text{ and}\\ &\frac{\partial P}{\partial \nu_{A^*,\xi}}(x,\xi,t)=0, \quad x\in\Omega, \quad \xi\in\Gamma,\end{aligned}\right\} \tag{5.17}$$

$$P(x,\xi,T)=0. \tag{5.18}$$

Proof. 1. We first note that

$$P(s)\in\mathscr{L}(L^2(\Omega), H^1(\Omega)). \tag{5.19}$$

Indeed, let β be given by (5.7); $\beta\in L^2(0,T;H^1(\Omega))$. Let γ be the solution of

$$-\frac{\partial\gamma}{\partial t}+A^*\gamma=\beta, \qquad \frac{\partial\gamma}{\partial\nu_{A^*}}=0 \quad \text{on } \Gamma\times\,]s,T[,$$

$$\gamma(x,T)=0.$$

Due to the regularity hypotheses, we have (cf. analogous to the Dirichlet problem of Section 3.1, Example 3.1)

$$\gamma\in H^{2,1}(\Omega\times\,]s,T[) \qquad \text{(cf. definition (3.5))}$$

whence $\gamma(x,s)\in H^1(\Omega)$, from which we deduce (5.19) (the mapping is also continuous which may be verified directly[30]).

We may also verify that the mapping $s\to P(s)h$ of $[0,T]\to H^1(\Omega)$ is continuous. Thus $P(t)y(t)\in L^2(0,T;V)$ $(V=H^1(\Omega))$ and hence in (4.27) we have,

$$r=p-Py\in L^2(0,T;V).$$

Let us make the change of variable $t=s+\tau(T-s)$ as in 2. of Lemma 4.4 and verify that

$$\text{the mapping } s\to P(s)h \text{ of } [0,T]\to V \text{ is continuously differentiable.} \tag{5.20}$$

[30] Or from the closed graph theorem.

2. The formal calculations of Section 4.3 can therefore be justified. We shall carry out these calculations explicitly[31]. The system of equations (3.25) may be written in variational form as

$$\left.\begin{aligned} &(y',\varphi)+a(t;y,\varphi)+(N^{-1}p,\varphi)_{L^2(\Gamma)}=(f,\varphi)\ \forall\varphi\in H^1(\Omega),\\ &-(\varphi',\psi)+a^*(t;\varphi,\psi)-(y,\psi)=-(z_d,\psi)\ \forall\psi\in H^1(\Omega),\\ &y(0)=y_0,\quad \varphi(T)=0.\end{aligned}\right\}\tag{5.21}$$

We have the identity: $p=Py+r$. Substituting this in the second equation of (5.21), we get

$$(-P'y+Py'+r',\psi)+a^*(t;Py+r,\psi)-(y,\psi)=-(z_d,\psi).\tag{5.22}$$

But

$$-(Py',\psi)=-(y',P\psi)$$

and from the first equation in (5.21) this is equal to

$$a(t;y,P\psi)+(N^{-1}p,P\psi)_{L^2(\Gamma)}-(f,P\psi);$$

hence (5.22) gives us

$$\left.\begin{aligned} &-(P'y,\psi)+a(t;y,P\psi)+a^*(t;Py,\psi)+(N^{-1}p,P\psi)_{L^2(\Gamma)}-(f,P\psi)\\ &-(r',\psi)+a^*(t;r,\psi)-(y,\psi)=-(z_d,\psi).\end{aligned}\right\}\tag{5.23}$$

In (5.23) replacing p by $Py+r$, we obtain

$$\left.\begin{aligned} &[-(P'y,\psi)+a(t;y,P\psi)+a^*(t;Py,\psi)+(N^{-1}Py,P\psi)_{L^2(\Gamma)}-(y,\psi)]\\ &+[-(r',\psi)+a^*(t;r,\psi)-(Pf-z_d,\psi)]=0\ \forall\psi\in H^1(\Omega).\end{aligned}\right\}\tag{5.24}$$

Let us fix $t=s$ (which has meaning almost everywhere). Since $y(s)$ may be chosen arbitrarily a priori, from (5.24) we deduce (5.11) and (5.13).

3. The kernel $P(x,\xi,t)$ of $P(t)$ satisfies (5.16) since $P(t)^*=P(t)$. Let us now carry out a formal calculation (with $N=\nu$ (identity)) in terms of the kernel:

$$a(t;\varphi,P\psi)+a^*(t;P\varphi,\psi)+\nu^{-1}(P\varphi,P\psi)_{L^2(\Gamma)}$$

$$=\sum_{i,j}\int_{\Omega\times\Omega} a_{ij}(x,t)\frac{\partial\varphi}{\partial x_i}(x)\frac{\partial P}{\partial x_j}(x,\xi,t)\psi(\xi)\,dx\,d\xi$$

[31] In examples, it is easier to carry out the calculations of Section 4.3 directly rather than identify the operators; in Section 4 the methods are more important than the results.

$$+\sum_{i,j} \int_{\Omega\times\Omega} a_{ji}(x,t)\frac{\partial P}{\partial x_i}(x,\xi,t)\varphi(\xi)\frac{\partial \psi}{\partial x_j}(x)\,dx\,d\xi$$

$$+\nu^{-1}\int_{\Omega\times\Omega}\left(\int_{\Gamma} P(x,\zeta,t)P(\zeta,\xi,t)\,d\Gamma_\zeta\right)\varphi(x)\psi(\xi)\,dx\,d\xi$$

= (by virtue of the fact that $P(x,\xi,t)=P(\xi,x,t)$)

$$\sum \int_{\Omega\times\Omega} a_{ij}(x,t)\frac{\partial \varphi}{\partial x_i}(x)\frac{\partial P}{\partial x_j}(x,\xi,t)\psi(\xi)\,dx\,d\xi$$

$$+\sum \int_{\Omega\times\Omega} a_{ij}(\xi,t)\frac{\partial P}{\partial \xi_j}(x,\xi,t)\varphi(x)\frac{\partial \psi}{\partial \xi_j}(\xi)\,dx\,d\xi$$

$$+\nu^{-1}\int_{\Omega\times\Omega}\left(\int_{\Gamma} P(x,\zeta,t)P(\zeta,\xi,t)\,d\Gamma_\zeta\right)\varphi(x)\psi(\xi)\,dx\,d\xi$$

= (utilising Green's formula)

$$\int_{\Gamma_x\times\Omega_\xi} \frac{\partial P}{\partial \nu_{A^*,x}}(x,\xi,t)\varphi(x)\psi(\xi)\,d\Gamma_x\,d\xi + \int_{\Omega\times\Omega}(A_x^* P)\varphi(x)\psi(\xi)\,dx\,d\xi$$

$$+\int_{\Omega_x\times\Gamma_\xi} \frac{\partial P}{\partial \nu_{A^*,\xi}}(x,\xi,t)\varphi(x)\psi(\xi)\,dx\,d\Gamma_\xi + \int_{\Omega\times\Omega}(A_\xi^* P)\varphi(x)\psi(\xi)\,dx\,d\xi$$

$$+\nu^{-1}\int_{\Omega\times\Omega}\left(\int_{\Gamma} P(\zeta,\xi,t)\,d\Gamma_\zeta\right)\varphi(x)\psi(\xi)\,dx\,d\xi$$

which when substituted in (5.11) gives us (5.15) and (5.17). □

Remark 5.4. As in Remark 5.2, a *direct* solution of the system of equations of the type (5.15)–(5.18) will be interesting. It will also be of interest to obtain results on the regularity properties of the kernel.

5.3. Example of a Mixed Neumann Problem with Observation of the Final State

We now consider a problem as in Section 3.2, subsection 3.2.2, Example 3.7. The system of equations determining $\{y,p\}$ is (3.55). *We shall directly adapt the reasoning of Section* 4 *to this example.*

The operator $P(s)$. We solve the system of equations

$$\left.\begin{aligned}
&\frac{\partial\beta}{\partial t}+A\beta=0 \quad \text{in } \Omega\times]s,T[,\\
&-\frac{\partial\gamma}{\partial t}+A^*\gamma=0 \quad \text{in } \Omega\times]s,T[,\\
&\frac{\partial\beta}{\partial\nu_A}+N^{-1}\gamma=0 \quad \text{on } \Gamma\times]s,T[, \qquad \frac{\partial y}{\partial\nu_{A^*}}=0,\\
&\beta(x,s)=h(x), \qquad x\in\Omega,\\
&\gamma(x,T)=\beta(x,T), \qquad x\in\Omega;
\end{aligned}\right\} \tag{5.25}$$

then

$$\gamma(0,s)=P(s)h. \tag{5.26}$$

The Function $r(s)$. We solve the system of equations

$$\left.\begin{aligned}
&\frac{\partial\eta}{\partial t}+A\eta=f \quad \text{in } \Omega\times]s,T[,\\
&-\frac{\partial\xi}{\partial t}+A^*\xi=0 \quad \text{in } \Omega\times]s,T[,\\
&\frac{\partial\eta}{\partial\nu_A}+N^{-1}\xi=0 \quad \text{on } \Gamma\times[s,T[,\\
&\frac{\partial\xi}{\partial\nu_{A^*}}=0 \quad \text{on } \Gamma\times]s,T[,\\
&\eta(x,s)=0, \qquad \xi(x,T)=\eta(x,T), \qquad x\in\Omega;
\end{aligned}\right\} \tag{5.27}$$

then

$$r(.,s)=\xi(.,s). \tag{5.28}$$

According to the regularity properties of the solution of a parabolic problem, we have (under the hypotheses of Theorem 5.2):

$$r(s)\in H^1(\Omega),$$

and $s\to r(s)$ is a function from $[0,T]\to H^1(\Omega)$ which is continuously differentiable. Therefore we may perform calculations analogous to that of Section 4.3, the point of departure being the identity

$$p(t)=P(t)\,y(t)+r(t). \tag{5.29}$$

Let us remark that the system of equations (3.25) may be written in the following variational form:

$$\left.\begin{array}{r}(y',\varphi)+a(t;y,\varphi)+(N^{-1}p,\varphi)_{L^2(\Gamma)}=(f,\varphi)\ \forall\varphi\in H^1(\Omega),\\ -(p',\psi)+a^*(t;p,\psi)=0\ \forall\psi\in H^1(\Omega),\\ y(0)=y_0,\quad p(T)-y(T)=-z_d.\end{array}\right\}\tag{5.30}$$

Hence as in Theorem 5.2, we deduce,

Theorem 5.3. *Let the hypothesis of Theorem 5.2 hold. The operator* $P(t)$ *(given by* (5.26)*) satisfies:*

$$\begin{array}{l}-(P'\varphi,\psi)+a(t;\varphi,P\psi)+a^*(t;P\varphi,\psi)+(N^{-1}P\varphi,P\psi)_{L^2(\Gamma)}=0\\ \forall\varphi,\psi\in H^1(\Omega),\end{array}\tag{5.31}$$

$$P(T)=\text{identity in } H,\tag{5.32}$$

$$P(t)^*=P(t),\tag{5.33}$$

and the function $r(t)$ *satisfies*

$$-(r',\psi)+a^*(t;r,\psi)+(N^{-1}r,P\psi)_{L^2(\Gamma)}=(f,P\psi)\ \forall\psi\in H^1(\Omega),\tag{5.34}$$

$$r(T)=-z_d.\tag{5.35}$$

The kernel $P(x,\xi,t)$ *of the operator* $P(t)$ [32] *satisfies* ($N=\nu$ *(identity)*)

$$\left.\begin{array}{l}-\dfrac{\partial P}{dt}+(A_x^*+A_\xi^*)P+\nu^{-1}\displaystyle\int_\Gamma P(x,\zeta,t)P(\zeta,\xi,t)\,d\Gamma_\zeta=0,\\ \text{in } \Omega_x\times\Omega_\xi\times\,]0,T[,\end{array}\right\}\tag{5.36}$$

$$P(x,\xi,T)=\delta(x-\xi),\qquad x,\xi\in\Omega_x\times\Omega_\xi,\tag{5.37}$$

$$P(x,\xi,t)=P(\xi,x,t),\tag{5.38}$$

$$\left.\begin{array}{l}\dfrac{\partial P}{\partial\nu_{A^*,x}}(x,\xi,t)=0,\qquad x\in\Gamma,\qquad \xi\in\Omega,\qquad t\in\,]0,T[,\ \text{and}\\ \dfrac{\partial P}{\partial\nu_{A^*,\xi}}(x,\xi,t)=0,\qquad x\in\Omega,\qquad \xi\in\Gamma,\qquad t\in\,]0,T[.\end{array}\right\}\tag{5.39}$$

Remark 5.5. The *method* of proof is general and is applicable to all the examples considered up to now in this chapter.

---

[32] See Section 5.1 for the utilization of the Schwartz-Kernel Theorem.

5.4. Mixed Neumann Problem, Observation of the Final State and Constraints in a Vector Space

We consider the case analysed in Section 3.2 with the cost function given by (3.31) and we assume that

$$\mathscr{U}_{ad} = \{v \mid v \text{ with support in } \Sigma_1, \Sigma_1 \subset \Sigma\}. \tag{5.40}$$

Hence $\mathscr{U}_{ad}$ is a subspace of $\mathscr{U}$.

The optimal control is given by the simultaneous solution of

$$\left.\begin{aligned}
&\frac{\partial y}{\partial t} + Ay = f, \qquad -\frac{\partial p}{\partial t} + A^* p = 0 \quad \text{in } Q,\\
&\frac{\partial p}{\partial \nu_{A^*}} = 0 \quad \text{on } \Sigma,\\
&\frac{\partial y}{\partial \nu_A} = 0 \quad \text{on } \Sigma - \Sigma_1,\\
&p + N\frac{\partial y}{\partial \nu_A} = 0 \quad \text{on } \Sigma_1,\\
&y(x,0) = y_0(x), \quad p(x,T) = y(x,T) - z_d(x), \qquad x \in \Omega,
\end{aligned}\right\} \tag{5.41}$$

and hence $u = \dfrac{\partial y}{\partial \nu_A}$ on Σ_1.

The decoupling of (5.41) is possible but extremely complicated, unless we assume

$$\Sigma_1 = \Gamma_1 \times]0, T[, \qquad \Gamma_1 \subset \Gamma. \tag{5.42}$$

Then assuming $N = \nu$(identity), we may show that the kernel $P(x, \xi, t)$ satisfies the equation (compare with (5.36)):

$$-\frac{\partial P}{\partial t} + (A_x^* + A_\xi^*)P + \nu^{-1} \int_{\Gamma_1} P(x,\zeta,t) P(\zeta,\xi,t)\, d\Gamma_\zeta = 0$$

and the equations (5.37), (5.38), (5.39) without any modification.

5.5. Remarks on Decoupling in the Presence of Constraints

We now return to the hypotheses of Section 4.1, with

$$\left.\begin{aligned}
&\mathscr{U}_{ad} = \{v \mid v \in \mathscr{U},\ v(t) \in E_{ad} \text{ a.e}\},\\
&E_{ad} = \text{closed convex subset of } E.
\end{aligned}\right\} \tag{5.44}$$

The optimal control u is then given by Theorem 2.1 and from Theorem 2.1 of Chapter 2, condition (2.21) is equivalent to

$$(\Lambda_{\mathscr{U}}^{-1} B^* p(t;u) + N u(t), e - u(t))_E \geq 0 \quad \forall e \in E_{ad}, \quad \text{for almost all } t. \tag{5.45}$$

We add the following hypothesis (on E_{ad}):

$$\left.\begin{array}{l} \text{for all } \xi \in E,\ \xi \neq 0, \text{ there exists } \Phi(\xi) \in E_{ad}, \text{ unique} \\ \text{such that } (\xi, e - \Phi(\xi))_E \geq 0 \ \forall e \in E_{ad}. \end{array}\right\} \tag{5.46}$$

Example 5.1. $E_{ad} = \{e \mid \|e\|_E \leq M\}$.
Then

$$\Phi(\xi) = -\xi \frac{M}{\|\xi\|_E}.$$

Hence (5.45) is equivalent to

$$\left.\begin{array}{l} u(t) = \Phi(\Lambda_{\mathscr{U}}^{-1} B^* p(t,u) + N u(t)), \\ \text{if } \Lambda_{\mathscr{U}}^{-1} B^* p(t,u) + N u(t) \neq 0, \quad \text{or} \\ \Lambda_{\mathscr{U}}^{-1} B^* p(t,u) + N u(t) = 0. \end{array}\right\} \tag{5.47}$$

We may now decouple the problem (2.19), (2.20), (5.47) in the following manner: in the interval $]s, T[$, $(0 < s < T)$, we consider (compare with Lemma 4.1) the non-linear system of equations in φ, ψ, u:

$$\left.\begin{array}{l} \dfrac{d\varphi}{dt} + A\varphi = f + B(t)u \quad \text{in }]s, T[, \\[2ex] -\dfrac{d\psi}{dt} + A^*\psi = C^* \Lambda_E (C\varphi - z_d) \quad \text{in }]s, T[, \\[2ex] \varphi(s) = h, \qquad \psi(T) = 0, \end{array}\right\} \tag{5.48}$$

u satisfies (5.47) almost everywhere in $]s, T[$.

This system of equations admits a unique solution from which we may define the mapping

$$h \to \psi(s) \quad \text{of } H \to H. \tag{5.49}$$

But due to (5.47), this mapping *is not affine* and there is no simple result analogous to that of "the Riccati integro-differential equation". Cf. also Section 14.

6. Behaviour as $T \to +\infty$

6.1. Orientation and Hypotheses

Upto now we have always assumed that $T < \infty$. We shall now see how the optimal control, the operator $P(t)$ etc., depend on T and study their behaviour as $T \to +\infty$.

We shall place ourselves in the framework of Section 4.1. We assume that

$$a(t;\varphi,\varphi)\geqslant\alpha\|\varphi\|^2 \quad \forall t\geqslant 0, \qquad \forall\varphi\in V, \qquad \alpha>0.^{33} \tag{6.1}$$

Further the operators $B(t)$, $C(t)$, $N(t)$ (cf. Section 4.1) are assumed to be given for all $t\geqslant 0$ with $\|B(t)\|_{\mathscr{L}(E;V')}\leqslant C$,

$$\|C(t)\|_{\mathscr{L}(V;F)}\leqslant C, \qquad \|N(t)\|_{\mathscr{L}(E;V')}\leqslant C, \qquad \forall t\geqslant 0.$$

The following notation will be used:
the state of the system will be denoted by $y_T(v)$ (to emphasize the dependence on T), that is, $y_T(v)=$solution of (2.2), (2.3), (2.4) considered in $]0,T[$; clearly if $T<T'$, $y_T(v)=y_{T'}(v)$ on $]0,T[$. We may therefore write $y(v)$ instead of $Y_T(v)$. The cost function is given by

$$J_T(v)=\int_0^T \|C(t)y(t,v)-z_d(t)\|_F^2\,dt+\int_0^T (N(t)v,v)_E\,dt. \tag{6.2}$$

The function z_d is given, with

$$z_d\in L^2(0,\infty;F). \tag{6.3}$$

Then, for each T, there exists a unique optimal control u_T (we consider the case where there are no constraints, but we clearly have the same result when constraints are present).

Let $\{y_T,p_T\}$ be the solution of (4.9), (4.10), (4.11) in $]0,T[$ [34] and let $P_T(t)$ and $r_T(t)$ be the corresponding operator and function $P(t)$ and $r(t)$ respectively (cf. Section 4).

6.2. The Case $T=\infty$

With the same notation as in Section 6.1, consider the minimization without constraints of $J_\infty(v)$.

We again have existence and uniqueness of the optimal control u_∞. The *adjoint state* is given by the solution of

$$-\frac{d}{dt}p(v)+A^*p(v)=D_2y(v)-g \qquad (g \text{ given as in (4.10)}), \tag{6.4}$$

$$p(v)\in L^2(0,\infty;V). \tag{6.5}$$

---

[33] We therefore assume that (1.2) holds with $\lambda=0$ and for all $t\geqslant 0$ (cf. also (1.13)).

[34] $y_T=y(u_T)$.

We then have

Proposition 6.1. *The adjoint state* $p(v)$ *is defined in a unique manner by* (6.4), (6.5).

Proof. 1. Let us first simplify the notation by defining $\tilde{f} = D_2 y(v) - g \in L^2(0, \infty; V')$. We are concerned with showing the existence and uniqueness of $p(v) = p \in L^2(0, \infty; V)$ satisfying

$$-\frac{dp}{dt} + A^* p = \tilde{f}. \tag{6.6}$$

It follows from (6.6) that $\frac{dp}{dt} \in L^2(0, \infty; V')$ which implies that $p(t)$ is a continuous function from $t \geqslant 0 \to H$ and *zero at infinity.*

2. *Uniqueness.* Suppose (6.6) holds with $\tilde{f} = 0$. Then

$$\int_0^\infty \left(-\frac{dp}{dt}, p\right) dt + \int_0^\infty a^*(t; p, p)\, dt = 0$$

and since from 1. above

$$\int_0^\infty \left(-\frac{dp}{dt}, p\right) dt = \tfrac{1}{2}|p(0)|^2,$$

we deduce that

$$\tfrac{1}{2}|v(0)|^2 + \int_0^\infty \|p(t)\|^2\, dt \leqslant 0, \quad \text{hence } p = 0.$$

3. *Existence.* Let q_T be the solution in $L^2(0, T; V)$ of

$$\left.\begin{aligned} -\frac{dq_T}{dt} + A^* q_T = \tilde{f} \quad \text{in }]0, T[, \\ q_T(T) = 0. \end{aligned}\right\} \tag{6.7}$$

From (6.7) we deduce that

$$\tfrac{1}{2}|q_T(0)|^2 + \alpha \int_0^T \|q_T(t)\|^2\, dt \leqslant \int_0^T \|\tilde{f}(t)\|_{V'}^2\, \|q_T(t)\|\, dt;$$

whence

$$\int_0^T \|q_T(t)\|^2\, dt \leqslant c_1 \int_0^T \|f(t)\|_{V'}^2 \leqslant c_2 \quad \text{(independent of } T). \tag{6.8}$$

Now, if

$$\tilde{q}_T = \text{extension of } q_T \text{ by } 0 \text{ for } t>T$$

$$\tilde{f}_T = \begin{cases} f & \text{in }]0,T[, \\ 0 & \text{for } t>T, \end{cases}$$

we have,

$$-\frac{d}{dt}\tilde{q}_T + A^*(\tilde{q}_T) = \tilde{f}_T \quad \text{on }]0,\infty[, \tag{6.9}$$

and hence from (6.8):

$$\int_0^\infty \|\tilde{q}_T(t)\|^2 dt \leqslant c_2. \tag{6.10}$$

We may then extract a sequence $T_n \to +\infty$ such that

$$\tilde{q}_{T_n} \to q \quad \text{weakly in } L^2(0,\infty; V)$$

and we may pass to the limit in (6.9) (for $T=T_n$). We thus obtain

$$-\frac{dq}{dt} + A^* q = \tilde{f}$$

and hence we may take $p=q$. ☐

The optimal control u_∞ is then given by the solution of the following system of equations in $\{y_\infty, p_\infty\}$:

$$\left.\begin{aligned} &\frac{dy_\infty}{dt} + A y_\infty + D_1 p_\infty = f \quad \text{in }]0,\infty[, \\ -&\frac{dp_\infty}{dt} + A^* p_\infty - D_2 y_\infty = g(= -C^* \Lambda_F z_d) \quad \text{in }]0,\infty[, \end{aligned}\right\} \tag{6.11}$$

with

$$\begin{aligned} &y_\infty, p_\infty \in L^2(0,\infty; V), \\ &y_\infty(0) = y_0. \end{aligned} \tag{6.12}$$

The Operator P_∞ and the Function r_∞. As in Section 4.2 (but now reasoning over the interval $]s,\infty[$ instead of $]s,T[$, we may derive the identity

$$p_\infty(t) = P_\infty(t) y_\infty(t) + r_\infty(t) \quad \text{on }]0,\infty[, \tag{6.13}$$

where P and r are defined as follows:

1. *Definition of* $P_\infty(s)$. We solve (compare with 4.28)

$$\left.\begin{aligned} &\frac{d\beta}{dt} + A\beta + D_1\gamma = 0 \quad \text{in }]s,\infty[, \\ -&\frac{d\gamma}{dt} + A^*\gamma - D_2\beta = 0 \quad \text{in }]s,\infty[, \\ &\beta(s) = h, \quad \beta,\gamma \in L^2(s,\infty; V) \end{aligned}\right\} \tag{6.14}$$

and then

$$\gamma(s) = P_\infty(s)h; \tag{6.15}$$

2. *Definition of* $r_\infty(s)$. We solve (compare with 4.30)

$$\left.\begin{aligned} &\frac{d\eta}{dt} + A\eta + D_1\xi = f \quad \text{in }]s,+\infty[, \\ -&\frac{d\xi}{dt} + A^*\xi - D_2\eta = g \quad \text{in }]s,+\infty[, \\ &\eta(s) = 0, \qquad \eta,\xi \in L^2(s,\infty; V), \end{aligned}\right\} \tag{6.16}$$

and then

$$\xi(s) = r(s). \tag{6.17}$$

We have the following result:

Theorem 6.1. *Let the hypotheses of Section* 6.1 *hold. Further let* $A(t) = A$, $B(t) = B$, $C(t) = C$, $N(t) = N$, *operators independent of* t. *Then*

$$P_\infty(s) = P_\infty \quad \text{not depending on } s, \tag{6.18}$$

$$P_\infty \in \mathscr{L}(H; H), \qquad P_\infty^* = P_\infty. \tag{6.19}$$

Under the hypotheses of Theorem 4.4 *(resp.* 5.2*) we also have,*

$$P_\infty \in \mathscr{L}(V; V) \cap \mathscr{L}(V'; V'), \tag{6.20}$$

$$(\text{resp. } P_\infty \in \mathscr{L}(H; V) \cap \mathscr{L}(V'; H)), \tag{6.21}$$

and

$$P_\infty A + A^* P_\infty + P_\infty D_1 P_\infty = D_2. \tag{6.22}$$

The function r_∞ *satisfies*

$$-r'_\infty + A^* r_\infty + P_\infty D_1 r_\infty = P_\infty f + g, \tag{6.23}$$

with

$$r_\infty \in L^2(0,\infty; V). \tag{6.24}$$

Proof. The only new element is (6.18), the remainder being consequences or variants of results which we have already established. Now if the operators $A(t), \dots$ are independent of t, problem (6.14) is independent of s

in the sense that if we translate the origin such that $t \to t-s$ then $P_\infty(s) = P_\infty(0)$. □

Remark 6.1. In cases where we are in a position to use the Schwartz-Kernel Theorem (cf. Section 5.1), the kernel $P_\infty(x,\xi)$ of P_∞ satisfies

$$\left.\begin{aligned}&(A_x^* + A_\xi^*) P_\infty(x,\xi) + \int_\Omega (D_1 P_\infty)(x,\zeta) P_\infty(\zeta,\xi)\, d\zeta = D_2(x,\xi),\\ &\text{in } \Omega_x \times \Omega_\xi \quad (\text{where } D_2(x,\xi) = \text{kernel of } D_2),\end{aligned}\right\} \tag{6.25}$$

$$P(x,\xi) = P(\xi,x), \tag{6.26}$$

with boundary conditions depending on the problem, for example:

$$\left.\begin{aligned}&P_\infty(x,\xi)=0 \ \text{ if } x\in\Gamma, \quad \xi\in\Omega \ (\text{and hence } P_\infty(x,\xi)=0 \ \text{ if}\\ &x\in\Omega \ \text{ and } \xi\in\Gamma), \text{in case we are in the framework of Section 5.1},\\ &\frac{\partial P_\infty}{\partial \nu_{A^*}}(x,\xi)=0 \ \text{ if } x\in\Gamma, \quad \xi\in\Omega \ \left(\text{and } \frac{\partial P_\infty}{\partial \nu_{A^*}}(x,\xi)=0 \ \text{ if}\right.\\ &\left.x\in\Gamma, \quad \xi\in\Omega\right) \quad \text{in the framework of Section 5.2}^{35}.\end{aligned}\right\} \tag{6.27}$$

The direct study of non-linear elliptic equations of the preceeding type appears to be an open interesting problem.

We shall now prove that in appropriate topologies

$$u_T \to u_\infty, \quad y_T \to y_\infty \quad \text{etc., as } T \to \infty.$$

6.3. Passage to the Limit as $T \to +\infty$

We shall prove

Theorem 6.2. *Let the hypotheses of Section 6.1 hold. Let $\tilde{u}_T$ (resp. $\tilde{y}_T$, resp. $\tilde{p}_T$) be the extension of u_T (resp. y_T, resp. p_T) in $]0,\infty[$ by 0 outside $]0,T[$. Then as $T \to \infty$,*

$$\tilde{u}_T \to u_\infty \quad \textit{weakly in } L^2(0,\infty;E), \tag{6.28}$$

$$\tilde{y}_T \to y_\infty, \quad \tilde{p}_T \to p_\infty \quad \textit{weakly in } L^2(0,\infty;V), \tag{6.29}$$

$$P_T(s)h \to P_\infty(s)h \quad \textit{weakly in } H, \quad \forall s \textit{ finite}, \quad \forall h \in H. \tag{6.30}$$

Proof. 1. Let $j_T = \inf J_T(v)$, $j_\infty = \inf J_\infty(v)$. For $v \in L^2(0,\infty;E)$ we have, $J_T(v) \leqslant J_\infty(v)$ and hence $j_T \leqslant j_\infty$. Thus $j_T = J_T(u_T) \geqslant \nu \int_0^T \|u_T\|_E^2\, dt$ and if $\tilde{u}_T$ is defined as in the statement of the theorem, we have

$$\|\tilde{u}_T\|_{L^2(0,\infty;E)} \leqslant C.^{36} \tag{6.31}$$

---

[35] Clearly there are other possibilities.

[36] The C's denoting constants independent of T.

But then

$$\|\tilde{y}_T\|_{L^2(0,\infty;V)} \leqslant C, \tag{6.32}$$

and

$$\|\tilde{p}_T\|_{L^2(0,\infty;V)} \leqslant C. \tag{6.33}$$

Now since

$$\frac{d}{dt}(\tilde{p}_T) = \left(\frac{dp_T}{dt}\right)^{\sim} \quad (\text{since } p_T(T)=0),$$

we have

$$\begin{gathered} -\frac{d}{dt}(\tilde{p}_T) + A^*(\tilde{p}_T) = D_2 \tilde{y}_T + \tilde{g}_T, \\ (\tilde{g}_T = g \text{ in }]0,T[, \quad 0 \text{ in }]T,\infty[), \end{gathered} \tag{6.34}$$

and hence

$$\left\|\frac{d}{dt} p_T\right\|_{L^2(0,\infty;V')} \leqslant C. \tag{6.35}$$

Since

$$\left\|\frac{dy_T}{dt}\right\|_{L^2(0,T;V')} \leqslant C,$$

we have

$$|y_T(T)| \leqslant C. \tag{6.36}$$

We may then find a sequence $T_n \to \infty$ such that

$$\left.\begin{aligned} &\tilde{u}_{T_n} \to w \quad \textit{weakly} \text{ in } L^2(0,\infty;E), \\ &\tilde{y}_{T_n} \to \bar{y} \quad \textit{weakly} \text{ in } L^2(0,\infty;V), \quad \tilde{p}_T \to \bar{p} \text{ in the same sense}, \\ &\frac{d\tilde{p}_{T_n}}{dt} \to \frac{d\bar{p}}{dt} \quad \textit{weakly} \text{ in } L^2(0,\infty;V'). \end{aligned}\right\} \tag{6.37}$$

In the limit, equation (6.34) gives us

$$-\frac{d\bar{p}}{dt} + A^*\bar{p} - D_2\bar{y} = g. \tag{6.38}$$

But we also have

$$\frac{d}{dt}\tilde{y}_{T_n} + A\tilde{y}_{T_n} + D_1\tilde{p}_{T_n} = \tilde{f}_{T_n} + y_{T_n}(T_n)\delta(t-T_n)$$

and by virtue of (6.36), $y_{T_n}(T_n)\delta(t-T_n) \to 0$ in the sense of $\mathscr{D}'(]0,\infty[;H)$ and hence in the limit

$$\frac{d}{dt}\bar{y} + A\bar{y} + D_1\bar{p} = f. \tag{6.39}$$

Since $y(0)=y_0$, we deduce by comparing with (6.11), (6.12) that $\bar{y}=y_\infty$, $\bar{p}=p_0$. The relation $N u_T+\Lambda_E^{-1} B^*(t) p_T=0$ gives us in the limit (by taking $T=T_n$): $N w+\Lambda_E^{-1} B^* p_\infty=0$, hence $w=u_\infty$ and we have thus already proved (6.28), (6.29)[37].

2. To prove (6.30), we observe that $P_T(s)h$ is defined by

$$\left.\begin{aligned}
&\frac{d\beta_T}{dt}+A\beta_T+D_1\gamma_T=0 \quad \text{in }]s,T[,\\
&-\frac{d}{dt}\gamma_T+A^*\gamma_T-D_2\beta_T=0 \quad \text{in }]s,T[,\\
&\beta_T(s)=h, \quad \gamma_T(T)=0
\end{aligned}\right\} \tag{6.40}$$

and then $P_T(s)h=\gamma_T(s)$.

But (cf. proof of Lemma 4.5), $\{\beta_T,\gamma_T\}$ corresponds to the optimal control of a system whose state is given by

$$\frac{d\varphi}{dt}+A\varphi=Bv \quad \text{in }]s,T[, \quad \varphi(s)=h$$

and whose cost function is given by

$$J_{s,T}^h(v)=\int_s^T \|C\varphi(v)\|_E^2\,dt+\int_s^T (Nv,v)\,dt.$$

We have (see proof of Lemma 4.5)

$$\inf J_{s,T}^h(v)=(P_T(s)h,h)\leqslant J_{s,T}^h(0)\leqslant C|h|^2,$$

whence

$$|P_T(s)h|\leqslant C|h|, \quad C=\text{constant independent of } s \text{ and } T. \tag{6.41}$$

Now if w_T is the optimal control of this problem, we obtain as in 1. of this proof

$$\int_s^T \|w_T(t)\|_E^2\,dt\leqslant C,$$

and extending w_T by 0 for $t>T$, we deduce that $\tilde{w}_T$ (resp. $\tilde{\beta}_T$, $\tilde{\gamma}_T$) ranges in a bounded set of $L^2(0,\infty;E)$ (resp. $L^2(0,\infty;V)$). We may then find a sequence $T_n\to\infty$ such that $\tilde{w}_{T_n}\to w$, $\tilde{\beta}_{T_n}\to\bar{\beta}$, $\tilde{\gamma}_{T_n}\to\bar{\gamma}$ in the corresponding weak topologies and hence (note that $|\beta_T(T)|\leqslant C$) $\{\bar{\beta},\bar{\gamma}\}$ satisfies (6.14). Hence $\bar{\beta}=\beta$, $\bar{\gamma}=\gamma$ and $\gamma_{T_n}(s)\to\gamma(s)$ weakly in H—whence the result. ☐

---

[37] Since $J_T(u_T)\to J(u_\infty)$ we have also proved that $u_T\to\tilde{u}$ strongly in $L^2(0,\infty;E)$.

7. Problems which are not Necessarily Coercive

7.1. Distributed Observation

Let us consider the problem in Section 4.1; the state $y(t;v)$ is given by

$$y'(t;v)+A(t)y(t;v)=F+B(t)v, \quad y(0;v)=y_0, \tag{7.1}$$

and the cost function is given by (case where "$N=0$")

$$J(v)=\int_0^t \|C(t)y(t;v)-z_d\|^2\,dt. \tag{7.2}$$

we then have

Theorem 7.1. *Let us assume that* (7.1), (7.2) *hold*[38], *the cost function being given by* (7.2). *We assume* $\mathscr{U}_{\text{ad}}\subset\mathscr{U}=L^2(0,T;E)$ *is bounded and that the operators B and C are injective. Then the optimal control exists and it is unique.*

Proof. We shall apply Remark 5.5, Chapter 1. For the moment, let us set,

$$y(v)-y(0)=\mathscr{L}v, \quad \mathscr{L}\in\mathscr{L}(L^2(0,T;E);W(0,T)). \tag{7.3}$$

Then if u is an optimal control (which exists since $\mathscr{U}_{\text{ad}}$ is bounded), the set of optimal controls is the set

$$(\mathscr{U}+\operatorname{Ker}(C\mathscr{L}))\cap\mathscr{U}_{\text{ad}}.$$

Let $V\in\operatorname{Ker}(C\mathscr{L})$. Since C is injective, $\mathscr{L}v=0$. Hence

$$\frac{d}{dt}(\mathscr{L}v)+A(\mathscr{L}v)=0, \quad \text{and therefore,}$$

$$\frac{d}{dt}(\mathscr{L}v)+A(\mathscr{L}v)=Bv,$$

and B being injective implies $v=0$.

Example 7.1. The hypotheses of Theorem 7.1 are valid for the problem considered in subsection 3.1.1. The same is true for problems considered in subsections 3.1.2 and 3.2.1.

Remark 7.1. If we observe only the final state, the mapping $v\to y(t;v)-y(t;0)$ is not injective and there is no analogue of Theorem 7.1. Cf. Section 7.2 for this case.

---

[38] With hypotheses (1.1), (1.2) on $A(t)$.

7.2. Observation of the Final State

We again consider a system whose state is given by (7.1) but the cost function is now given by

$$J(v) = |D y(T; v) - z_d|^2, \qquad D \in \mathscr{L}(H; H). \tag{7.4}$$

We need to make use of the following concept:

Definition 7.1. The operator (parabolic) $-\frac{d}{dt} + A^*(t)$ is said *to possess the backward uniqueness property if the following conditions hold:*

$$\psi \in L^2(0, T; V), \qquad \frac{d\psi}{dt} \in L^2(0, T; V'), \tag{7.5}$$

$$-\frac{d\psi}{dt} + A^*(t)\psi = 0 \quad \text{in }]0, T[, \tag{7.6}$$

$$\psi(s) = 0, \qquad s \in]0, T[\text{ fixed}, \tag{7.7}$$

imply $\psi = 0$ in $]s, T[$.

Remark 7.2. The natural boundary conditions connected with the problem (7.5), (7.6) are:

$$\psi(T) = \text{given}.$$

The problem with $\psi(s)$ given is in general *ill-posed*, but it may have the uniqueness property.

By *reversing the direction of time*, it is clear that this amounts to assuming

$$\theta \in W(0, T), \qquad \frac{d\theta}{dt} + A^*(T-t)\theta = 0 \quad \text{in }]0, T[, \qquad \theta(T) = 0 \Rightarrow \theta \equiv 0.$$

This explains the terminology "backward uniqueness".

Remark 7.3. In Lions-Malgrange [1] a result exhibiting the backward uniqueness property is proved provided (7.5)[39] is satisfied. Here we present a more elementary example where this property obtains: assume that $A(t) = A = A^*$, and that the injection of V into H is compact. Let $w_1, \ldots, w_m$ be the eigen-functions of A:

$$A w_i = \lambda_i w_i, \qquad i = 1, 2, \ldots, \lambda_i > 0, \qquad (w_i, w_j) = \delta_i^j.$$

---

[39] This is not particularly distressing when concrete examples are considered: for parabolic equations if the coefficients are sufficiently regular this property holds.

Then all ψ satisfying (7.5), (7.6) may be written as

$$\left.\begin{aligned} &\psi(t)=\sum_{j=1}^{\infty}\psi_j(t)w_j, \qquad \psi_j\in L^2(0,T)\ \forall_j,\\ &\sum\lambda_j\int_0^t|\psi_j(t)|^2\,dt<\infty, \qquad \psi_j(t)=c_je^{\lambda_jt}.\end{aligned}\right\} \tag{7.8}$$

Then if $\psi(s)=0$, we necessarily have $\psi_j(s)=0\ \forall_j$ and hence $c_j=0$. Therefore $\psi\equiv 0$ in $]s,T[$.[40]

We shall now prove

Theorem 7.2. *We assume that the state is given by* (7.1) *and the cost function by* (7.4). *We assume that the backward uniqueness property holds, that*

$$\operatorname*{Inf}_{v\in\mathscr{U}_{ad}} J(v)>0, \tag{7.9}$$

and that

$$B^*(t) \text{ is injective (for almost all } t\text{)}, \ D^* \text{ is injective.} \tag{7.10}$$

Finally assume that

$$\left.\begin{aligned}&\mathscr{U}_{ad}=\{v\,|\,v\in L^2(0,T;E),\ v(t)\in E_{ad}\},\\ &E_{ad}=\text{closed, bounded, convex subset of } E\}.\end{aligned}\right\} \tag{7.11}$$

Then the optimal control u exists and satisfies

$$u(t)\in\partial E_{ad}=\text{boundary of } E_{ad}, \text{ for almost all } t.^{41} \tag{7.12}$$

Proof. Since $\mathscr{U}_{ad}$ is bounded, an optimal control u exists (the question of uniqueness of u is not considered). The adjoint state $p(u)$ is defined by

$$\begin{aligned}&-\frac{dp}{dt}(u)+A^*(t)p(u)=0,\\ &p(u)\in L^2(0,T;V),\\ &p(T;u)=D^*(Dy(T;u)-z_d),\end{aligned}$$

the optimality of u being characterized by $(p(t)=p(t;u(t)))$,

$$\int_0^T(B^*(t)p(t),v(t)-u(t))_E\,dt\geqslant 0.$$

---

[40] By using the spectral decomposition of A, a similar proof may be given when the injection of V into H is not compact.

[41] This is a "bang-bang" type property—in the terminology of finite dimensional control problems.

From Theorem 2.1, Chapter 2, this condition is equivalent to

$$(B^*(t)p(t), e-u(t))_E \geqslant 0 \quad \forall e \in E_{ad}, \quad \text{almost all } t. \tag{7.13}$$

But we have

$$B^*(t)p(t) \neq 0 \quad \text{almost everywhere (everywhere if } B(t) \text{ is defined everywhere).} \tag{7.14}$$

This is true, since if $B^*(s)p(s)=0$ then from (7.10) $p(s)=0$ and hence the backward uniqueness property implies $p=0$ in $]s, T[$. Hence $p(t)=0$ and hence

$$D^*(D y(T; u) - Z_d) = 0.$$

But since from (7.10), D^* is injective, we deduce that $D y(T; u) - z_d = 0$ and hence

$$\inf_{v \in \mathscr{U}_{ad}} J(v) = J(u) = |D y(T; u) - z_d|^2 = 0,$$

which contradicts (7.9). Hence (7.14).

Finally from (7.13) we obtain (7.12). □

Corollary 2.1. *Let the hypotheses of Theorem* 7.2 *hold. If further* E_{ad} *is strictly convex, the optimal control* u *is unique.*

Proof. If u_1 and u_2 are two optimal controls, $\dfrac{u_1+u_2}{2}$ is also an optimal control[42] and hence from (7.12) and the strict convexity of E_{ad} we obtain $u_1 = u_2$ almost everywhere. □

Example 7.2. Theorem 7.2 is applicable to the case considered in Section 3.2.2. For example, suppose that the state is given by

$$\left.\begin{aligned} &\frac{\partial y(v)}{\partial t} + A(t) y(v) = f \quad \text{in } Q = \Omega \times]0, T[, \\ &\frac{\partial y(v)}{\partial \nu_A} = v \quad \text{on } \Sigma, \\ &y(x, 0; v) = y_0(x) \quad \text{on } \Omega, \end{aligned}\right\} \tag{7.15}$$

and $D =$ identity map, hence

$$J(v) = \int_\Omega (y(x, T; v) - z_d(x))^2 \, dx. \tag{7.16}$$

We assume that

$$\mathscr{U} = L^2(0, T; L^2(\Gamma)) \; (= L^2(\Sigma)), \quad \text{hence } E = L^2(\Gamma),$$

---

[42] The set of optimal controls is convex. Cf. Chapter 1, Section 5.1.

and that

$$\mathscr{U}_{\text{ad}} = \{v \mid \xi_0(x) \leqslant v(x,t) \leqslant \xi_1(x) \text{ almost everywhere on } \Sigma, \xi_0, \xi_1 \text{ given in } L^\infty(\Gamma)\}. \tag{7.17}$$

Hence

$$E_{\text{ad}} = \{e \mid e \in L^2(\Gamma), \xi_0(x) \leqslant e(x) \leqslant \xi_1(x) \text{ almost everywhere on } \Gamma\} \tag{7.18}$$

and we may apply corollary 7.1. We thus have:

$$\left.\begin{array}{l} \text{If } \operatorname{Inf} J(v) > 0 \text{ for } v \in \mathscr{U}_{\text{ad}}, \text{ the optimal control } u \text{ satisfies} \\ u(x,t) = \xi_0(x) \text{ or } u(x,t) = \xi_1(x) \text{ almost everywhere,} \\ \text{according as } p(x,t) > 0 \text{ or } < 0. \end{array}\right\} \tag{7.19}$$

It would be extremely interesting to obtain results on the topology of the "switching surface" of $u(x,t)$.

Remark 7.4. Condition (7.9) signifies that there exists no $u \in \mathscr{U}_{\text{ad}}$ such that $J(u)=0$ provided the image of $\mathscr{U}_{\text{ad}}$ by $v \to Dy(T;v)$ is closed. Care must be taken here, since this image is in general not closed. For example, in the case of example 7.2 the image is not closed.

Remark 7.5. Let $\mathscr{U}_{\text{ad}}$ be given by (7.17), (7.18) and let us assume that there exists a u satisfying

(i) $y(T;u) = z_d$ (hence $\operatorname{Inf} J(v) = 0$),

(ii) $\xi_0(x) < u(x,t) < \xi_1(x)$ on Σ_0, Σ_0 non-empty open subset of Σ. (7.20)

Then there exists *an infinity* of optimal controls. Indeed, we may show that there exists an infinite number of functions w on Σ such that

$$w \text{ has its support in } \Sigma_0, \quad w \in L^2(\Sigma), \tag{7.21}$$

and

$$\left.\begin{array}{l} \text{the solution } \psi(w) \text{ of } \dfrac{\partial \psi(w)}{\partial t} + A(t)\psi(w) = 0 \text{ in } \Omega \times]0,T[, \\ \dfrac{\partial \psi(w)}{\partial \nu_A} = w \text{ on } \Sigma, \quad \psi(x,0;w) = 0 \text{ satisfies} \\ \psi(x,T;w) = 0. \end{array}\right\} \tag{7.22}$$

Then $y(u+\varepsilon w) = y(u) + \varepsilon\psi(w)$ satisfies

$$y(T; u+\varepsilon w) = z_d.$$

and $u + \varepsilon w \in \mathscr{U}_{\text{ad}}$ for $|\varepsilon| \leqslant \varepsilon_0$.

For an explicit example, consult Yu. V. Egorov [1].

7.3. Examples where $N=0$ and $\mathcal{U}_{ad}$ is not Bounded [43]

Let us consider the system whose state $y(v)$ is given by

$$\left.\begin{aligned} &\frac{dy}{dt}(v)+A(t)y(v)=f+v,\\ &y(0;v)=y_0,\\ &y(v)\in L^2(0,T;V),\end{aligned}\right\} \tag{7.23}$$

$$v\in\mathcal{U}=L^2(0,T;V'). \tag{7.24}$$

Let the observation be:

$$Cy(v)=\left\{y(v),\frac{dy(v)}{dt}\right\}\in L^2(0,T;V)\times L^2(0,T;V'),^{44} \tag{7.25}$$

and let

$$\left.\begin{aligned} J(v)&=\|Cy(v)-Z_d\|^2_{W(0,T)}=(\text{if } z_d=\{z_d^0,z_d^1\})\\ &=\int_0^T\|y(v)-z_d^0\|_V^2\,dt+\int_0^T\left\|\frac{dy(v)}{dt}-z_d^1\right\|_{V'}^2 dt.\end{aligned}\right\} \tag{7.26}$$

We then have:

Theorem 7.3. *Let hypotheses* (7.23) – (7.26) *hold. Let* $\mathcal{U}_{ad}\subset\mathcal{U}$ *be any closed, bounded, convex set. Then a unique optimal control exists.*

Proof. Using the notation of Chapter 2, Section 1, we have

$$\pi(v,v)=\|y(v)-y(0)\|^2_{W(0,T)} \tag{7.27}$$

and since the mapping

$$v\to y(v)-y(0)=\mathcal{L}v \tag{7.28}$$

is an *isomorphism* of $L^2(0,T;V')=\mathcal{U}\to W_0(0,T)$, where

$$W_0(0,T)=\{\psi\,|\,\psi\in W(0,T),\ \psi(0)=0\}, \tag{7.29}$$

we see that

$$\pi(v,v)\geq C\|v\|^2_{\mathcal{U}},\qquad C>0,\qquad \forall v\in\mathcal{U},$$

whence the result. □

---

[43] This is the analogue in the evolution case of the state of affairs considered in Chapter 2, Section 3.1.

[44] The observation may also be taken in $W(0,T)$.

We have to transpose the isomorphism (7.28)[45] in order to obtain the equations characterizing the optimal control. This is the same as transposing its inverse. The mapping

$$\psi \to \frac{d\psi}{dt} + A\psi \; (=\mathscr{L}^{-1}(\psi))$$

is an isomorphism of $W_0(0,T) \to \mathscr{U}$. Hence if σ is a continuous linear form on $W_0(0,T)$, there exists a unique $p \in L^2(0,T;V)$ such that

$$\int_0^T \left(p, \frac{d\psi}{dt} + A\psi\right) dt = \sigma(\psi) \quad \forall \psi \in W_0(0,T). \tag{7.30}$$

Consequently, there *exists a unique* $p(u) \in L^2(0,T;V)$ *such that*

$$\begin{aligned}\int_0^T \left(p(u), \frac{d\psi}{dt} + A\psi\right) dt = &\int_0^T (y(u) - z_d^0, \psi(t))_V \, dt \\ &+ \int_0^T \left(\frac{dy(u)}{dt} - z_d^1, \frac{d\psi}{dt}(t)\right)_{V'} dt \quad \forall \psi \in W_0(0,T).\end{aligned} \tag{7.31}$$

Equation (7.31) defines the adjoint state uniquely. Hence the optimal control u is characterized by (7.23) (with $v=u$), (7.31) and

$$\int_0^T (p(u), v-u)\, dt \geq 0 \quad \forall v \in \mathscr{U}_{ad}. \tag{7.32}$$

Example 7.3. Let us consider the case of Section 3.1, (with $V = H_0^1(\Omega)$) and

$$\mathscr{U}_{ad} = \{v \mid v \geq 0 \text{ almost everywhere in } Q \times]0,T[\}. \tag{7.33}$$

Equation (7.33) may then be interpreted as[46]:

$$-\frac{\partial p}{dt}(u) + A^* p(u) = \Lambda_V(y(u) - z_d^0) - \frac{d}{dt}\Lambda_V^{-1}\left(\frac{dy(u)}{dt} - z_d^1\right), \tag{7.34}$$

$$p(u)|_\Sigma = 0, \tag{7.35}$$

$$p(T;u) = \Lambda_V^{-1}\left[\frac{d}{dt}y(u) - z_d^1\right]_T, \tag{7.36}$$

---

[45] This procedure forms the basis of the study of non-homogeneous boundary value problems, cf. Chapter 2, Section 4.2 and Lions-Magenes [1].

[46] Λ_V = canonical isomorphism of V onto V' is here the operator $\Lambda_V = (-\Delta_x + \text{Identity})$ of $H_0^1(\Omega) \to H^{-1}(\Omega)$, if

$$(\varphi, \psi)_{H_0^1(\Omega)} = \int_\Omega (\varphi\psi + \operatorname{grad}\varphi \operatorname{grad}\psi)\, dx.$$

and (7.32) is equivalent to

$$\left.\begin{aligned} p(u)&\geqslant 0 \quad \text{in } Q,\\ u&\geqslant 0 \quad \text{in } Q,\\ u(p(u))&=0 \quad \text{almost everywhere.}\end{aligned}\right\} \tag{7.37}$$

8. Other Observations of the State and other Types of Control

8.1. Pointwise Observation of the State

Consider the following situation: the operator $A=A\left(x,t,\dfrac{\partial}{\partial x}\right)$ is given as in Section 3, and we place ourselves in the situation of Section 3.1; hence the state $y(v)$ is given by

$$\left.\begin{aligned} &\frac{\partial y}{\partial t}(v)+A\,y(v)=f+v, \qquad f\in L^2(Q), \qquad v\in L^2(Q)=\mathscr{U},\\ &y(v)|_\Sigma=0,\\ &y(x,0;v)=y_0(x), \qquad x\in\Omega.\end{aligned}\right\} \tag{8.1}$$

Let $x^1,\dots,x^\mu$ be points of Ω; we assume that the observation is $\{y(x^j,t;v)\}$, $1\leqslant j\leqslant\mu$—provided we can attach a meaning to this. If we now assume that the coefficients of A are sufficiently regular, the solution of $y(v)$ of (8.1) satisfies (cf. (3.4), 3.5)):

$$y(v)\in H^{2,1}(Q). \tag{8.2}$$

Hence $y(v)\in L^2(0,T;H^2(\Omega))$ and $y(x^j,t)$ has meaning (and "$t\to y(x^j,t)$" $\in L^2(0,T)$) if

$$H^2(\Omega)\subset C^0(\Omega), \tag{8.3}$$

which is true if (and only if)

$$\frac{1}{2}-\frac{2}{n}<0 \quad \text{i.e. } n\leqslant 3.$$

Hence we make the standing hypothesis that the dimension $n\leqslant 3$. Then

$$C\,y(v)=\{y(x^j,t;v)\}\in(L^2(0,T))^\mu. \tag{8.4}$$

The cost function is now given by

$$J(v)=\|C\,y(v)-z_d\|^2_{L^2(0,T)^\mu}+(N\,v,v)_{\mathscr{U}}. \tag{8.5}$$

If $z_d = \{z_{d_1}, \ldots, z_{d_\mu}\}$,

$$J(v) = \sum_{j=1}^{\mu} \int_0^T |y(x^j, t; v) - z_{d_j}(t)|^2 \, dt + \int_Q (N v) v \, dx \, dt. \tag{8.6}$$

If the observation is given by (8.4) we say that we make local observations.

If $\mathcal{U}_{\text{ad}} \subset \mathcal{U}$, $\mathcal{U}_{\text{ad}}$ = closed, convex set, and if $N \geqslant \nu$ (Identity), $\nu > 0$, a unique optimal control exists and is characterized by

$$\sum_{j=1}^{\mu} \int_0^T (y(x^j, t; u) - z_{d_j})(y(x^j, t; v) - y(x^j, t; u)) \, dt + (N u, v - u) \geqslant 0 \quad \forall v \in \mathcal{U}_{\text{ad}}. \tag{8.7}$$

We introduce the *adjoint state* by

$$-\frac{\partial}{\partial t} p(u) + A^* p(u) = \sum_{j=1}^{\mu} (y(x^j, t; u) - z_{d_j}(t)) \otimes \delta(x - x^j), \text{[47]} \tag{8.8}$$

$$p(u)|_\Sigma = 0, \tag{8.9}$$

$$p(x, T; u) = 0, \qquad x \in \Omega. \tag{8.10}$$

Remark 8.1. The problem (8.8), (8.9), (8.10) admits a unique solution in $L^2(Q)$ (for example), defined by transposition as in Section 7.3:

$$\left.\begin{aligned} &\int_Q p(u) \left(\frac{\partial \psi}{\partial t} + A \psi \right) dx \, dt = \sum_{j=1}^{\mu} \int_0^T (y(x^j, t; u) - z_{d_j}(t)) \psi(x^j, t) \, dt, \\ &\forall \psi \in H^{2,1}(Q), \quad \psi|_\Sigma = 0, \quad \psi(x, T) = 0.\text{[48]} \end{aligned}\right\} \tag{8.11}$$

Condition (8.7) then becomes

$$\int_Q (p(u) + N u)(v - u) \, dx \, dt \geqslant 0 \quad \forall v \in \mathcal{U}_{\text{ad}}, \qquad u \in \mathcal{U}_{\text{ad}}. \tag{8.12}$$

Example 8.1. $\mathcal{U}_{\text{ad}} = \{v \mid v \geqslant 0 \text{ in } Q\}$.

We have considered this example several times; (8.12) reduces to $u \geqslant 0$, $p(u) \geqslant 0$, $u \cdot p(u) = 0$ in Q.

---

[47] $g(t) \otimes \delta(x - x^j)$ is the distribution,

$$\psi \to \int_0^T g(t) \psi(x^j, t) \, dt, \qquad \psi \in \mathcal{D}(Q).$$

[48] The right hand side of (8.11) is a continuous linear form on $H^{2,1}(Q)$ if $n \leqslant 3$.

Example 8.2. The case of no constraints, $\mathscr{U}_{ad}=\mathscr{U}$.

Then the optimal control is determined by the simultaneous solution of:

$$\left.\begin{aligned} &\frac{\partial y}{\partial t}+Ay+N^{-1}p=f,\\ &-\frac{\partial p}{\partial t}+A^*p=\sum_{j=i}^{\mu}(y(x^j,t)-z_{d_j}(t))\otimes\delta(x-x^j),\\ &y|_\Sigma=0,\quad p|_\Sigma=0,\\ &y(x,0)=y_0(x),\quad p(x,T)=0.\end{aligned}\right\}\tag{8.13}$$

The properties relative to decoupling (Sections 4, 5, and 6) are valid here. We present the following result: we have the identity

$$p(t)=P(t)y(t)+r(t),\tag{8.14}$$

where the kernel $P(x,\xi,t)$ of $P(t)$ satisfies

$$\left.\begin{aligned} &-\frac{\partial P}{\partial t}+(A_x^*+A_\xi^*)P(x,\xi,t)\\ &+\iint_{\Omega\times\Omega}P(x,\xi_1,t)N^{-1}(\xi_1,\xi_2)P(\xi_2,\xi,t)\,d\xi_1\,d\xi_2\\ &=\sum_{j=1}^{\mu}\delta(x-x^j)\otimes\delta(\xi-x^j),\\ &\text{in }\Omega_x\times\Omega_\xi\times\,]0,T[,\end{aligned}\right\}\tag{8.15}$$

with

$$P(x,\xi,t)=P(\xi,x,t),\tag{8.16}$$

$$P(x,\xi,t)=0\quad\text{if } x\in\Gamma,\quad \xi\in\Omega,\quad t\in\,]0,T[,\tag{8.17}$$

$$P(x,\xi,T)=0,\ x,\ \xi\in\Omega\times\Omega,\tag{8.18}$$

and r satisfies

$$\left.\begin{aligned} &-r'+A^*r+PN^{-1}r=Pf-\sum_{j=1}^{\mu}z_{d_j}\otimes\delta(x-x^j),\\ &r|_\Sigma=0,\\ &r(x,T)=0,\quad x\in\Omega.\end{aligned}\right\}\tag{8.19}$$

Example 8.3. Let us consider an analogous problem but with $N=0$ and $\mathscr{U}_{ad}$ bounded. Then an optimal control u exists and all otpimal u's satisfy

$$\int_Q p(v-u)\,dx\geqslant 0\quad\forall\, v\in\mathscr{U}_{ad}.\tag{8.20}$$

More precisely, assume that

$$\mathscr{U}_{\mathrm{ad}}=\{v|\xi_0(x,t)\leqslant v(x,t)\leqslant \xi_1(x,t),\xi_0,\xi_1\in L^\infty(Q)\}. \tag{8.21}$$

Then (8.20) is equivalent to

$$\left.\begin{array}{l} p(x,t;u)u(x,t)=\inf[p(x,t;u)\cdot\xi], \quad \text{almost everywhere in } Q, \\ \xi_0(x,t)\leqslant\xi\leqslant\xi_1(x,t). \end{array}\right\} \tag{8.22}$$

Let us assume that the coefficients of A are analytic in Q and the boundary Γ is analytic. Then $p=0$ almost everywhere in Q. To see this, we note that p is analytic inside Q other than that in $\bigcup_{j=1}^{\mu}\{x^j\}\times]0,T[$ and if $p=0$ on a set of measure >0, p then has support in $\bigcup_{j=1}^{\mu}\{x^j\}\times]0,T[$. But then

$$-\frac{\partial p}{\partial t}+A^*p=\sum_{j=1}^{\mu}(y(x^j,t)-z_{d_j}(t))\otimes\delta(x-x^j)$$

is impossible unless $y(x^j,t;u)=z_{d_j}(t), j=i,\dots,\mu$. Hence:
If $\operatorname{Inf} J(v)>0$, then $p\neq 0$ almost everywhere and

$$u(x,t)=\xi_0(x,t) \quad \text{if } p(x,t;u)>0,$$
$$u(x,t)=\xi_1(x,t) \quad \text{if } p(x,t;u)<0,$$

and the optimal control is unique.

Remark 8.2. Analogous considerations can be made for systems whose state is given as in Section 3.2.

Remark 8.3. The same remarks are true for systems described by parabolic operators

$$\frac{\partial}{\partial t}+A\left(x,t,\frac{\partial}{\partial x}\right), \quad A \text{ of order } 2m.\text{ [49]}$$

Remark 8.4. If the observation is $\mathscr{M}_j y(x^j,t)$, $\mathscr{M}_j\in\mathscr{L}(L^2(0,T);L^2(0,T))$ the same remarks apply with p defined by

$$-\frac{\partial p}{\partial t}+A^*p=\sum_{j=1}^{\mu}\mathscr{M}_j^*(\mathscr{M}_j y(x^j,t)-z_{d_j}(t))\otimes\delta(x-x^j), \tag{8.23}$$

the other conditions remaining the same.

---

[49] Then $y(x^j,t;v)$ has meaning if $n\leqslant 4m-1$.

8.2. Pointwise Control

In applications another situation, which is a variant of the preceeding, is important. Let A be as in Section 8.1, the state $y(v)$ being given by

$$\frac{\partial y}{\partial t}(v)+A\,y(v)=\sum_{j=1}^{\mu} v_j(t)\otimes\delta(x-x^j), \quad x^j\in Q, \tag{8.24}$$

where

$$v=\{v_j\}\in(L^2(0,T))^{\mu}. \tag{8.25}$$

The boundary conditions are (for example)

$$y(v)|_{\Sigma}=0 \tag{8.26}$$

and

$$y(x,0;v)=y_0(x), \quad x\in\Omega. \tag{8.27}$$

It is clear that we must attach some meaning to this problem (termed pointwise control).

Let us proceed by transposition (cf. Section 7.3 and Remark 8.1). We define $y(v)$ as the unique element in $L^2(Q)$ such that

$$\left.\begin{aligned}\int_Q y(v)\left(-\frac{\partial\psi}{\partial t}+A^*\psi\right)dx\,dt=\sum_{j=1}^{\mu}\int_0^T \psi(x^j,t)\,v_j(t)\,dt\\ +\int_Q \psi(x^j,0)\,y_0(x)\,dx,\\ \forall\psi\in H^{2,1}(Q), \quad \psi|_{\Sigma}=0, \quad \psi(x,T)=0,\end{aligned}\right\} \tag{8.28}$$

provided we assume $n\leqslant 3$ (since then the right hand side is continuous on $H^{2,1}(Q)$).

If the cost function is given by

$$\left.\begin{aligned}J(v)=\int_Q (y(v)-z_d)^2\,dx\,dt+(N\,v,v)_{\mathcal{U}},\\ \mathcal{U}=(L^2(0,T))^{\mu}, \quad N\in\mathscr{L}(\mathcal{U},\mathcal{U}), \quad N\geqslant\nu\,(\text{Identity}), \quad \nu>0,\end{aligned}\right\} \tag{8.29}$$

a unique optimal control exists; the adjoint state $p(v)$ is defined by

$$\left.\begin{aligned}-\frac{\partial p(v)}{\partial t}+A^*p(v)=y(v)-z_d,\\ p(v)|_{\Sigma}=0, \quad p(x,T;v)=0, \quad x\in\Omega,\end{aligned}\right\} \tag{8.30}$$

and u is characterized by (8.24), (8.27), (8.30) (where $v=u$) and

$$\sum_{j=1}^{\mu} \int_0^T p(x^j,t;u)(v_j(t)-u_j(t))\,dt + (N u, v-u)_{\mathscr{U}} \geqslant 0 \quad \forall v \in \mathscr{U}_{\text{ad}}. \tag{8.31}$$

Remark 8.5. We note, that in this section and in the preceeding one, y and p are in duality (as it must be) in the sense that if y is regular (case 8.1) (resp. irregular (case 8.2)) then p is irregular (resp. regular) and Green's formula is applicable in each case.

8.3. Control and Observation on the Boundary

We shall now study the first example (important for applications) where *both the control and the observation is on the boundary.*

There are an infinite number of possibilities. In this chapter we shall examine two typical cases. The first example is considered in this section. The second (which introduces serious additional technical difficulties) will be studied in the next section.

We consider the same situation as in Section 3.2. Hence the state is defined by

$$\left.\begin{aligned} &\frac{\partial}{\partial t} y(v) + A y(v) = f \quad \text{in } Q, \\ &\frac{\partial y}{\partial \nu_A}(v) = v \quad \text{on } \Sigma, \quad (v \in L^2(\Sigma) = \mathscr{U}), \\ &y(x,0;v) = y_0(x), \quad x \in Q. \end{aligned}\right\} \tag{8.32}$$

The observation is

$$\left.\begin{aligned} &C y(v) = \mathscr{M}(y(v)|_\Sigma), \\ &\mathscr{M} \in \mathscr{L}(L^2(\Sigma); L^2(\Sigma)). \end{aligned}\right\} \tag{8.33}$$

This is interpreted as follows: we take the trace of $y(v)$ on Σ, which is in particular[50] in $L^2(\Sigma)$. Let this be denoted by $y(v)|_\Sigma$; (8.33) is then clear.

Example 8.4. If $\mathscr{M} =$ identity, we observe $y(v)$ on Σ.

If $\mathscr{M} =$ multiplication by the characteristic function of $\Sigma_0 \subset \Sigma$, we observe $y(v)$ on Σ_0.

[50] We have a better result: $y(v)|_\Sigma \in H^{\frac{1}{2},\frac{1}{4}}(\Sigma)$; cf. Lions-Magenes [1].

The *cost function* is

$$\left.\begin{aligned} &J(v)=\int_\Sigma (\mathcal{M}(y(v)|_\Sigma)-z_d)^2\,d\Sigma+(N v,v)_{L^2(\Sigma)},\\ &N\in\mathcal{L}(\mathcal{U};\mathcal{U}),\qquad N\geqslant \nu\text{ (identity)},\qquad \nu>0.\end{aligned}\right\}\tag{8.34}$$

The optimal control exists and it is unique. It is characterized by

$$\left.\begin{aligned} &\int_\Sigma (\mathcal{M}(y(u)|_\Sigma-z_d)\mathcal{M}(y(v)|_\Sigma-y(u)|_\Sigma)\,d\Sigma+(Nu,v-u)_{L^2(\Sigma)}\geqslant 0,\\ &\forall v\in\mathcal{U}_{\mathrm{ad}}\subset L^2(\Sigma).\end{aligned}\right\}\tag{8.35}$$

The *adjoint state* is defined by

$$\left.\begin{aligned} -\frac{\partial p(u)}{\partial \nu_{A^*}}+A^*p(u)&=0\quad\text{in } Q,\\ \frac{\partial p(u)}{\partial \nu_{A^*}}&=\mathcal{M}^*(\mathcal{M}(y(u)|_\Sigma)-z_d)\quad\text{on } \Sigma,\\ p(x,T;u)&=0,\qquad x\in Q.\end{aligned}\right\}\tag{8.36}$$

Let us note that problem (8.36) admits a unique solution in $L^2(0,T;H^1(\Omega))$ defined by

$$\left.\begin{aligned} &\int_Q\left[\left(p(u),\frac{\partial\psi}{\partial t}\right)+a^*(t;p(u),\psi)\right]dx\,dt=\int_\Sigma \mathcal{M}^*(\mathcal{M}(y(u)|_\Sigma)-z_d)\psi\,d\Sigma\\ &\forall\psi\in L^2(0,T;H^1(\Omega)),\qquad \frac{\partial\psi}{\partial t}\in L^2(Q),\qquad \psi(x,0)=0.\end{aligned}\right\}\tag{8.37}$$

Then condition (8.35) is equivalent to

$$\int_\Sigma (p(u)+Nu)(v-u)\,d\Sigma\geqslant 0\quad\forall v\in\mathcal{U}_{\mathrm{ad}}.\tag{8.38}$$

Thus:

Theorem 8.1. *The state being defined by* (8.32) *and the cost function by* (8.34), *the optimal control* u *is characterized by the system of equations* (8.32) (*where* $v=u$), (8.36) *and the inequality* (8.38).

Example 8.5. We take

$$\mathcal{U}_{\mathrm{ad}}=\{v\,|\,v\in L^2(\Sigma),\ v\geqslant 0\text{ on }\Sigma\},$$

and

$$\mathcal{M}=\text{identity map}.$$

Then we must solve the unilateral problem

$$\left.\begin{aligned}
&\frac{\partial y}{\partial t}+Ay=f, \quad -\frac{\partial p}{\partial t}+A^*p=0 \quad \text{in } Q,\\
&\frac{\partial p}{\partial \nu_{A^*}}=y-z_d \quad \text{on } \Sigma, \quad \frac{\partial y}{\partial \nu_A}\geqslant 0 \quad \text{on } \Sigma,\\
&p+N\frac{\partial y}{\partial \nu_A}\geqslant 0 \quad \text{on } \Sigma, \quad \frac{\partial y}{\partial \nu_A}\left(p+N\frac{\partial y}{\partial \nu_A}\right)=0 \quad \text{on } \Sigma,\\
&y(x,0)=y_0(x), \quad p(x,T)=0.
\end{aligned}\right\} \tag{8.39}$$

Then the optimal control is given by

$$u=\frac{\partial y}{\partial \nu_A}. \tag{8.40}$$

We again encounter an open problem which he have encountered several times before: what is the topological nature of the "lines of switching" on Σ separating the domains where $\frac{\partial y}{\partial \nu_A}=0$ from that of where $p+N\frac{\partial y}{\partial v_A}=0$?

Example 8.6. We take

$\mathscr{U}_{\rm ad}$ as in example 8.5,
$\mathscr{M}=$ Multiplication by the characteristic function of the measurable set $\Sigma_0\subset\Sigma$.

Then the boundary condition on Σ in (8.36) is

$$\frac{\partial}{\partial \nu_{A^*}}p(u)=\begin{cases} y(u)-z_d & \text{on } \Sigma_0,\\ 0 \text{ outside on } \Sigma.\end{cases} \tag{8.41}$$

It now suffices to make the necessary modifications in the unilateral problem (8.39).

Example 8.7. $\mathscr{U}_{\rm ad}=\mathscr{U}$ (case of no constraints).
The optimal control is determined by the solution of

$$\left.\begin{aligned}
&\frac{\partial y}{\partial t}+Ay=f, \quad -\frac{\partial p}{\partial t}+A^*p=0,\\
&\frac{\partial y}{\partial \nu_A}+N^{-1}p=0 \quad \text{on } \Sigma, \quad \frac{\partial p}{\partial \nu_{A^*}}=\mathscr{M}^*(\mathscr{M}(y|_\Sigma)-z_d),\\
&y(x,0)=y_0(x), \quad p(x,T)=0,
\end{aligned}\right\} \tag{8.42}$$

and then

$$u=\frac{\partial y}{\partial \nu_A} \quad \text{on } \Sigma. \tag{8.43}$$

We shall now indicate how we may "decouple" problem (8.42). To simplify somewhat the calculation let $\mathcal{M}$ = identity map. Let us rewrite the system in variational form:

$$\left.\begin{aligned}
&\left(\frac{dy(t)}{dt},\varphi\right)+a(t;y,\varphi)+(N^{-1}p,\varphi)_{L^2(\Gamma)}=(f,\varphi)\ \ \forall\varphi\in H^1(\Omega),\\
&-\left(\frac{dp(t)}{dt},\psi\right)+a^*(t;p,\psi)-(y,\psi)_{L^2(\Gamma)}=-(z_d,\psi)_{L^2(\Gamma)}\ \ \forall\psi\in H^1(\Omega),\\
&\qquad y(0)=y_0,\quad p(T)=0.
\end{aligned}\right\} \tag{8.44}$$

As in Sections 4 and 5, we prove the identity

$$p(t)=P(t)y(t)+r(t), \quad P^*(t)=P(t) \tag{8.45}$$

and as in Section 5 the following calculations are valid: carrying (8.45) into the second equation of (8.44), we obtain

$$\left.\begin{aligned}
&\left(\frac{dP(t)}{dt}y(t),\psi\right)-\left(P(t)\frac{dy(t)}{dt},\psi\right)-\left(\frac{dr(t)}{dt},\psi\right)\\
&\qquad+a^*(t;Py+r,\psi)-(y,\psi)_{L^2(\Gamma)}=-(z_d,\psi)_{L^2(\Gamma)}.
\end{aligned}\right\} \tag{8.46}$$

But

$$\left(P(t)\frac{dy}{dt},\psi\right)=\left(\frac{dy}{dt},P(t)\psi\right)$$

and utilising the first equation (8.44), this equals

$$=-a(t;y,P(t)\psi)-(N^{-1}p,(P\psi))_{L^2(\Gamma)}+(f,P\psi)$$

and substituting this in (8.46) (using (8.45) once more) we obtain:

$$\begin{aligned}
-\left(\frac{dP(t)y}{dt},\psi\right)-\left(\frac{dr(t)}{dt},\psi\right)+a(t;y,P(t)\psi)+(N^{-1}(Py+r),P\psi)_{L^2(\Gamma)}\\
-(f,P\psi)+a^*(t;Py+r,\psi)-(y,\psi)_{L^2(\Gamma)}=-(z_d,\psi)_{L^2(\Gamma)}
\end{aligned}$$

from which we deduce that

$$\left.\begin{aligned}
&-\left(\frac{d}{dt}P(t)\varphi,\psi\right)+a(t;\varphi,P\psi)+a^*(t;P\varphi,\psi)+(N^{-1}P\varphi,P\psi)_{L^2(\Gamma)}\\
&\qquad=(\varphi,\psi)_{L^2(\Gamma)},\ \ \forall\varphi,\psi\in H^1(\Omega),
\end{aligned}\right\} \tag{8.47}$$

and

$$P(T)=0. \tag{8.48}$$

Further $r(t)$ is characterized by

$$-\left(\frac{dr(t)}{dt},\psi\right)+a^*(t;r,\psi)=(Pf,\psi)-(z_d,\psi)_{L^2(\Gamma)} \tag{8.49}$$

and

$$r(T)=0. \tag{8.50}$$

We now obtain the equation for the kernel $P(x,\xi,t)$ of $P(t)$ (cf. Section 5.1 for the application of the Schwartz Kernel-Theorem).

From (8.47) we deduce that (assuming $N=\nu\times$ identity to simplify the calculations)

$$-\frac{\partial P}{\partial t}(x,\xi,t)+(A_x^*+A_\xi^*)P(x,\xi,t)+\frac{1}{\nu}\int_\Gamma P(x,\zeta,t)P(\zeta,\xi,t)d\Gamma=0$$

in $\Omega_x\times\Omega_\xi\times]0,T[$, (8.51)

$$P(x,\xi,t)=P(\xi,x,t), \tag{8.52}$$

$$\left.\begin{aligned}&\frac{\partial P}{\partial\nu_{A^*,x}}(x,\xi,t)=0 \quad \text{if } x\in\Gamma_x,\quad \xi\in\Omega_\xi,\\ &\frac{\partial P}{\partial\nu_{A^*,\xi}}(x,\xi,t)=0 \quad \text{if } x\in\Omega_x,\quad \xi\in\Gamma_\xi,\\ &\frac{\partial P}{\partial\nu_{A^*,x}}(x,\xi,t)=\frac{\partial P}{\partial\nu_{A^*,\xi}}(x,\xi,t)=\tfrac{1}{2}\delta(x-\xi) \quad \text{if } x\in\Gamma_x,\quad \xi\in\Gamma_\xi\end{aligned}\right\} \tag{8.53}$$

[51]

and

$$P(x,\xi,T)=0,\qquad x,\xi\in\Omega\times\Omega. \tag{8.54}$$

Remark 8.6. Here again, a *direct study* of problems (non-linear) of the type (8.51)–(8.54), in particular for the regularity of P, appears to us to be extremely interesting.

---

[51] $\delta(x-\xi)$ is the distribution on Γ_x,Γ_ξ:

$$\theta\to\int_{\Gamma\times\Gamma}\delta(x-\xi)\theta(x,\xi)d\Gamma_x d\Gamma_\xi=\int_\Gamma\theta(x,x)d\Gamma_x.$$

Analytically (8.53) may be expressed as

$$\int_{\Gamma_x\times\Omega_\xi}\frac{\partial P}{\partial\nu_{A^*,x}}\varphi(x)\psi(\xi)d\Gamma_x d\Gamma_\xi+\int_{\Omega_x\times\Gamma_\xi}\frac{\partial P}{\partial\nu_{A^*,x}}\varphi(x)\psi(\xi)d\Gamma_x d\Gamma_\xi$$

$$=-\int_{\Gamma\times\Gamma}\varphi(x)\psi(\xi)d\Gamma_x d\Gamma_\xi\quad\forall\varphi,\psi.$$

Remark 8.7. Expansion in Eigenfunctions. Let us reconsider (8.32) with the assumption:

$$A=A^*, \quad A \text{ independent of } t,$$

Ω bounded with a sufficiently regular boundary.

Let $w_1,\dots,w_m,\dots$ be the eigenfunction of A, for the Neumann problem:

$$\begin{aligned} &A w_j=\lambda_j w_j, \quad \frac{\partial w_j}{\partial \nu_A}=0 \quad \text{on } \Gamma, \quad j=1,2,\dots \\ &(w_j, w_k)=\delta_j^k. \end{aligned} \tag{8.55}$$

The solution of (8.32) may be represented with the aid of w_j. In order not to commit any errors in sign, it is better to rewrite (8.32) in variational form. This is[52]

$$\begin{aligned} &\frac{d}{dt}(y,\psi)+a(y,\psi)=(f,\psi)+(v,\psi)_{L^2(\Gamma)} \quad \forall \psi\in H^1(\Omega), \\ &y(0)=y_0. \end{aligned} \tag{8.56}$$

We may represent y in the form

$$\left.\begin{aligned} &y=\sum_{j=1}^{\infty} y_j(t) w_j, \\ &y_j\in L^2(0,T), \quad \sum_{j=1}^{\infty} |\lambda_j| \int_0^T |y_j(t)|^2\,dt<\infty. \end{aligned}\right\} \tag{8.57}$$

If we set

$$\begin{aligned} &f_j(t)=(f(t), w_j), \\ &v_j(t)=(v(t), w_j)_{L^2(\Gamma)}=\int_\Gamma v(x,t)\, w_j(x)\, d\Gamma_x, \end{aligned}$$

and $y_{0j}=(y_0, w_j)$, we obtain, substituting (8.57) in (8.56) and taking $\psi=w_j$:

$$\begin{aligned} &y_j'(t)+\lambda_j y_j(t)=f_j(t)+v_j(t), \\ &y_j(0)=y_{0j}. \end{aligned} \tag{8.58}$$

We note that the "boundary term" $v(t)$ gives rise to the expression $v_j(t)$ in (8.58). Further the condition "$\frac{\partial y}{\partial \nu_A}=v$ on Σ" in (8.32) can be extended to have meaning in an "average" sense and hence there is no contradiction in representing $\frac{\partial y}{\partial \nu_A}$ on Σ with the aid of $\frac{\partial w_j}{\partial \nu_A}$ which are zero.

---

[52] We write y instead of $y(v)$.

With the aid of the basis w_j, $P(t)$ may clearly be represented as

$$P(t)=\sum_{j,k} P_{jk}(t)\, w_j \otimes w_k \tag{8.59}$$

$$\left(\text{where } P(x,\xi,t)=\sum_{j,k} P_{jk}(t)\, w_j(x)\, w_k(\xi)\right)$$

where

$$P_{jk}=P_{kj}\ \forall j,k; \tag{8.60}$$

taking $\varphi=w_j$, $\psi=w_k$, in (8.47), we find that $P_{jk}(t)$ is given by the infinite system of equations of "*Riccati type*":

$$\left.\begin{aligned}
&-\frac{d}{dt}P_{jk}(t)+(\lambda_j+\lambda_k)P_{jk}(t)+\frac{1}{\nu}\sum_{r,s=1}^{\infty}P_{rj}(t)P_{sk}(t)\int_\Gamma w_r w_s\, d\Gamma\\
&=\int_\Gamma w_j w_k\, d\Gamma \quad \forall j,k, \qquad t\in\,]0,T[,\\
&\qquad P_{jk}(T)=0.
\end{aligned}\right\} \tag{8.61}$$

Let us remark that if we truncate the system (8.61) by taking terms up to order m, $j,k\leqslant m$, we are assured that the truncated system of order m will converge towards the infinite system (using arguments of Section 4.5).

9. Boundary Control and Observation on the Boundary or of the Final State for a System Governed by a Mixed Dirichlet Problem

9.1. Orientation and Problem Statement

In Section 8.3 we have presented the first example of a problem where both the control and observation are on the boundary. We shall now study the following case[53]: the operator A is given as in Section 8[54], and the state $y(v)$ is given by

$$\frac{\partial}{\partial t}y(v)+A\,y(v)=f \quad \text{in } Q, \tag{9.1}$$

$$y(v)|_\Sigma=v \quad \text{on } \Sigma, \tag{9.2}$$

$$y(x,0;v)=y_0(x), \qquad x\in\Omega. \tag{9.3}$$

---

[53] In the sequel we shall indicate in what sense these boundary value problems are to be considered (which requires some additional development).

[54] That is, second order elliptic operator with coefficients which are sufficiently smooth functions of x and t.

The *observation* is given by

$$z(v) = C\,y(v) = \mathcal{M}\left(\frac{\partial y(v)}{\partial \nu_A}\right) \tag{9.4}$$

where $\mathcal{M}$ is an "operator defined on Σ".[55]

We now encounter the following difficulties:

(i) If we take $v \in L^2(\Sigma)$, for example, we must first solve (9.1), (9.2), (9.3) (in some appropriate sense) and then $\dfrac{\partial y(v)}{\partial \nu_A}$ is in a space which is larger than $L^2(\Sigma)$ (due to loss of regularity). We then have to define $\dfrac{\partial y(v)}{\partial \nu_A}$ and consequently define the cost function.

(ii) The difficulty indicated in (i) can be avoided if we take v to be *more regular*[56], for example, if $v \in H_0^1(\Sigma)$. However, then the term $(N\,v, v)_{\mathcal{U}}$ becomes more complicated and the determination of the optimal control becomes correspondingly more difficult.

We shall successively examine these two points of view.

9.2. Non Homogeneous Mixed Dirichlet Problem[57]

As in Section 7.3 and Remark 8.1 we proceed by transposition. We consider the isomorphism

$$\varphi \xrightarrow{\mathcal{L}} -\frac{\partial \varphi}{\partial t} + A^* \varphi$$

of $\Phi \to L^2(Q)$, where

$$\Phi = \{\varphi \mid \varphi \in H^{2,1}(Q),\ \varphi|_{\Sigma} = 0,\ \varphi(x, T) = 0\}.$$

By transposition, we deduce:

$\varphi \to M(\varphi)$ being a continuous linear form on Φ (endowed with the topology induced by $H^{2,1}(Q)$, cf. (3.5)), there exists a unique $y \in L^2(Q)$ such that

$$\int_Q y\left(-\frac{\partial \varphi}{dt} + A^* \varphi\right) dx\,dt = M(\varphi)\ \ \forall \varphi \in \Omega, \tag{9.5}$$

and

the mapping $M \to y$ is a continuous linear mapping of $\Phi' \to L^2(Q)$. (9.6)

---

[55] This will be indicated later.

[56] This is not very realistic from an applications point of view.

[57] Compare with Chapter 2, Section 5.

Let us choose $M(\varphi)$ by

$$M(\varphi)=\int_Q f\varphi\,dx\,dt+\int_\Omega y_0\,\varphi(x,0)\,dx-\int_\Sigma v\frac{\partial\varphi}{\partial\nu_{A^*}}\,d\Sigma \tag{9.7}$$

where

$$f\in L^2(Q),\quad y_0\in L^2(\Omega),\quad v\in L^2(\Sigma). \tag{9.8}$$

Then (9.7) defines a continuous linear form on Φ.[58] Therefore

Lemma 9.1. *For given f, y_0, v as in* (9.8), *with M given by* (9.7), *there exists a unique $y=y(v)$ in $L^2(Q)$ such that* (9.5) *is satisfied.*

Let us now formally verify that it is reasonable to define $y(v)$, the solution of (9.1), (9.2), (9.3) with the aid of (9.5) and (9.7). To see this, let us first take $\varphi\in\mathscr{D}(\Omega)$ in (9.5); $M(\varphi)$ then reduces to

$$\int_Q f\varphi\,dx\,dt$$

and hence

$$\int_Q y\left(-\frac{\partial\varphi}{\partial t}+A^*\right)dx\,dt=\int_Q f\varphi\,dx\,dt\quad\forall\varphi\in\mathscr{D}(Q),$$

and thus we have (9.1) $(y=y(v))$; now multiplying (9.1) by $\varphi\in\Phi$ and applying *Green's Formula*, we obtain

$$\int_Q f\varphi\,dx\,dt=-\int_\Omega y(x,0)\,\varphi(x,0)\,dx+\int_\Sigma y\frac{\partial\varphi}{\partial\nu_{A^*}}\,d\Sigma+\int_Q y\left(-\frac{\partial\varphi}{dt}+A^*\varphi\right)dx\,dt$$

whence comparing with (9.5), (9.7):

$$\int_\Omega y(x,0)\,\varphi(x,0)\,dx-\int_\Sigma y\frac{\partial\varphi}{\partial\nu_{A^*}}\,d\Sigma=\int_\Omega y_0\,\varphi(x,0)\,dx-\int_\Sigma v\frac{\partial\varphi}{\partial\nu_{A^*}}\,d\Sigma\quad\forall\varphi\in\Phi.$$

From this we deduce (9.2), (9.3).

---

[58] In fact, $\varphi(x,0)\in H^1(\Omega)$ and hence in particular $\in L^2(\Omega)$ and $\frac{\partial\varphi}{\partial\nu_{A^*}}$ is in particular in $L^2(\Sigma)$ (actually $\frac{\partial\varphi}{\partial\nu_{A^*}}\in H^{\frac{1}{2},\frac{1}{4}}(\Sigma)$, cf. Lions-Magenes [1]).

We may summarize the above in

Theorem 9.1. *There exists a unique* $y(v) \in L^2(Q)$ *such that*

$$\left.\begin{aligned}\int_Q y(v)\left(-\frac{\partial\varphi}{\partial t}A^*\varphi\right)dx\,dt &= \int_Q f\varphi\,dx\,dt + \int_\Omega y_0(x)\varphi(x,0)\,dx\\ &\quad -\int_\Sigma v\frac{\partial\varphi}{\partial\nu_{A^*}}d\Sigma \quad \forall\varphi\in\Phi,\end{aligned}\right\}\tag{9.9}$$

f, y_0, v *satisfying* (9.8).

Remark 9.1. The state of the system is therefore defined by (9.9).

Remark 9.2. (Compare with Remark 8.7). Assume that $A=A^*$ independent of t; let $w_1,\dots,w_m,\dots$ be the eigenfunctions of A for the Dirichlet problem:

$$A w_j = \lambda_j w_j, \qquad w_j|_\Gamma = 0, \qquad (w_j, w_k) = \delta_j^k. \tag{9.10}$$

Then the solution $y(v)=y$ in $L^2(Q)$ may be represented in a unique manner by

$$\left.\begin{aligned}&y = \sum_{j=1}^{\infty} y_j(t)w_j,\\ &y_j\in L^2(0,T), \qquad \sum_{j=1}^{\infty}\int_0^T |y_j(t)|^2\,dt < \infty.\end{aligned}\right\}\tag{9.11}$$

To determine $y_j(t)$, let us take in (9.9)

$$\varphi(x,t) = \psi(t)w_j(x), \qquad \psi\in C^1([0,T]), \qquad \psi(T)=0$$

(so that $\varphi\in\Phi$). We obtain:

$$\left.\begin{aligned}\int_0^T y_j(t)\left(-\frac{d\psi}{dt}+\lambda_j\psi\right)dt &= \int_0^T f_j(t)\psi(t)\,dt + y_{0j}\psi(0)\\ &\quad -\int_0^T v_j(t)\psi(t)\,dt \quad \forall\psi,\end{aligned}\right\}\tag{9.12}$$

where

$$\left.\begin{aligned}&f_j(t) = \int_\Omega f(x,t)w_j(x)\,dx, \qquad y_{0j} = \int_\Omega y_0(x)w_j(x)\,dx,\\ &v_j(t) = \int_\Gamma v(x,t)\frac{\partial w_j}{\partial\nu_{A^*}}(x)\,d\Gamma_x.\end{aligned}\right\}\tag{9.13}$$

But (9.12) is equivalent to

$$\left.\begin{aligned} \frac{dy_j}{dt} + \lambda_j y_j = f_j - v_j, \\ y_j(0) = y_{0j}. \end{aligned}\right\} \tag{9.14}$$

Same comments as in Remark 8.7.

9.3. Definition of $\frac{\partial y}{\partial \nu_A}$; Observation

Let us now assume that in the calculation of $\int_Q \left(\frac{\partial y}{\partial t} + A y\right) \varphi \, dx \, dt$ we take

$$\varphi \in H^{2,1}(Q), \quad \psi(x,T) = 0, \text{ but } \psi|_\Sigma = 0.$$

We obtain (utilizing *Green's Formula* formally)

$$\int_Q y \varphi \, dx \, dt = -\int_\Omega y(x,0) \varphi(x,0) \, dx - \int_\Sigma \frac{\partial y}{\partial \nu_A} \varphi \, d\Sigma + \int_\Sigma v \frac{\partial \varphi}{\partial \nu_{A^*}} d\Sigma$$
$$+ \int_Q y \left(-\frac{\partial \varphi}{\partial t} + A^* \varphi\right) dx \, dt$$

and hence

$$\left.\begin{aligned} \int_\Sigma \frac{\partial y}{\partial \nu_A} \varphi \, d\Sigma = \int_Q y \left(-\frac{\partial \varphi}{\partial t} + A^* \varphi\right) dx \, dt + \int_\Sigma v \frac{\partial \varphi}{\partial \nu_{A^*}} d\Sigma \\ - \int_\Omega y_0(x) \varphi(x,0) \, dx - \int_Q f \varphi \, dx \, dt, \\ \forall \varphi \in H^{2,1}(Q), \quad \varphi(x,T) = 0. \end{aligned}\right\} \tag{9.15}$$

We may then propose to define $\frac{\partial y}{\partial \nu_A}$ with the aid of (9.15): we take an "appropriate" function ψ on Σ, we define φ (non unique) in such a way that $\varphi|_\Sigma = \psi$, and then we define $\frac{\partial y}{\partial \nu_A}$ by

$$\left.\begin{aligned}\int_\Sigma \frac{\partial y}{\partial \nu_A}\psi\, d\Sigma = \int_Q y\left(-\frac{\partial \varphi}{\partial t} + A^*\varphi\right) dx\, dt + \int_\Sigma v \frac{\partial \varphi}{\partial \nu_{A^*}} d\Sigma \\ - \int_\Omega y_0(x)\varphi(x,0)\, dx - \int_Q f\varphi\, dx\, dt.^{59}\end{aligned}\right\} \tag{9.16}$$

In this way we may show[60]

$$\frac{\partial y}{\partial \nu_A} \in H^{-1}(\Sigma),^{61} \tag{9.17}$$

where $H^{-1}(\Sigma)$ = dual of $H_0^1(\Sigma)$,

$$H_0^1(\Sigma) = \{\psi \mid \psi \in H^1(\Sigma),\ \psi(x,0)=0, \psi(x,T)=0\}.$$

The observation now is

$$\left.\begin{aligned}&z(v) = \frac{\partial y}{\partial \nu_A}(v)|_{\Sigma_0}, \text{ restriction of } \frac{\partial y}{\partial \nu_A}(v) \text{ to } \Sigma_0 = \text{open subset of } \Sigma, \\ &\text{with the possibility of } \Sigma_0=\Sigma); \quad z(v) \in H^{-1}(\Sigma_0).\end{aligned}\right\} \tag{9.18}$$

9.4. Cost Function; Equations for Optimal Control

We take

$$\left.\begin{aligned}&J(v) = \left\| \frac{\partial y}{\partial \nu_A}(v) - z_d \right\|^2_{H^{-1}(\Sigma_0)} + (Nv, v)_{\mathscr{U}}, \\ &\mathscr{U} = L^2(\Sigma), \quad N \in \mathscr{L}(\mathscr{U};\mathscr{U}), \quad N \geqslant \nu(\text{identity}), \quad \nu>0.\end{aligned}\right\} \tag{9.19}$$

The norm in $H^{-1}(\Sigma_0)$ is fixed in the following manner: let $-\Delta_\Sigma$ be the Laplace-Beltrami operator on Σ (hence $-\Delta_\Sigma = -\Delta_\Gamma - \frac{\partial^2}{\partial t^2}$, where $-\Delta_\Gamma$ is the Laplace-Beltrami operator on Γ); then if $f, g \in H^{-1}(\Sigma_0)$, we take

$$(f,g)_{H^{-1}(\Sigma_0)} = \int_{\Sigma_0} ((-\Delta_\Sigma)^{-1} f) g\, d\Sigma = \int_{\Sigma_0} f((-\Delta_\Sigma)^{-1} g)\, d\Sigma,$$

---

[59] φ is not unique, once ψ is given; the right hand side of (9.16) is independent of the choice of φ due to the fact that $\varphi|_\Sigma = \psi$.

[60] The "technical details" may be found in Lions-Magenes [1], Chapter 4.

[61] Compare with Chapter 2, Section 5.2.

where $(-\Delta_\Sigma)^{-1} f=\psi$ is the solution in $H_0^1(\Sigma_0)$ of $(-\Delta_\Sigma)\psi=f$, $\psi=0$ on the boundary of Σ_0; then

$$\|f\|^2_{H^{-1}(\Sigma_0)}=(f,f)_{H^{-1}(\Sigma_0)}$$

and (9.19) is defined without any ambiguity.

If $\mathscr{U}_{\mathrm{ad}}$ = closed, convex subset of $L^2(\Sigma)$, there exists a unique optimal control which is characterized by

$$\left(\frac{\partial y}{\partial v_A}(u)-z_d,\ \frac{\partial y}{\partial v_A}(v)-\frac{\partial y}{\partial v_A}(u)\right)_{H^{-1}(\Sigma_0)}+(Nu,v-u)_{\mathscr{U}}\geqslant 0\quad \forall v\in\mathscr{U}_{\mathrm{ad}}. \tag{9.20}$$

The adjoint state is defined by

$$-\frac{\partial p(u)}{\partial t}+A^*p(u)=0 \quad \text{in } Q, \tag{9.21}$$

$$p(u)=\begin{cases}(-\Delta_\Sigma)^{-1}\left(\dfrac{\partial y}{\partial v_A}(u)-z_d\right) & \text{on } \Sigma_0,^{62}\\[2ex] 0 \text{ on } \Sigma-\Sigma_0,\end{cases} \tag{9.22}$$

$$p(x,T;u)=0,\quad x\in\Omega. \tag{9.23}$$

The solution p is now "regular"[63]: $p(u)|_\Sigma\in H_0^1(\Sigma)$ such that $p(u)$ is in particular in $L^2(0,T;H^1(\Omega))$.

To transform (9.20), we utilize *Green's formula*. Multiplying (9.21) by $y(v)-y(u)$ and integrating over Q, we get

$$0=-\int_\Sigma \frac{\partial p(u)}{\partial v_{A^*}}(y(v)-y(u))\,d\Sigma+\int_\Sigma p(u)\left(\frac{\partial y(v)}{\partial v_A}-\frac{\partial y(u)}{\partial v_A}\right)d\Sigma$$
$$+\int_Q p(u)\left[\frac{\partial y(v)}{\partial t}+Ay(v)-\left(\frac{\partial y(u)}{\partial t}+Ay(u)\right)\right]dx\,dt$$

whence utilizing (9.22):

$$\left(\frac{\partial y(u)}{\partial v_A}-z_d,\ \frac{\partial y(v)}{\partial v_A}-\frac{\partial y}{\partial v_A}(u)\right)_{H^{-1}(\Sigma_0)}=\int_\Sigma \frac{\partial p(u)}{\partial v_{A^*}}(v-u)\,d\Sigma.^{64} \tag{9.24}$$

---

[62] For the Dirichlet problem $(-\Delta_\Sigma)^{-1}$ is taken in Σ_0. Hence $-\Delta_\Sigma p=\dfrac{\partial y(u)}{\partial v_A}z_d$ on Σ_0, $p=0$ on the boundary of Σ_0.

[63] This is the same duality phenomenon which he have indicated in Remark 8.5.

[64] We may show that $\dfrac{\partial p(u)}{\partial v_{A^*}}\in L^2(\Sigma)$.

We then have

Theorem 9.2. *We assume that the state is given by* (9.9). *Then, if the cost function is given by* (9.19), *the optimal control u is given by the simultaneous solution of* (9.9) *(where $v=u$),* (9.21), (9.22), (9.23) *and of*

$$\int_{\Sigma}\left(\frac{\partial p(u)}{\partial \nu_{A^*}}+Nu\right)(v-u)\,d\Sigma\geqslant 0 \quad \forall v\in\mathcal{U}_{ad}. \tag{9.25}$$

Example 9.1. Case where there are no constraints: $\mathcal{U}_{ad}=\mathcal{U}$.
Then the optimal control is obtained by simultaneously solving

$$\left.\begin{aligned}
&\frac{\partial y}{\partial t}+Ay=f, \qquad -\frac{\partial p}{\partial t}+A^*p=0 \quad \text{in } Q,\\
&y+N^{-1}\frac{\partial p}{\partial \nu_{A^*}}=0 \quad \text{on } \Sigma,\\
&p=\begin{cases}(-\Delta_\Sigma)^{-1}\left(\dfrac{\partial y}{\partial \nu_{A^*}}-z_d\right) & \text{on } \Sigma_0,\\ 0 \text{ on } \Sigma-\Sigma_0,\end{cases}\\
&y(x,0)=y_0(x), \quad p(x,T)=0, \quad x\in\Omega,
\end{aligned}\right\} \tag{9.26}$$

and then we have

$$u=y|_{\Sigma}. \tag{9.27}$$

Problem (9.26) may be decoupled in the manner indicated in the previous section but we were unable to justify the formal calculations.

Example 9.2. If we take $\mathcal{U}_{ad}=\{v \mid v\geqslant 0$ almost everywhere on $\Sigma\}$, then we obtain the following unilateral problem:

$$\left.\begin{aligned}
&\frac{\partial y}{\partial t}+Ay=f, \quad -\frac{\partial p}{\partial t}+A^*p=0 \quad \text{in } Q,\\
&p=(-\Delta_\Sigma)^{-1}\left(\frac{\partial y}{\partial \nu_A}-z_d\right) \quad \text{on } \Sigma, \quad (\text{if } \Sigma_0=\Sigma),\\
&y\geqslant 0 \quad \text{on } \Sigma, \quad \frac{\partial p}{\partial \nu_{A^*}}+Ny\geqslant 0 \quad \text{on } \Sigma,\\
&y\left(\frac{\partial p}{\partial \nu_{A^*}}\right)+Ny=0 \quad \text{on } \Sigma,\\
&y(x,0)=y_0(x), \quad p(x,T)=0, \quad x\in\Omega.
\end{aligned}\right\} \tag{9.28}$$

Then the optimal control is given by $u=y|_{\Sigma}$.

9.5. Regular Control

Let us now examine the second possibility mentioned at the end of Section 9.1. We then take

$$\mathscr{U} = H_0^1(\Sigma).$$

The *state* $y(v)$ is given by (9.1), (9.2), (9.3) and we may then show that

$$\frac{\partial y(v)}{\partial \nu_A} \in L^2(\Sigma). \tag{9.29}$$

If now we take the observation to be $\left(\frac{\partial y(v)}{\partial \nu_A}\right)$, we may consider the cost function

$$\left.\begin{aligned} J(v) &= \int_\Sigma \left|\frac{\partial y(v)}{\partial \nu_A} - z_d\right|^2 d\Sigma + (Nv, v)_{\mathscr{U}}, \\ N &\in \mathscr{L}(\mathscr{U};\mathscr{U}), \quad N > \nu \text{ (identity)}, \quad \nu > 0. \end{aligned}\right\} \tag{9.30}$$

The *adjoint state* is now defined more simply than in (9.21), (9.22), (9.23):

$$\left.\begin{aligned} -\frac{\partial}{\partial t} p(u) + A^* p(u) &= 0 \quad \text{in } Q, \\ p(u) &= \frac{\partial y}{\partial \nu_A}(u) - z_d \quad \text{on } \Sigma, \\ p(x, T; u) &= 0, \quad x \in \Omega. \end{aligned}\right\} \tag{9.31}$$

The control u is optimal if and only if

$$\int_\Sigma \left(\frac{\partial y(u)}{\partial \nu_A} - z_d\right)\left(\frac{\partial y(v)}{\partial \nu_A} - \frac{\partial y(u)}{\partial \nu_A}\right) d\Sigma + (Nu, v-u)_{\mathscr{U}} \geqslant 0$$

from which utilizing the expression for $p(u)$, we get

$$\int_\Sigma \frac{\partial p(u)}{\partial \nu_{A^*}}(v-u)\, d\Sigma + (Nu, v-u)_{\mathscr{U}} \geqslant 0 \;\; \forall v \in \mathscr{U}_{\text{ad}}. \tag{9.32}$$

But

$$(Nu, v-u)_{\mathscr{U}} = \int_\Sigma (-\Delta_\Sigma(Nu))(v-u)\, d\Sigma,$$

whence

$$\int_\Sigma \left(\frac{\partial p(u)}{\partial \nu_{A^*}} + (-\Delta_\Sigma) N u\right)(v-u)\, d\Sigma \geqslant 0 \;\; \forall v \in \mathscr{U}_{\text{ad}}. \tag{9.33}$$

We see that the simplification found in the definition of the adjoint state is nullified by the complications introduced in (9.33). This contains a term involving $-\Delta_\Sigma$, whereas (9.25) did not.

9.6. Observation of the Final State

Let us again take equations (9.1), (9.2), (9.3) to define the state and let us now observe the *final state*, that is,

$$z(v) = y(.\,, T; v). \tag{9.34}$$

If $v \in L^2(\Sigma)$, the state $y(v) \in L^2(Q) = L^2(0,T; L^2(\Omega))$ (and is defined by (9.9)). Since $A\,y(v) \in L^2(0,T; H^{-2}(\Omega))$, we deduce from (9.1) that

$$\frac{d}{dt} y(v) \in L^2(0,T; H^{-2}(\Omega))$$

from which we may deduce (cf. Lions-Magenes [1], Chapter 1) that

$$t \to y(t; v) \text{ is a continuous function of } [0,T] \to H^{-1}(\Omega). \tag{9.35}$$

Hence (9.34) has meaning and *the observation is in* $H^{-1}(\Omega)$.

Remark 9.3. (9.35) is not the best possible result: we may replace $H^{-1}(\Omega)$ by $H^{-\frac{1}{2}}(\Omega)$ [65]—but $y(x,T;v)$ *is not in* $L^2(\Omega)$.

Let us take for example $\Omega =]0,1[$, the state being defined by

$$\frac{\partial y}{\partial t} - \frac{\partial^2 y}{\partial x^2} = 0, \quad y(x,0) = 0, \quad y(0,t) = 0, \quad y(1,t) = v(t).\text{[66]} \tag{9.36}$$

Then

$$\left.\begin{aligned} y(x,t) &= \sum_{j=1}^{\infty} y_j(t)\, w_j(x), \qquad w_j(x) = \sqrt{2}\sin(j\pi x),\\ y_j(t) &= -w_j'(1)\int_0^t e^{-\pi^2 j^2 (t-\sigma)}\, v(\sigma)\, d\sigma. \end{aligned}\right\} \tag{9.37}$$

If we take

$$v(\sigma) = \frac{1}{(T-\sigma)^\alpha}, \quad \frac{1}{4} \leqslant \alpha < \frac{1}{2}, \tag{9.38}$$

then $v \in L^2(0,T)$ and $|y_j(T)|^2 \geqslant C(1)|j^{2-4\alpha}$ and hence $\Sigma |y_j(T)|^2 = \infty$ which shows that $y(T; v) \notin L^2(\Omega)$.

---

[65] Since the norm in $H^{-\frac{1}{2}}(\Omega)$ is complicated it is preferable to consider the observation as an element of $H^{-1}(\Omega)$.

[66] Hence the control function has support in the subset $\{1\} \times]0,T[$ of Σ.

Remark 9.4. The previous considerations may be extended to the case where the parabolic operator is of arbitrary order. If, for example, A is a fourth order elliptic operator (cf. Section 3.3.2) and if we assume that the state is defined by

$$\left.\begin{aligned} &\frac{\partial y(v)}{\partial t}+A\,y(v)=f \quad \text{in } Q,\\ &y(v)|_\Sigma = v\in L^2(\Sigma),\\ &\frac{\partial}{\partial \nu}y(v)|_\Sigma=0 \quad \left(\frac{\partial}{\partial \nu}=\text{normal derivative}\right),\\ &y(x,0;v)=y_0(x), \qquad x\in\Omega, \end{aligned}\right\} \tag{9.39}$$

then $y(v)\in L^2(0,T;L^2(\Omega))$ (defined again by transposition) and from (9.39) $(dy(v)/dt\in L^2(0,T;H^{-4}(\Omega))$ which implies that $y(.,T;v)\in H^{-2}(\Omega)$.

In the special case where the parabolic operator is of second order, we may obtain additional information on $y(.,T;v)$ when v, for example, belongs to $L^\infty(\Sigma)$; cf. Section 9.7 in the sequel.

The cost function is now defined by

$$\left.\begin{aligned} &J(v)=\|y(T;v)-z_d\|^2_{H^{-1}(\Omega)}+(N\,v,v)_{L^2(\Sigma)},\\ &N\in\mathscr{L}(L^2(\Sigma);L^2(\Sigma)), \qquad N\geqslant \nu \text{ (identity)}, \qquad \nu>0 \end{aligned}\right\} \tag{9.40}$$

and the norm in $H^{-1}(\Omega)$ is defined by

$$\|f\|^2_{H^{-1}(\Omega)}=\int_\Omega((-\Delta)^{-1}f)f\,dx \quad \text{where } (-\Delta)^{-1}f=\varphi \quad \text{satisfies}$$

$$-\Delta\varphi=f \quad \text{in } \Omega, \qquad \varphi=0 \quad \text{on } \Gamma.$$

The optimal control u is characterized by

$$(y(T;u)-z_d,\,y(T;v)-y(T;u))_{H^{-1}(\Omega)}+(N\,u,v-u)_{L^2(\Sigma)}\geqslant 0 \quad \forall v\in\mathscr{U}_{\mathrm{ad}}. \tag{9.41}$$

The adjoint state is defined by

$$\left.\begin{aligned} &-\frac{\partial p(u)}{\partial t}+A^*p(u)=0 \quad \text{in } Q,\\ &p(u)|_\Sigma=0,\\ &p(x,T;u)=-(-\Delta)^{-1}(y(T;u)-z_d).\,^{67} \end{aligned}\right\} \tag{9.42}$$

(9.42) is now transformed by multiplying the first equation in (9.42) by $y(v)-y(u)$ and using Green's Formula. We then get

---

[67] That is, $\Delta p(x,T;u)=y(x,T;u)-z_d$ in Ω, $p(x,T;u)=0$ if $x\in\Gamma$.

Theorem 9.3. *The state is defined by* (9.1), (9.2), (9.3) (*or* (9.9)). *The optimal control corresponding to the cost function* (9.40) *is determined by* (9.1), (9.2), (9.3) (*where* $v=u$), (9.42) *and by*

$$\int_\Sigma \left(\frac{\partial p(u)}{\partial \nu_{A^*}} + N u\right)(v-u)\, d\Sigma \geqslant 0 \quad \forall v \in \mathscr{U}_{\text{ad}}. \tag{9.43}$$

Example 9.3. Case where $\mathscr{U}_{\text{ad}} = \mathscr{U}$.

Then the optimal control is obtained by the simultaneous solution of

$$\left.\begin{aligned} &\frac{\partial y}{\partial t} + A y = f, \quad -\frac{\partial p}{\partial t} + A^* p = 0 \quad \text{in } Q,\\ &p = 0, \quad \frac{\partial p}{\partial \nu_{A^*}} + N y = 0 \quad \text{on } \Sigma,\\ &y(x,0) = y_0(x), \quad p(x,T) = -(-\Delta)^{-1}(y(x,T) - z_d) \quad \text{in } \Omega, \end{aligned}\right\} \tag{9.44}$$

and then the optimal control $u = y|_\Sigma$.

Example 9.4. Case where $\mathscr{U}_{\text{ad}} = \{v | v \geqslant 0$ almost everywhere on $\Sigma\}$. We obtain the *unilateral problem*

$$\left.\begin{aligned} &\frac{\partial y}{\partial t} + A y = f, \quad -\frac{\partial p}{\partial t} + A^* p = 0 \quad \text{in } Q,\\ &p = 0 \text{ on } \Sigma, \quad y \geqslant 0 \text{ on } \Sigma, \quad \frac{\partial p}{\partial \nu_{A^*}} + N y \geqslant 0 \quad \text{on } \Sigma,\\ &y\left(\frac{\partial p}{\partial \nu_{A^*}} + N y\right) = 0 \quad \text{on } \Sigma,\\ &y(x,0) = y_0(x), \quad p(x,T) = -(-\Delta)^{-1}(y(x,T) - z_d) \quad \text{in } \Omega, \end{aligned}\right\} \tag{9.45}$$

and then $u = y|_\Sigma$.

Example 9.5. We take $N=0$ (compare with Section 7.2) and

$$\begin{aligned} &\mathscr{U}_{\text{ad}} = \{v | v(t) \in E_{\text{ad}} \text{ almost everywhere}\},\\ &E_{\text{ad}} = \text{closed, bounded, convex subset of } L^2(\Gamma). \end{aligned} \tag{9.46}$$

Then (result analogous to Theorem 7.2) if the coefficients of A are sufficiently regular and if $\operatorname*{Inf}_{v \in \mathscr{U}_{\text{ad}}} J(v) > 0$, the optimal control u satisfies

$$u(t) \in \partial E_{\text{ad}} = \text{boundary of } E_{\text{ad}}, \text{ almost everywhere.} \tag{9.47}$$

To see this, if u is optimal, we have (9.43) with $N=0$ which gives us (cf. Chapter 2, Theorem 2.1) almost everywhere in t:

$$\int_{\Gamma} \frac{\partial p(t;u)}{\partial \nu_{A^*}} u(t)\, d\Gamma \leqslant \int_{\Gamma} \frac{\partial p(t;u)}{\partial \nu_{A^*}} e\, d\Gamma \quad \forall e \in E_{\mathrm{ad}}. \tag{9.48}$$

But from the *backward uniqueness property* (cf. Section 7.2) which is valid here—we have,

$$\frac{\partial p(t;u)}{\partial \nu_{A^*}} \neq 0 \quad \text{almost everywhere}$$

and (9.48) implies (9.47).

Note that if E_{ad} is strictly convex, the optimal control may be shown to be unique.

Application. If E_{ad} is chosen as in (7.18), we conclude that

$$\left.\begin{aligned} u(x,t) &= \xi_0(x) \quad \text{if } \frac{\partial p}{\partial \nu_{A^*}}(x,t;u) > 0, \\ u(x,t) &= \xi_1(x) \quad \text{if } \frac{\partial p}{\partial \nu_{A^*}}(x,t;u) < 0. \end{aligned}\right\} \tag{9.49}$$

Here again, it would be interesting to obtain additional information on the topology of the switching surface.

Remark 9.5. We may avoid using the norm of $H^{-1}(\Omega)$ if we assume v to be "regular" (as in Section 9.5). But then we have the additional difficulty in the term $(N v, v)_{\mathscr{U}}$—as in Section 9.5.

Remark 9.6. In examples corresponding to (7.18), (9.49) (utilizing the fact that A is of second order) we may avoid using the norm $H^{-1}(\Omega)$, cf. Section 9.7.

9.7. Observation of the Final State, Second Order Parabolic Operator

Let the state be given by

$$\left.\begin{aligned} \frac{\partial y(v)}{\partial t} + A y(v) &= 0 \quad \text{in } Q, \\ y(v) &= v \quad \text{on } \Sigma, \\ y(x,0;v) = y_0(x), &\quad y_0 \in L^\infty(\Omega), \end{aligned}\right\} \tag{9.50}$$

and assume that

$$\mathscr{U}_{\mathrm{ad}} = \text{closed convex set in } L^\infty(\Sigma) \tag{9.51}$$

(which implies that $\mathscr{U}_{\mathrm{ad}}$ is closed in $L^2(\Sigma)$).

Then from the *maximum principle for parabolic equations* we have: (cf. Ladyzenskaya-Solonnikov-Uralts'eva [1], Theorem 7.2, Chapter 3)

$$y(v) \in L^\infty(Q),$$

the mapping $v \to y(v)$ being an (affine) continuous map of $L^\infty(\Sigma) \to L^\infty(Q)$.
Then

$$y(T; v) \in L^\infty(\Omega)$$

and hence in particular $y(T; v) \in L^2(\Omega)$ and we may consider as the cost function

$$\left.\begin{aligned} &J(v) = \int_\Omega [y(x, T; v) - z_d(x)]^2 \, dx + (N v, v)_{\mathscr{U}}, \\ &\mathscr{U} = L^2(\Sigma), \quad N \in \mathscr{L}(\mathscr{U}; \mathscr{U}), \quad N \geqslant \nu \text{ (identity)}, \quad \nu > 0. \end{aligned}\right\} \tag{9.52}$$

According to Remark 1.1, Chapter 1, Section 1.2, there exists a unique $u \in \mathscr{U}_{\text{ad}}$ such that $J(u) \leqslant J(v) \;\; \forall v \in \mathscr{U}_{\text{ad}}$, and this is again characterized by

$$\begin{aligned} &\int_\Omega (y(x, T; u) - z_d)\,(y(x, T; v) - y(x, T; u))\, dx \\ &\qquad + \int_\Sigma N u (v - u)\, d\Sigma \geqslant 0 \quad \forall v \in \mathscr{U}_{\text{ad}}. \end{aligned} \tag{9.53}$$

We now introduce the adjoint state as in Section 9.6—but without having to use $(-\Delta)^{-1}$:

$$\left.\begin{aligned} -\frac{\partial p(u)}{\partial t} + A^* p(u) = 0 \quad &\text{in } Q, \\ p(u) = 0 \quad &\text{on } \Sigma, \\ p(x, T; u) = -(y(x, T; u) - z_d(x)), \quad &x \in \Omega. \end{aligned}\right\} \tag{9.54}$$

Then *u is an optimal control if and only if* (9.50) (where $v = u$), (9.54) and

$$\int_\Sigma \left(\frac{\partial p(u)}{\partial \nu_{A^*}} - N u \right)(v - u)\, d\Sigma \geqslant 0 \quad \forall v \in \mathscr{U}_{\text{ad}} \tag{9.55}$$

are satisfied.

Example 9.6. Let us take $\mathscr{U}_{\text{ad}}$, E_{ad}, as in (9.46), (7.18).
Then the preceeding result is applicable. We get

$$u(x, t) = \xi_0(x, t) \quad \text{if} \quad \frac{\partial p(x, t; u)}{\partial \nu_{A^*}} + (N u)(x, t) > 0,$$

$$u(x, t) = \xi_1(x, t) \quad \text{if} \quad \frac{\partial p(x, t; u)}{\partial \nu_{A^*}} + (N u)(x, t) < 0,$$

and

$$\frac{\partial p(x,t;u)}{\partial \nu_{A^*}} + (N u)(x,t) = 0 \quad \text{otherwise on } \Sigma.$$

Remarks analogous to Example 9.5 if $N = 0$.

10. Controllability

10.1. Problem Statement

Definition 10.1. The system whose state is defined by (2.2), (2.3), (2.4) is said to be *controllable* if as v is varied without any constraints, the observation $C y(v)$ generates a dense (affine) subspace of the space of observations.

Remark 10.1. Definition 10.1 is an adaptation of the usual definition in finite dimensions when (observation of final state)

$$C y(v) = y(T; v).$$

Remark 10.2. In infinite dimensions it is clear that we must distinguish between a dense subspace and the whole space. In applications, the space (affine) generated by $C y(v)$ is *not closed* in the *parabolic* case; it may be closed in the hyperbolic case (cf. Chapter 4).

Remark 10.3. Definition 10.1 is not restricted in any way to (2.2), (2.3), (2.4) only and may be clearly adapted to cover all the systems we have studied upto now.

By means of examples we shall demonstrate how the question of controllability can be reduced to a question of uniqueness.

10.2. Controllability and Uniqueness

Example 10.1. Let us first consider the case of Section 3.1.1. Hence the state of the system is given by

$$\left.\begin{aligned} \frac{\partial}{\partial t} y(v) + A y(v) &= f + v \quad \text{in } Q, \\ y(v) &= 0 \quad \text{on } \Sigma, \\ y(x,0;v) &= y_0(x) \quad \text{in } \Omega, \end{aligned}\right\} \tag{10.1}$$

and the observation $y(v)$ is in $L^2(Q)$.

As v ranges over $L^2(Q)$, $y(v)$ generates a dense (affine) subspace of $L^2(Q)$; hence the *system is controllable.*

To see this, let us first remark that by translation we may always reduce the problem to the case where $f=0$, $y_0=0$.

Let $\psi \in L^2(Q)$ be orthogonal to the subspace generated by $y(v)$:

$$\int_Q y(v)\psi \, dx\, dt = 0 \quad \forall v. \tag{10.2}$$

We introduce ξ as the solution of

$$\left.\begin{aligned} -\frac{\partial \xi}{\partial t} + A^* \xi = \psi \quad \text{in } Q, \\ \xi = 0 \quad \text{on } \Sigma, \qquad \xi(x,T) = 0 \quad \text{on } \Omega. \end{aligned}\right\} \tag{10.3}$$

Then

$$\begin{aligned} \int_Q \psi\, y(v)\, dx\, dt &= \int_Q \left(-\frac{\partial \xi}{\partial t} + A^* \xi\right) y(v)\, dx\, dt \\ &= \int_Q \xi \left(\frac{\partial y(v)}{\partial t} + A\, y(v)\right) dx\, dt = \int_Q \xi\, v\, dx\, dt = 0 \quad \forall v; \end{aligned}$$

hence $\xi=0$ and hence $\psi=0$. We now get the required result from the Hahn-Banach Theorem.

Remark 10.4. The preceeding method is quite general. In the sequel we shall give other examples illustrating the method.

Example 10.2. We now place ourselves in the framework of Section 3.2.2. The state is given by

$$\left.\begin{aligned} \frac{\partial}{\partial t} y(v) + A\, y(v) = 0 \quad \text{in } Q, \\ \frac{\partial y(v)}{\partial \nu_A} = v \quad \text{on } \Sigma, \qquad v \in \mathcal{U} = L^2(\Sigma), \\ y(x,0;v) = 0 \quad \text{in } \Omega. \end{aligned}\right\} \tag{10.4}$$

The observation is $y(x,T;v)$ in $L^2(\Omega)$. The system is controllable. In fact let $\psi \in L^2(\Omega)$ with

$$\int_Q y(x,T;v)\psi(x)\, dx = 0 \quad \forall v \in \mathcal{U}. \tag{10.5}$$

We introduce ξ as the solution of

$$\left.\begin{aligned} -\frac{\partial \xi}{\partial t}+A^*\xi=0 \quad &\text{in } Q,\\ \frac{\partial \xi}{\partial \nu_{A^*}}=0 \quad &\text{on } \Sigma,\\ \xi(x,T)=\psi(x). & \end{aligned}\right\} \tag{10.6}$$

Then

$$\int_Q \left(-\frac{\partial \xi}{\partial t}+A^*\xi\right) y(v)\,dx\,dt=0=\int_\Sigma \xi \frac{\partial}{\partial \nu_A} y(v)\,d\Sigma-\int_\Omega \psi(x)\,y(x,T;v)\,dx,$$

and hence from (10.5)

$$\int_\Sigma \xi v\,d\Sigma=0 \quad \forall v;$$

hence $\xi=0$ on Σ. The Cauchy data of ξ on Σ being zero, we conclude (S. Mizohata [1]) $\xi=0$ and hence $\psi=0$.

Example 10.3. Let us now consider a system governed by a fourth order parabolic equation:

$$\frac{\partial y(v)}{\partial t}+\Delta^2 y(v)=0 \quad \text{in } Q, \tag{10.7}$$

$$y(v)|_\Sigma=v_1, \qquad v_1\in L^2(\Sigma), \tag{10.8}$$

$$\frac{\partial}{\partial \nu} y(v)|_\Sigma=v_2, \qquad v_2\in L^2(\Sigma), \tag{10.9}$$

$$y(x,0;v)=0 \quad \text{in } \Omega; \tag{10.10}$$

$$v=\{v_1,v_2\}\in\mathscr{U}=L^2(\Sigma)\times L^2(\Sigma).$$

The solution of (10.7)...(10.10) is defined by transposition: there exists a unique $y(v)\in L^2(Q)$ such that

$$\int_Q y(v)\left(-\frac{\partial \varphi}{\partial t}+\Delta^2\varphi\right)dx\,dt=\int_\Sigma v_1\frac{\partial \Delta\varphi}{\partial \nu}\,d\Sigma-\int_\Sigma v_2\,\Delta\varphi\,d\Sigma \quad \forall \varphi\in\Phi, \tag{10.11}$$

where

$$\Phi=\left\{\varphi \,|\, D_x^p\varphi\in L^2(Q),\ |p|\leqslant 4,\ \frac{\partial \varphi}{\partial t}\in L^2(Q),\ \varphi(x,t)=0,\ \frac{\partial \varphi}{\partial \nu}(x,t)=0 \text{ on } \Sigma,\ \varphi(x,T)=0 \text{ on } \Omega\right\}. \tag{10.12}$$

Let us suppose that the observation is $y(x,T;v)$, $x\in\Omega$. Then (see Remark 9.4).

$$y(.,T;v)\in H^{-2}(\Omega). \tag{10.13}$$

The system is *controllable.*
To see this, let $\psi\in H_0^2(\Omega)$ such that

$$(y(.,T;v),\psi)=0 \quad \forall v\in\mathscr{U}.\text{[68]} \tag{10.14}$$

Let us define ξ by

$$\left.\begin{aligned} &\frac{\partial\xi}{\partial t}+\Delta^2\xi=0 \quad \text{in } Q,\\ &\xi(x,T)=\psi(x) \quad \text{in } \Omega,\\ &\xi=0, \quad \frac{\partial\xi}{\partial\nu}=0 \quad \text{on } \Sigma. \end{aligned}\right\} \tag{10.15}$$

Then

$$\begin{aligned} \int_Q\left(-\frac{\partial\xi}{\partial t}+\Delta^2\xi\right)y(v)\,dx\,dt&=0=-(y(.,T;v),\psi)\\ &+\int_\Sigma\frac{\partial\Delta\xi}{\partial\nu}v_1\,d\Sigma-\int_\Sigma\Delta\xi v_2\,d\Sigma\\ &+\int_Q\xi\left(\frac{\partial}{\partial t}y(v)+\Delta^2 y(v)\right)dx\,dt, \end{aligned}$$

from which we deduce that

$$\int_\Sigma\frac{\partial\Delta\xi}{\partial\nu}v_1\,d\Sigma-\int_\Sigma\Delta\xi\cdot v_2\,d\Sigma=0 \quad \forall v\in\mathscr{U}. \tag{10.16}$$

Hence

$$\Delta\xi=0, \quad \frac{\partial\Delta\xi}{\partial\nu}=0 \quad \text{on } \Sigma. \tag{10.17}$$

Comparing with the last condition in (10.15) we see that the *Cauchy data* on Σ is zero and hence (S. Mizohata [1] and H. Tanabe [1]) $\xi=0$ and $\psi=0$ from which the result follows.

---

[68] The bracket denotes duality between $H^{-2}(\Omega)$ and $H_0^2(\Omega)$.

10.3. Super-Controllability and Super-Uniqueness

Let us again consider Example 10.3; we assume that v ranges over the subspace $\{0, L^2(\Sigma)\}$ of $L^2(\Sigma) \times L^2(\Sigma)$. We say that the system is super-controllable if the space generated by $y(x,T;v)$ is also dense in $H^{-2}(\Omega)$. Hence the state $y(v)$ is given by (we write v instead of v_2)

$$\left.\begin{aligned} \frac{\partial}{\partial t} y(v) + \Delta^2 y(v) = 0, \\ y(v)|_\Sigma = 0, \\ \frac{\partial}{\partial \nu} y(v)|_\Sigma = v \in L^2(\Sigma), \\ y(x,0;v) = 0. \end{aligned}\right\} \tag{10.18}$$

To see if $y(x,T;v)$ is dense, consider $\psi \in H_0^2(\Omega)$ satisfying (10.14). We introduce ξ by means of (10.15) and in this case (10.16) becomes

$$\int_\Sigma \Delta \xi v \, d\Sigma = 0 \quad \forall v \quad \text{and hence } \Delta \xi = 0 \quad \text{on } \Sigma.$$

Hence ξ satisfies

$$\left.\begin{aligned} \frac{\partial \xi}{\partial t} + \Delta^2 \xi = 0 \quad \text{in } Q, \\ \xi, \frac{\partial \xi}{\partial \nu}, \frac{\partial^2 \xi}{\partial \nu^2} = 0 \quad \text{on } \Sigma. \end{aligned}\right\} \tag{10.19}$$

The problem now is to see if we have *super-uniqueness:* do the hypotheses (10.19) imply that $\xi = 0$?

In the sequel we present a positive result in this sense due to H. O. Fattorini [5]:

Theorem 10.1. *If ξ satisfies* (10.19) *and $\xi \in L^2(0,T; H^2(\Omega))$ where the dimension $n=1$ and $\Omega =]0,\infty[$, then $\xi = 0$.*

Proof. Let us introduce

$$\hat{\xi}(\mu,t) = \frac{2}{\pi} \int_0^\infty \sin(x\mu)\xi(x,t)\,dx.$$

Since $\xi(0,t) = 0$, $\dfrac{\partial^2 \xi}{\partial x^2}(0,t) = 0$ and from the equation

$$-\frac{\partial \xi}{\partial t} + \frac{\partial^4 \xi}{\partial x^4} = 0$$

we deduce that

$$-\frac{d\hat{\xi}}{dt}+\mu^4\hat{\xi}=0$$

and hence

$$\hat{\xi}(\mu,t)=\exp(-\mu^4(T-t))\hat{\xi}(\mu,T).$$

Hence

$$\begin{aligned}\frac{\partial\xi}{\partial x}(0,t)&=\frac{\partial}{\partial x}\left(\int_0^\infty \sin(x\mu)\hat{\xi}(\mu,t)\,d\mu\right)_{x=0}\\&=\int_0^\infty \mu\exp(-\mu^4(T-t))\hat{\xi}(\mu,T)\,d\mu\\&=\frac{1}{4}\int_0^\infty \exp(-\lambda\tau)\lambda^{-\frac{1}{2}}\hat{\xi}(\lambda^{\frac{1}{4}},T)\,d\lambda=0,\qquad 0<\tau<T,\end{aligned}$$

and hence the Laplace Transform of $\lambda^{-\frac{1}{2}}\hat{\xi}(\lambda^{\frac{1}{4}}T)$ is zero and hence $\hat{\xi}(\lambda^{\frac{1}{4}}T)=0$. Thus $\xi(x,T)=0$ and hence $\xi=0$. □

Remark 10.5. We say that the system is *null-controllable* (resp. *null-controllable under constraints* "$v\in\mathscr{U}_{ad}$") if 0 belongs to the closure of the set described by $Cy(v)$ as v spans $\mathscr{U}$ (resp. $\mathscr{U}_{ad}$). One can give (see Fattorini [8]) sufficient conditions for these two notions to be *equivalent*, when $\mathscr{U}_{ad}$ is defined by "soft constraints"; these conditions are satisfied, in particular, for systems governed by wave operators (of course the above definitions readily extend to "hyperbolic systems" such as those considered in Chapter 4).

Remark 10.6. An important problem in practice is the problem of controllability in the presence of an *additive disturbance*. We look for an admissible control which steers the state into a given target set in the presence of the *worst* disturbance. This is therefore a problem related to *game theory*. For a geometrical approach to the problem (and, in particular, an extension of the method of Antosiewicz [1]) we refer to M. C. Delfour and S. K. Mitter [1].

Remark 10.7. Another notion related to controllability is that of *modal controllability* which, roughly speaking, means: *find a control which "improves" the spectrum* (for example to improve stability). *Modal controllability with respect to a given spectrum* is the problem of choosing a control so that the desired spectrum is obtained.

This idea, introduced by Rosenbrock [1] has been studied in detail in J. D. Simon-S. K. Mitter [1] and W. M. Wonham [2] for the finite dimensional case.

We would like to mention here the interesting analogy between this problem and problems of the following type arising in numerical analysis of linear systems (the so-called "inverse problem"): given a matrix A (hermitian or real) find a real diagonal matrix B such that $A+B$ has a given real spectrum—see A. C. Downing, A. S. Householder [1], K. P. Hadeler [1], F. Laborde [1].

11. Control via Initial Conditions; Estimation

11.1. Problem Statement. General Results

Let us consider the situation of Section 2, with

$$\mathscr{U} = H. \tag{11.1}$$

The state $y(v)$ is given in $L^2(0,T;V)$ by

$$\frac{dy(v)}{dt} + A(t)y(v) = f, \quad f \text{ given in } L^2(0,T;V), \tag{11.2}$$

$$y(0;v) = v. \tag{11.3}$$

The "control" here is therefore the *initial state.* To fix ideas[69], let the observation be $y(T;v)$ and let the cost function be given by

$$\begin{aligned} &J(v) = |y(T;v) - z_d|^2 + (Nv, v), \\ &N \in \mathscr{L}(H;H), \quad N \geqslant \nu \text{ (identity)}, \quad \nu > 0. \end{aligned} \tag{11.4}$$

The *optimal control* u is characterized by

$$(y(T;u) - z_d, y(T;v) - y(T;u)) + (Nu, v-u) \geqslant 0 \ \ \forall v \in \mathscr{U}_{\text{ad}}. \tag{11.5}$$

We introduce the adjoint state by

$$\left.\begin{aligned} &-\frac{dp(u)}{dt} + A^*(t)p(u) = 0, \\ &\qquad p(T;u) = y(T;u) - z_d, \\ &p(t;u) \in L^2(0,T;V). \end{aligned}\right\} \tag{11.6}$$

[69] We leave it to the reader to modify these arguments using the results of the previous section to cover the case of other observations.

Then (11.5) may be written in the equivalent form

$$(p(0,u)+Nu, v-u) \geqslant 0 \quad \forall v \in \mathscr{U}_{\mathrm{ad}}. \tag{11.7}$$

Therefore

Theorem 11.1. *Let us suppose that the state and the cost function are given by* (11.2), (11.3), *and* (11.4). *Then the optimal control u is characterized by* (11.2), (11.3), *(with $v=u$)*, (11.6) *and* (11.7).

11.2. Examples

Example 11.1. In the case where there are no constraints ($\mathscr{U}=H$), we have

$$p(0;u)+Nu=0.$$

Hence u is obtained by the simultaneous solution of

$$\left.\begin{aligned} &\frac{dy}{dt}+A(t)y=f, \qquad -\frac{dp}{dt}+A^*(t)p=0,\\ &y(0)+N^{-1}p(0)=0, \qquad p(T)=y(T)-z_d,\\ &y,p\in L^2(0,T;V). \end{aligned}\right\} \tag{11.8}$$

Example 11.2. Let us consider, for example, the situation of Section 3.1, $\frac{\partial}{\partial t}+A$ being a second order parabolic operator, and let us take

$$\mathscr{U}_{\mathrm{ad}}=\{v \mid v\geqslant 0 \text{ almost everywhere in } \Omega\}. \tag{11.9}$$

Then the optimal control is determined by the solution of the *unilateral problem*

$$\left.\begin{aligned} &\frac{\partial y}{\partial t}+Ay=f, \qquad -\frac{\partial p}{\partial t}+A^*p=0 \quad \text{in } Q,\\ &y=0, \qquad p=0 \quad \text{on } \Sigma,\\ &y(x,0)\geqslant 0, \qquad p(x,0)+Ny(x,0)\geqslant 0,\\ &y(x,0)(p(x,0)+Ny(x,0))=0 \quad \text{in } \Omega,\\ &p(x,T)=y(x,T)-z_d(x) \quad \text{in } \Omega, \end{aligned}\right\} \tag{11.10}$$

and then $u(x)=p(x,0)$.

Example 11.3. (Case where $N=0$.)
Suppose that in the example of Section 11.1 we have $N=0$ and

$$\mathscr{U}_{\mathrm{ad}} = \text{closed, bounded, convex subset of } H(=\mathscr{U}). \tag{11.11}$$

Then an optimal u exists and all optimal u are characterized by (taking $N=0$ in (11.7)):

$$(p(0;u),v-u)\geqslant 0 \quad \forall v\in\mathscr{U}_{ad}. \tag{11.12}$$

But if we assume that the backward uniqueness property holds (cf. Section 7.2, Definition 7.1), then

$$\text{if} \quad \operatorname*{Inf}_{v\in\mathscr{U}_{ad}} J(v)>0, \quad \text{we have } p(0;u)=0. \tag{11.13}$$

Then

$$u\in\partial\mathscr{U}_{ad} = \text{boundary of } \mathscr{U}_{ad} \tag{11.14}$$

and if $\mathscr{U}_{ad}$ is strictly convex then the optimal control is unique.

11.3. Controllability

Definition 10.1 may be easily adapted to the problem considered in this section: the system whose state is given by (11.2), (11.3) is controllable (at the instant T) if the space (affine) generated by $y(T;v)$ as v ranges over $\mathscr{U}=H$ is dense in H.

We shall see (by using the same arguments as in Section 10) that this property may be reduced to the backward uniqueness property.

By translation, we assume $f=0$ in (11.2). Let $h\in H$ be such that

$$(y(T;v),h)=0 \quad \forall v\in H. \tag{11.15}$$

Let us introduce ξ as the solution of

$$\left.\begin{aligned} -\frac{\partial\xi}{dt}+A^*(t)\xi=0,\\ \xi(T)=h.\end{aligned}\right\} \tag{11.16}$$

Then

$$\int_0^T\left(-\frac{d\xi}{dt}+A^*(t)\xi,y(t;v)\right)dt=0=-(h,y(T;v))+(\xi(0),y(0;v))$$

and hence from (11.15) and (11.3):

$$(\xi(0),v)=0 \quad \forall v,$$

hence $\xi(0)=0$. But from the backward uniqueness property, $\xi\equiv 0$ and hence $h=0$ whence the following result:

Theorem 11.2. *If the backward uniqueness property holds, the system whose state is given by* (11.2), (11.3) *is controllable.*

Remark 11.1. Under the hypotheses of Theorem 11.2, the space spanned by $y(T; v)$ as v varies is not closed in the *parabolic case*. In fact, if it were, the mapping $v \to y(T; v)$ would be surjective. Hence $\forall h \in H$, there would exist a solution y in $L^2(0, T; V)$ of the backward equation

$$\frac{dy}{dt} + A(t)y = 0, \quad y(T) = h$$

which is not true ("irreversibility of parabolic equations").

11.4. An Estimation Problem

We consider a system whose state is described by a second order parabolic operator[70]:

$$\frac{\partial y}{\partial t} + Ay = 0 \quad \text{in } Q. \tag{11.17}$$

We are interested in *estimating the initial state* (unknown $y(0)$), from the measurements (or observations)

$$y|_{\Sigma} = g_0, \tag{11.18}$$

$$\left.\frac{\partial y}{\partial \nu_A}\right|_{\Sigma} = g_1. \tag{11.19}$$

To be precise, assume that g_0 and g_1 belong to $L^2(\Sigma)$ and assume that there exists a y satisfying (11.17), (11.18) and (11.19). Then y is unique; otherwise if $\bar{y}$ were another solution, the difference $\hat{y} = y - \bar{y}$ satisfies

$$\frac{\partial \hat{y}}{dt} + A\hat{y} = 0 \quad \text{in } Q, \qquad \hat{y} = 0 \quad \text{on } \Sigma, \qquad \frac{\partial \hat{y}}{\partial \nu_A} = 0 \quad \text{in } \Sigma,$$

and hence, according to S. Mizohata [1], $\hat{y} = 0$.

Therefore there exists a unique $y(0)$ which we are interested in finding—more precisely we want to approximate $y(0)$ since the Cauchy problem (11.17), (11.18), (11.19) is ill-posed (unstable).

Transformation of the Problem. We shall consider $y(0) = u$ as the control variable which we shall determine in an optimal manner.

---

[70] The extension to parabolic operators of arbitrary order is easy. For this we may apply the theorem of S. Mizohata [1].

Let then $v \in L^2(\Omega) = \mathscr{U}$; we introduce two systems whose states $y^i(v)$, $i=1,2$, are given by

$$\left.\begin{aligned} \frac{\partial y^1}{\partial t} + A y^1(v) &= 0 \quad \text{in } Q, \\ y^1(v) &= g_0 \quad \text{on } \Sigma, \\ y^1(0; v) &= v \quad \text{in } \Omega, \end{aligned}\right\} \tag{11.20}$$

and

$$\left.\begin{aligned} \frac{\partial y^2(v)}{\partial t} + A y^2(v) &= 0 \quad \text{in } Q, \\ \frac{\partial y^2(v)}{\partial \nu_A} &= g_1 \quad \text{on } \Sigma, \\ y^2(0; v) &= v \quad \text{in } \Omega, \end{aligned}\right\} \tag{11.21}$$

Then, when $v=u$ we have $y^1(u) = y^2(u)$. We introduce

$$J(v) = \int_Q (y^1(v) - y^2(v))^2 \, dx\, dt, \tag{11.22}$$

and we have

$$\inf_{v \in \mathscr{U}} J(v) = 0; \quad \text{there exists a unique } u \text{ such that } J(u) = 0. \tag{11.23}$$

It remains to find u using the fact that (11.23) is true.

Remark 11.2. Since (11.23) is equivalent to the initial problem, the *instability* of the problem is conserved. In the form of (11.23), $J(v)$ is not *coercive* on $\mathscr{U}$ (cf. Chapter 1, Section 5). Hence we shall *regularize* $J(v)$ (as in Chapter 1, Section 5).

Application of Theorem 5.3, Chapter 1. We introduce, for $\varepsilon > 0$:

$$J_\varepsilon(v) = J(v) + \varepsilon |v|^2 \qquad \left(|v|^2 = \int_\Omega v(x)^2 \, dx\right). \tag{11.24}$$

Then from Theorem 5.3, Chapter 1, we have

$$\left.\begin{aligned} &\text{the unique } u_\varepsilon \in \mathscr{U} = H \text{ which minimizes } J_\varepsilon(v) \text{ satisfies} \\ &u_\varepsilon \to u \text{ in } H \text{ as } \varepsilon \to 0 \end{aligned}\right. \tag{11.25}$$

We have thus reduced the problem to a search (or approximation) for u_ε. This problem is however *stable.*

(11.23) is only true for particular g_0, g_1 (satisfying the compatibility relations), because u does not depend continuously on g_i in the form (11.23). However u_ε does depend continuously on g_i.

Search for u_ε. More generally, consider the problem

$$\text{Find Inf } J_\varepsilon(v), \quad v \in \mathscr{U}_{ad} \subset \mathscr{U} = H. \tag{11.26}$$

Let u_ε be the optimal control. It is characterized by

$$\begin{aligned}&\int_Q (y^1(u_\varepsilon)-y^2(u_\varepsilon))(y^1(v)-y^2(v)-y^1(u_\varepsilon)+y^2(u_\varepsilon))\,dx\,dt\\&\quad+\varepsilon(u,v-u_\varepsilon)\geqslant 0 \ \forall v\in\mathscr{U}_{ad}.\end{aligned} \tag{11.27}$$

The *adjoint state* $p=\{p^1,p^2\}$ is defined by

$$\left.\begin{aligned}-\frac{\partial p^1}{\partial t}+A^*p^1&=y^1(u_\varepsilon)-y^2(u_\varepsilon) \quad \text{in } Q,\\ p^1&=0 \quad \text{on } \Sigma,\\ p^1(x,T)&=0 \quad \text{in } \Omega,\end{aligned}\right\} \tag{11.28}$$

$$\left.\begin{aligned}-\frac{\partial p^2}{\partial t}+A^*p^2&=y^1(u_\varepsilon)-y^2(u_\varepsilon) \quad \text{in } Q,\\ \frac{\partial p^2}{\partial \nu_{A^*}}&=0 \quad \text{on } \Sigma,\\ p^2(x,T)&=0 \quad \text{in } \Omega.\end{aligned}\right\} \tag{11.29}$$

We now transform (11.27). Multiplying the 1st equation in (11.28) (resp. (11.29)) by $y^1(v)-y^1(u_\varepsilon)$ (resp. $y^2(v)-y^2(u_\varepsilon)$), integrating over Q, and substracting, we get

$$\begin{aligned}&\int_Q\left(-\frac{\partial p^1}{\partial t}+A^*p^1\right)(y^1(v)-y^1(u_\varepsilon))\,dx\,dt\\&-\int_Q\left(-\frac{\partial p^2}{\partial t}+A^*p^2\right)(y^2(v)-y^2(u_\varepsilon))\,dx\,dt=\int_\Omega(p^1(x,0)-p^2(x,0))(v-u_\varepsilon)\,dx.\end{aligned}$$

Hence,

Theorem 11.3. *The control* u_ε *in* (11.25) *is determined by the simultaneous solution of the following system of equations:*

$$\left.\begin{aligned}&\frac{\partial y^1}{\partial t}+Ay^1=0, \qquad \frac{\partial y^2}{\partial t}+Ay^2=0,\\ &-\frac{\partial p^1}{\partial t}+A^*p^1=y^1-y^2, \qquad -\frac{\partial p^2}{\partial t}+A^*p^2=y^1-y^2\end{aligned}\right\} \text{ in } Q, \tag{11.30}$$

$$\left.\begin{aligned} y^1 &= g_0, & \frac{\partial y^2}{\partial \nu_A} &= g_1 \\ p^1 &= 0, & \frac{\partial p^2}{\partial \nu_{A^*}} &= 0 \end{aligned}\right\} \text{ on } \Sigma, \tag{11.31}$$

$$\left.\begin{aligned} y^1(x,0) &= y^2(x,0) \\ p^1(x,0)-p^2(x,0)+\varepsilon y^1(x,0) &= 0 \\ p^1(x,T)=0, \quad p^2(x,T) &= 0 \end{aligned}\right\} \text{in } \Omega. \tag{11.32}$$

12. Duality

12.1. General Remarks

Let $J(v)$ be a quadratic[71] functional on $\mathscr{U}$. We may write:

$$\begin{aligned} &J(v) = \pi(v,v) - 2L(v) + J(0), \\ &\pi(u,v) \text{ a continuous bi-linear form on } \mathscr{U}, \\ &L(v) \text{ a continuous linear form on } \mathscr{U}. \end{aligned} \tag{12.1}$$

We assume that

$$\pi(v,v) \geqslant c\,\|v\|_{\mathscr{U}}^2 \quad \forall v \in \mathscr{U}, \qquad c>0. \tag{12.2}$$

Let $\mathscr{U}_{\text{ad}}$ be a closed, convex subset of $\mathscr{U}$. We seek:

$$\operatorname*{Inf}_{v\in\mathscr{U}_{\text{ad}}} J(v). \tag{12.3}$$

Let us introduce the function $v \to F(v)$ defined on $\mathscr{U}$ by

$$F(v) = \begin{cases} -L(v) & \text{if } v \in \mathscr{U}_{\text{ad}}, \\ +\infty & \text{if } v \notin \mathscr{U}_{\text{ad}}. \end{cases} \tag{12.4}$$

Then (12.3) is equivalent to the search of

$$\operatorname*{Inf}_{v\in\mathscr{U}} \left[\pi(v,v) + 2F(v)\right]. \tag{12.5}$$

The function F has the following properties (these are the only properties we will use in this section):

$$\begin{aligned} &F \text{ is convex, lower semicontinuous, with values in} \\ &]-\infty; +\infty], \qquad F \not\equiv +\infty. \end{aligned} \tag{12.6}$$

---

[71] This is not absolutely essential; cf. bibliographical notes.

We introduce the *functional dual to F with respect to* π*:*

$$F^*(u) = \sup_{v\in\mathscr{U}} [\pi(u,v) - F(v)]. \tag{12.7}$$

The function F^* *satisfies* (12.6). In fact it is the upper envelope of continuous affine functions; hence it suffices to show that $F^* \not\equiv +\infty$. Now, there exists u_0 and $\gamma \in R$ such that

$$\pi(u_0, v) - \gamma \leqslant F(v) \tag{12.8}$$

and hence

$$F^*(u_0) \leqslant \gamma < \infty.$$

Remark 12.1. If we assume that we only have (instead of (12.2))

$$\pi(v,v) \geqslant 0 \quad \forall v \in \mathscr{U}, \tag{12.9}$$

then F^* may be identically $+\infty$ (taking $\pi = 0$), but clearly $F^* \not\equiv \infty$ if F is bounded below.

Remark 12.2. The function F^{**} (dual of F^* with respect to π) coincides with F. If only (12.9) holds this is false in general (cf. H. Brezis [3]). If π satisfies (12.2) and is not symmetric, the function dual to F^* with respect to π^* coincides with F.

We may now prove

Theorem 12.1. *Let us assume that* (12.2) *holds. Then*

$$\inf_{v\in\mathscr{U}} [\pi(v,v) + 2F(v)] + \inf_{v\in\mathscr{U}} [\pi(v,v) + 2F^*(v)] = 0. \tag{12.10}$$

If the first lower bound is attained for the element u*, then the second lower bound is attained for the element* $\{-u\}$*.*

Proof. 1. From Chapter 1, Theorem 1.3 we know that u is characterized by

$$\pi(u,u) - L(u) \leqslant \pi(u,v) - L(v) \quad \forall v \in \mathscr{U}_{ad},$$

that is,

$$\pi(u,u) + F(u) = \inf_{v\in\mathscr{U}} [\pi(u,v) + F(v)]. \tag{12.11}$$

This is also valid for F^* satisfying (12.6) (Chapter 1, Theorem 1.6). Therefore there exists a unique $w \in \mathscr{U}$ such that

$$\inf_{v\in\mathscr{U}} [\pi(v,v) + 2F^*(v)] = \pi(w,w) + 2F^*(w), \tag{12.12}$$

and (12.12) is equivalent to

$$\pi(w,w) + F^*(w) = \inf_{v\in\mathscr{U}} [\pi(w,v) + F^*(v)]. \tag{12.13}$$

2. But (12.11) may equally well be written as

$$\pi(u,u)+F(u)+\sup_{v\in\mathscr{U}}[\pi(-u,v)-F(v)]=0,$$

that is,

$$\pi(u,u)+F(u)+F^*(-u)=0. \tag{12.14}$$

Hence

$$\begin{aligned}\pi(-u,-u)+F^*(-u)&=-F(u)=-F^{**}(u)=\text{(from Remark 12.2)}\\&=-\sup_v[\pi(u,v)-F^*(v)]=\inf_v[\pi(-u,v)+F^*(v)]\end{aligned}$$

and comparing with (12.13) we obtain $w=-u$.

Then

$$\inf_{v\in\mathscr{U}}[\pi(v,v)+2F(v)]+\inf_{v\in\mathscr{U}}[\pi(v,v)+2F^*(v)]$$
$$=\pi(u,u)+2F(u)+\pi(-u,-u)+2F^*(-u)=2[\pi(u,u)+F(u)+F^*(-u)]=0$$

from (12.14). □

Remark 12.3. Let us assume that only (12.9) holds. Let X be the set (assumed non-empty) of optimal controls (X is a closed, convex subset of $\mathscr{U}_{\mathrm{ad}}$—cf. Chapter 1, Section 5). Then $F^*\not\equiv+\infty$ and the set X^* of w such that (12.13) holds is non-empty. More precisely

$$X^*\supset -X \tag{12.15}$$

and we have (12.10) (same type of proof; observe that $F^{**}\leqslant F$).

Remark 12.4. The result (12.10) may also be stated as

$$\inf_{v\in\mathscr{U}}[\pi(v,v)+2F(v)]=\sup_{v\in\mathscr{U}}[-\pi(v,v)-F^*(v)] \tag{12.16}$$

which allows us in applications to obtain bounds on the value of $\inf J(v)$.

Remark 12.5. The right hand side of (12.16) is also equal to

$$\sup_{v\in\mathscr{U}}[-\pi(v,v)-F^*(-v)].$$

But from definition (12.7):

$$F^*(-u)=\sup_{v\in\mathscr{U}}[\pi(u,-v)-F(v)]=\sup_{v\in\mathscr{U}_{\mathrm{ad}}}[\pi(u,-v)+L(v)]$$

and since (cf. Chapter 1, Section 1)

$$\pi(u,-v)+L(v)=\tfrac{1}{2}J'(u)(-v), \tag{12.17}$$

we have

$$F^*(-u)=\tfrac{1}{2}\sup_{v\in\mathscr{U}_{\mathrm{ad}}}(-J'(u),v),\qquad J'(u)\in\mathscr{U}'. \tag{12.18}$$

But if $\xi \in \mathscr{U}'$, we set

$$\mathscr{A}_{\mathscr{U}_{ad}}(\zeta) = \sup(\zeta, v) = \text{support function of } \mathscr{U}_{ad} \text{ at the point } \xi, \tag{12.19}$$

and hence

$$F^*(-u) = \tfrac{1}{2}\mathscr{A}_{\mathscr{U}_{ad}}(-J'(u)). \tag{12.20}$$

Thus, we may also write (12.16) in the form

$$\inf_{v \in \mathscr{U}} [\pi(v,v) + 2F(v)] = \sup_{v \in \mathscr{U}} [-\pi(v,v) - \tfrac{1}{2}\mathscr{A}_{\mathscr{U}_{ad}}(-J'(v))]. \tag{12.21}$$

12.2. Example

Let us suppose that

$$\left.\begin{aligned} &\mathscr{U}_{ad} = \{v \mid (v, g_i)_{\mathscr{U}} \leqslant \alpha_i,\ i = 1, \dots, m\}, \\ &g_i \in \mathscr{U}, \quad \alpha_i \in \mathbb{R}, \end{aligned}\right\} \tag{12.22}$$

and to fix ideas let $\mathscr{U} = L^2(\Sigma)$, and let the state $y(v)$ be given as in Section 8.3:

$$\left.\begin{aligned} \frac{\partial y(v)}{\partial t} + A y(v) &= f \quad \text{in } Q, \\ y(v) &= v \quad \text{on } \Sigma, \\ y(x,0;v) &= y_0(x), \quad x \in \Omega. \end{aligned}\right\} \tag{12.23}$$

Assume that

$$J(v) = \int_\Sigma (y(v) - z_d)^2 \, d\Sigma + (N v, v)_{L^2(\Sigma)}. \tag{12.24}$$

Then

$$\pi(u,v) = \int_\Sigma (y(u) - y(0))(y(v) - y(0)) \, d\Sigma + (N u, v)_{L^2(\Sigma)},$$

$$L(v) = \int_\Sigma (z_d - y(0))(y(v) - y(0)) \, d\Sigma.$$

If $\zeta \in L^2(\Sigma)$, we have:

$$\sup(\zeta, v)_{L^2(\Sigma)} = \begin{cases} \sum_{i=1}^m \xi_i \alpha_i & \text{if } \zeta = \sum_{i=1}^m \xi_i g_i, \quad \xi_i \geqslant 0, \\ +\infty & \text{otherwise.} \end{cases} \tag{12.25}$$

To see this, if

$$\zeta = \sum_{i=1}^m \xi_i g_i, \quad \xi_i \geqslant 0,$$

then the result is obvious. If ζ is not of this form, from the Hahn-Banach Theorem, there exists an element $h \in L^2(\Sigma)$ such that

$$(\zeta, h) > 0, \quad (h, g_i) < 0 \ \forall i = 1, \ldots, m.$$

Then $\lambda h \in \mathscr{U}_{ad}$ for $\lambda > 0$ sufficiently large and hence

$$\sup_{v \in \mathscr{U}_{ad}} (\zeta, v)_{L^2(\Sigma)} \geqslant \sup_{\lambda \to +\infty} (\zeta, \lambda h) = +\infty.$$

Hence we have (12.25).

Therefore the right hand side of (12.21) is equal to

$$\sup \left[-\pi(v,v) - \frac{1}{2} \sum_{i=1}^{m} \xi_i \alpha_i \right], \quad -J'(v) = \sum_{i=1}^{m} \xi_i g_i, \quad \xi_i \geqslant 0, \quad (12.26)$$

—a problem in *quadratic programming* (but requires preliminary calculation of $\pi((J')^{-1} g_i, (J')^{-1} g_j)$).

13. Constraints on the Control and the State

13.1. A General Result

Let $\mathscr{U}$ be the (Hilbert) space of controls; let Φ be a reflexive locally convex topological vector space on $\mathbb{R}$ and let

$$\rho \in \mathscr{L}(\mathscr{U}; \Phi). \quad (13.1)$$

Let

$$\mathscr{C} = \text{closed convex set in } \Phi \text{ with the vertex at the origin.} \quad (13.2)$$

We define the set of admissible controls by

$$\mathscr{U}_{ad} = \{ v \mid \varphi_0 + \rho v \in \mathscr{C} \}, \quad \varphi_0 \text{ given in } \Phi. \quad (13.3)$$

We assume that $\mathscr{U}_{ad} \neq \emptyset$. The set $\mathscr{U}_{ad}$ is closed and convex.

Remark 13.1. In Section 13.2 we shall see examples which show how (13.3) defines constraints on the control and the state.

We are given a function $v \to J(v)$ mapping $\mathscr{U}_{ad} \to \mathbb{R}$ which we assume to be convex and differentiable (cf. Chapter 1, Section 1) and as usual we look for $\inf_{v \in \mathscr{U}_{ad}} J(v)$. We know (cf. Chapter 1, Section 1) that if u is an optimal control then

$$J'(u)(v-u) \geqslant 0 \ \forall v \in \mathscr{U}_{ad} \text{ and conversely.} \quad (13.4)$$

Let us set, in general,

$$\mathscr{C}^+ = \{\varphi' \mid \varphi' \in \Phi', \ \langle \varphi', \varphi \rangle \geqslant 0 \ \forall \varphi \in \mathscr{C}\}, \tag{13.5}$$

(where $\langle \varphi', \varphi \rangle$ denotes the scalar product between $\varphi' \in \Phi'$ and $\varphi \in \Phi$).
Let us note that

$$(\mathscr{C}^+)^+ = \mathscr{C}. \tag{13.6}$$

To see this, first, it is clear that $\mathscr{C} \subset (\mathscr{C}^+)^+$. Let us suppose that the inclusion is strict. Then there exists a $\xi \in (\mathscr{C}^+)^+$, $\xi \notin \mathscr{C}$. But from the Hahn-Banach Theorem, there then exists $\varphi' \in \Phi'$, $\varphi' \geqslant 0$ on $\mathscr{C}$, $\langle \varphi', \xi \rangle < 0$; but "$\varphi' \geqslant 0$ on $\mathscr{C}$"

$$\Rightarrow \varphi' \in \mathscr{C}^+ \quad \text{and since } \xi \in (\mathscr{C}^+)^+, \quad \langle \varphi', \xi \rangle \geqslant 0,$$

which is a contradiction. Hence we have (13.6).

We now prove

Theorem 13.1. *Let* $\rho^* \in \mathscr{L}(\Phi'; \mathscr{U}')$ *be the conjugate transformation of* ρ. *We assume that*

$$\rho^*(\mathscr{C}^+) \quad \text{is closed in } \mathscr{U}', \tag{13.7}$$

and

$$-\varphi_0 \in \text{ closure of } \rho(\mathscr{U}_{ad}) \quad \text{in } \Phi. \tag{13.8}$$

Then if u is an optimal control (assumed to exist), there exists a $\lambda \in \mathscr{C}^+$ *such that*

$$J'(u) = \rho^* \lambda \tag{13.9}$$

and we have

$$\langle \lambda, \rho u + \varphi_0 \rangle = 0. \tag{13.10}$$

Conversely if u and λ *satisfy* (13.9), (13.10), $\lambda \in \mathscr{C}^+$, $u \in \mathscr{U}_{ad}$, *then u is optimal.*

Proof 1. For u fixed in $\mathscr{U}_{ad}$, define

$$\mathscr{U}_{ad}^u = \{v \mid v \in \mathscr{U}, u + \varepsilon v \in \mathscr{U}_{ad} \text{ for appropriate } \varepsilon > 0\}. \tag{13.11}$$

If $v \in \mathscr{U}_{ad}$, $v - u \in \mathscr{U}_{ad}^u$ since $u + (v-u) \in \mathscr{U}_{ad}$, and hence from (13.4):

$$J'(u) \in (\mathscr{U}_{ad}^u)^+ \tag{13.12}$$

(note that definition (13.5) has meaning irrespective of whether $\mathscr{C}$ is a cone or not).

But

$$(\rho^* \mathscr{C}^+)^+ \subset \mathscr{U}_{ad}^u; \tag{13.13}$$

in fact

$$w \in (\rho^* \mathscr{C}^+)^+ \Leftrightarrow (w, \rho^* c_+) \geqslant 0 \ \forall c_+ \in \mathscr{C}^+, \text{[72]}$$

[72] Duality between $\mathscr{U}$ and $\mathscr{U}'$.

that is $(\rho w, c_+) \geqslant 0$. But $\rho u + \varphi_0 \in \mathscr{C}$ (since $u \in \mathscr{U}_{\mathrm{ad}}$ by hypothesis) and hence from (13.6) $\rho u + \varphi_0 \in (\mathscr{C}^+)^+$ and hence

$$\langle \rho(u+\theta w) + \varphi_0, c_+ \rangle \geqslant 0 \quad \forall \theta \geqslant 0.$$

Therefore $u + \theta w \in \mathscr{U}_{\mathrm{ad}}$ $\forall \theta \geqslant 0$, implying $w \in \mathscr{U}_{\mathrm{ad}}^u$ whence (13.13).

But since from (13.7) $\rho^* \mathscr{C}^+$ is closed, we may apply (13.6) to $\rho^* \mathscr{C}^+$: $((\rho^* \mathscr{C}^+)^+)^+ = \rho^* \mathscr{C}^+$. Therefore from (13.7) we deduce that

$$(\mathscr{U}_{\mathrm{ad}}^u)^+ \subset \rho^* \mathscr{C}^+. \tag{13.14}$$

From (13.12), (13.14) we deduce that there exists a $\lambda \in \mathscr{C}^+$ such that (13.9) holds.

Then

$$\begin{aligned}(J'(u), u) = \langle \lambda, \rho u \rangle &= \inf_{v \in \mathscr{U}_{\mathrm{ad}}} (J'(u), v) = \inf_{v \in \mathscr{U}_{\mathrm{ad}}} \langle \lambda, \rho v \rangle \\ &= \inf_{v \in \mathscr{U}_{\mathrm{ad}}} \langle \lambda, \rho v + \varphi_0 \rangle - \langle \lambda, \varphi_0 \rangle;\end{aligned} \tag{13.15}$$

now $\lambda \in \mathscr{C}^+ \Rightarrow \langle \lambda, \rho v + \varphi_0 \rangle \geqslant 0$ and since we have (13.8), we also have

$$\inf_{v \in \mathscr{U}_{\mathrm{ad}}} \langle \lambda, \rho v + \varphi_0 \rangle = 0,$$

and therefore (13.5) gives us $\langle \lambda, \rho u + \varphi_0 \rangle = 0$.

2. Conversely let $\lambda \in \mathscr{C}^+$, $u \in \mathscr{U}_{\mathrm{ad}}$ and assume that (13.9), (13.10) hold. Then if $v \in \mathscr{U}_{\mathrm{ad}}$:

$$\begin{aligned}J'(u)(v-u) &= \langle \lambda, \rho v - \rho u \rangle = \langle \lambda, \rho v + \varphi_0 \rangle - \langle \lambda, \rho u + \varphi_0 \rangle \\ &= \langle \lambda, \rho v + \varphi_0 \rangle \geqslant 0\end{aligned}$$

since $\lambda \in \mathscr{C}^+$, whence $J'(u)(v-u) \geqslant 0$ $\forall v \in \mathscr{U}_{\mathrm{ad}}$ and u is optimal.

Example 13.1. $\mathscr{C} = \{0\}$.

Then $\mathscr{C}^+ = \Phi'$. By hypothesis $\mathscr{U}_{\mathrm{ad}} \neq \emptyset$, hence $\exists\, v$ with $\rho v + \varphi_0 = 0$. Then (13.8) holds and hence:

$$\left.\begin{array}{l}\textit{if } \mathscr{C} = \{0\}, \textit{ Theorem 13.1 is applicable if we assume that} \\ \rho^* \Phi' \textit{ is closed in } \mathscr{U}'.\end{array}\right\} \tag{13.16}$$

Remark 13.2. In general (this remark covers the case of Example 13.1)

$$\textit{if } \varphi_0 \textit{ is the form } \varphi_0 = \rho u_0, \quad u_0 \in \mathscr{U}, \textit{ then (13.8) holds.} \tag{13.17}$$

To see this, note that $w = u_0$ belongs to $\mathscr{U}_{\mathrm{ad}}$ and $\varphi_0 + \rho w = 0$.

13.2. Applications (I)

Let us consider the situation of Section 2.1 with the following data:

$$\left.\begin{aligned}&\mathscr{U}=L^2(0,T;V'),\\&\frac{dy(v)}{dt}+A(t)\,y(v)=f+v,\quad v\in\mathscr{U},\quad (f\in L^2(0,T;V')),\\&y(0;v)=y_0,\quad y_0 \text{ given in } H,\end{aligned}\right\}\tag{13.18}$$

$$\left.\begin{aligned}&J(v)=\int_0^T |y(t;v)-z_d|^2\,dt+(N\,v,v)_{\mathscr{U}},\quad z_d\in L^2(0,T;H),\\&N\in\mathscr{L}(\mathscr{U};\mathscr{U}),\quad N\geqslant\nu \text{ (Identity)},\quad \nu>0.\end{aligned}\right\}\tag{13.19}$$

We assume that $\mathscr{U}_{\text{ad}}$ is defined by

$$\left.\begin{aligned}&\mathscr{U}_{\text{ad}}=\{v\,|\,y(T;v)\in y_1+\mathscr{C}\},\\&y_1 \text{ given in } H,\\&\mathscr{C}=\text{closed convex cone in } H \text{ with vertex at the origin.}\end{aligned}\right\}\tag{13.20}$$

We assume that $\mathscr{U}_{\text{ad}}\neq\emptyset$.

Let us identify with the notation in Section 13.1:

$$\Phi=H,\quad \rho\,v=y(T;v)-y(T;0),\quad \varphi_0=y(T;0)-y_1.$$

The conjugate $\rho^*\in\mathscr{L}(H;L^2(0,T;V))$ is defined as follows: for $h\in H$, $\rho^*h=w$ is the solution of

$$\left.\begin{aligned}-\frac{dw}{dt}+A^*(t)\,w=0\quad &\text{in }]0,T[,\\ w(T)=h.&\end{aligned}\right\}\tag{13.21}$$

In fact,

$$\int_0^T\left(-\frac{dw}{dt}+A^*(t)\,w,\;y(t;v)-y(t;0)\right)dt=0=-(h,\rho\,v)+\int_0^T(w,v)\,dt$$

hence

$$w=\rho^*h.\tag{13.22}$$

We have:

Lemma 13.1. *The hypothesis* (13.7) *holds.*

Proof. Let $h_n \in \mathscr{C}^+$, $w_n = \rho^* h_n \to w$ in $L^2(0, T; V)$. Then

$$-\frac{dw_n}{dt} + A^*(t) w_n = 0, \qquad A^*(t) w_n \to A^*(t) w \quad \text{in } L^2(0, T; V'),$$

hence $\frac{dw_n}{dt} \to \frac{dw}{dt}$ in $L^2(0, T; V')$, hence $w_n(T) = h_n \to w(T)$ in H.

Set $h = w(T)$. Hence $h_n \to h$ in H and since $\mathscr{C}^*$ is closed, we have:

$$h \in \mathscr{C}^* \quad \text{and} \quad -\frac{dw}{dt} + A^* w = 0, \qquad w(T) = h, \quad \text{implying } w = \rho^* h. \quad \square$$

Lemma 13.2. *Hypothesis* (13.8) *holds.*

Proof. Actually more is true: there exists a v such that $\rho(v) + \varphi_0 = 0$, that is, $y(T; V) = y_1$. To see this, we note that there exists a $z \in L^2(0, T; V)$ such that $\frac{dz}{dt} \in L^2(0, T; V')$ and $z(0) = y_0$, $z(T) = y_1$ (for y_0, y_1 arbitrary given elements in H); hence it suffices to take $v = \frac{dz}{dt} + A(t) z - f$. (We have the same case as in Remark 13.2.) $\square$

Hence we may apply Theorem 13.1. We introduce $p(u)$ by

$$\left.\begin{aligned} -\frac{dp(u)}{dt} + A^*(t) p(u) &= y(u) - z_d, \\ p(T; u) &= 0, \end{aligned}\right\} \tag{13.23}$$

$$\left.\begin{aligned} &J'(u) = 2(p(u) + \Lambda_V^{-1} N u), \\ &\quad \Lambda_V = \text{canonical isomorphism of } V \text{ onto } V'. \end{aligned}\right\} \tag{13.24}$$

Therefore there exists a λ such that[73]

$$\lambda \in \mathscr{C}^+, \qquad p(u) + \Lambda_V^{-1} N u = \rho^* \lambda, \tag{13.25}$$

and

$$(\lambda, y(T; u) - y_1) = 0. \tag{13.26}$$

We now set $w = \rho^* \lambda$ and $q = p(u) - w$. Then (using (13.21))

$$\left.\begin{aligned} -\frac{dq}{dt} + A^*(t) q &= y(u) - z_d, \\ q(T) &= -\lambda, \end{aligned}\right\} \tag{13.27}$$

and eliminating u we arrive at

---

[73] We may always change λ to $\frac{\lambda}{2}$.

Theorem 13.2. *Assume that the state is given by* (13.18), *the cost function by* (13.19) *and* $\mathscr{U}_{ad}$ *by* (13.20), *with* $\mathscr{U}_{ad} \neq \emptyset$. *Then if* u *is an optimal control there exists a* $\lambda \in H$ *such that*

$$\left.\begin{aligned} &\frac{dy}{dt} + A(t)y + N^{-1}\Lambda_V q = f, \\ &-\frac{dq}{dt} + A^*(t)q = y - z_d, \\ &y(0) = y_0, \\ &q(T) = -\lambda, \\ &(\lambda, y(T) - y_1) = 0, \quad \lambda \in \mathscr{C}^+, \end{aligned}\right\} \tag{13.28}$$

and the optimal control is given by

$$\Lambda_V^{-1} N u + q = 0. \tag{13.29}$$

13.3. Applications (II)

We now place ourselves within the framework of Section 7.2: the state $y(v)$ is given by

$$\frac{\partial y(v)}{\partial t} + A(t)y(v) = f \quad \text{in } Q \quad (f \in L^2(Q)), \tag{13.30}$$

$$\frac{\partial}{\partial \nu_A} y(v) = v, \quad v \in \mathscr{U} = L^2(\Sigma), \tag{13.31}$$

$$y(x,0;v) = y_0(x), \quad x \in \Omega, \quad y_0 \in L^2(\Omega). \tag{13.32}$$

Assume that

$$\begin{aligned} &J(v) = \int_\Omega [y(x,T;v) - z_d(x)]^2\,dx + (Nv, v)_{L^2(\Sigma)}, \\ &N \in \mathscr{L}(L^2(\Sigma); L^2(\Sigma)), \quad N \geqslant \nu \text{ (identity) } \nu > 0, \end{aligned} \tag{13.33}$$

and that

$$\left.\begin{aligned} &\mathscr{U}_{ad} = \{v \mid y(T;v) \in y_1 + \mathscr{C}\}, \\ &y_1 \text{ given in } H = L^2(\Omega), \quad \mathscr{C} = \text{closed, convex cone in } H \text{ with} \\ &\text{vertex at the origin } \{0\}. \end{aligned}\right\} \tag{13.34}$$

Hence

$$\Phi = H = L^2(\Omega), \quad \rho v = y(T;v) - y(T;0),$$
$$\varphi_0 = y(T;0) - y_1, \quad \rho \in \mathscr{L}(L^2(\Sigma); L^2(\Omega)).$$

Hence $\rho^* \in \mathscr{L}(L^2(\Omega); L^2(\Sigma))$. The operator ρ^*, conjugate to ρ, is defined as follows: for h given in $L^2(\Omega)$, we solve

$$\left.\begin{aligned} -\frac{\partial w}{\partial t} + A^*(t)w = 0 \quad \text{in } Q, \\ w(x,T) = h(x), \quad x \in \Omega, \\ \frac{\partial w}{\partial \nu_{A^*}} = 0 \quad \text{on } \Sigma. \end{aligned}\right\} \tag{13.35}$$

Then

$$\rho^* h = w|_E. \tag{13.36}$$

In this case hypothesis (13.7) is in general not satisfied. In fact, the solution of (13.35) is in $L^2(0,T; H^1(\Omega))$ and hence in particular $w|_E \in L^2(0,T; H^{\frac{1}{2}}(\Gamma))$ which is not a closed subspace of $L^2(\Sigma)$.

But clearly

hypothesis (13.7) *holds if* $\mathscr{C}^+$ *is contained in a finite dimensional space.* (13.37)

Therefore we have

Theorem 13.3. *We assume that the state is given by* (13.30), (13.31), (13.32) *and the cost function by* (13.33). *We assume that* $\mathscr{U}_{ad}$ *is given by* (13.34) *with* $y_1 = y(T;0)$ *and* (cf. (13.37)) $\mathscr{C}^+$ *is contained in a finite dimensional space. Then if u is an optimal control, there exists a* $\lambda \in L^2(\Omega)$ *such that we have*

$$\left.\begin{aligned} &\frac{\partial y}{\partial t} + A(t)y = f, \quad -\frac{\partial q}{\partial t} + A^*(t)q = 0 \quad \text{in } Q, \\ &N\frac{\partial y}{\partial \nu_A} + q = 0, \quad \frac{\partial q}{\partial \nu_{A^*}} = 0 \quad \text{on } \Sigma, \\ &y(x,0) = y_0(x), \quad q(x,T) = y(x,T) - z_d(x) - \lambda(x) \quad \text{in } \Omega, \\ &\lambda \in \mathscr{C}^+, \\ &\int_\Omega \lambda(x)\, y(x,T)\, dx = 0. \end{aligned}\right\} \tag{13.38}$$

Then $u = -N^{-1}(q|_E)$.

Proof. In fact, if $y_1 = y(T;0)$ then $\varphi_0 = 0$ and we may apply Remark 13.2. Let us introduce $p(u)$ by

$$-\frac{\partial}{\partial t} p(u) + A^*(t)p(u) = 0, \qquad p(T;u) = y(T;u) - z_d.$$

Then $\frac{1}{2}J'(u)=p(u)|_\Sigma+Nu$ and hence (changing λ into $\frac{\lambda}{2}$) there exists a λ such that

$$p(u)|_\Sigma+Nu=\rho^*\lambda\pm w|_\Sigma.$$

The above is obtained by using (13.35) (with $h=\lambda$). Also $\lambda\in\mathscr{C}^+$ and satisfies according to (13.10):

$$\int_\Omega \lambda(x)\,y(x,T;u)\,dx=0.$$

Setting $q=p(u)-w$ we deduce the theorem. □

Remark 13.3. Similar examples may be constructed for systems governed by parabolic equations of arbitrary order or by a system of parabolic equations.

Remark 13.4. Clearly, Theorem 13.1 is also applicable to systems governed by *elliptic equations* (Chapter 2).

14. Non Quadratic Cost Functions

14.1. Orientation

Up to now we have only considered quadratic cost functions. Clearly, everything we have done so far (excepting for decoupling—cf. Section 14.3 following) can be extended to the case where $J(v)$ is a convex, non-quadratic differentiable function by applying Theorem 1.3, Chapter 1. We shall confine ourselves to considering a simple example (in Section 14.2) and we then show in Section 14.3 that considerable difficulties exist if we want to decouple the system.

14.2. An Example

We consider a system governed by a second order parabolic operator[74]:

$$\left.\begin{aligned} \frac{\partial y(v)}{\partial t}+A(t)y(v)&=f+v \quad \text{in } Q,\\ y(v)|_\Sigma&=0,\\ y(x,0;v)&=y_0(x) \quad \text{in } \Omega, \end{aligned}\right\} \tag{14.1}$$

[74] What follows may be extended to the general case.

with

$$\mathscr{U} = L^q(Q), \qquad 1 < q < \infty,^{75} \tag{14.2}$$

$$f \in L^q(Q), \qquad y_0 \in L^q(\Omega). \tag{14.3}$$

There exists a unique solution to problem (14.1). The solution satisfies

$$y(v), \qquad \frac{\partial}{\partial x_i} y(v) \in L^q(Q), \qquad i = 1, \ldots, n.^{76} \tag{14.4}$$

We may then consider the cost function:

$$\left.\begin{aligned} &J(v) = \int_Q |y(x,t;v) - z_d(x,t)|^q \, dx\,dt + \nu \|v\|^q_{L^q(Q)}, \\ &z_d \text{ given in } L^q(Q), \qquad \nu > 0. \end{aligned}\right\} \tag{14.5}$$

Remark 14.1. We may also consider a *non-quadratic* form on a Hilbert Space: take $\mathscr{U} = L^2(Q)$; then $y(v) \in L^q(Q)$ if $1 < q < 2$ (we may go further by applying—for example—Sobolev's Theorem) and hence, in particular, we may consider (14.5) with $1 < q < 2$.

Remark 14.2. A far more difficult step is to pass from a consideration of a linear system to that of a non-linear system. Some partial results in this direction will be presented in the following section.

From Theorem 1.3, Chapter 1 (in the case where $\mathscr{U}$ is a reflexive Banach space), there *exists a unique optimal control u which is characterized by*

$$J'(u)(v-u) \geqslant 0 \quad \forall v \in \mathscr{U}_{\mathrm{ad}} \qquad (u \in \mathscr{U}_{\mathrm{ad}}) \tag{14.6}$$

(if $\mathscr{U}_{\mathrm{ad}}$ = closed, convex set in $\mathscr{U}$).

But from (14.5), we have for $w \in \mathscr{U}$

$$\left.\begin{aligned} J'(u)\cdot w &= \frac{d}{d\xi} J(u + \xi w)|_{\xi=0} \\ &= q \int_Q |y(u) - z_d|^{q-2} (y(u) - z_d) \frac{\partial}{\partial u} y(u) \cdot w \, dx\,dt \\ &\quad + q\nu \int_Q |u|^{q-2} u w \, dx\,dt. \end{aligned}\right\} \tag{14.7}$$

---

[75] If $q \neq 2$, $\mathscr{U}$ is a reflexive Banach space (but not Hilbert); but as indicated in Chapter 1, results of the type of Theorem 1.3, Chapter 1 remain valid.

[76] For the solution of parabolic problems in L^q, $q \neq 2$, consult P. Grisvard [1], [2], V. A. Solonnikov [1].

But if we set

$$\frac{\partial}{\partial u} y(u) \cdot w = \psi(w), \tag{14.8}$$

we may verify that

$$\left.\begin{aligned} &\frac{\partial \psi(w)}{\partial t} + A\psi(w) = w \quad \text{in } Q, \\ &\psi(w)|_\Sigma = 0, \qquad \psi(x,0;w) = 0 \quad \text{in } \Omega. \end{aligned}\right\} \tag{14.9}$$

Hence condition (14.6) is equivalent to (dividing through by q):

$$\left.\begin{aligned} &\int_Q |y(u) - z_d|^{\,-2}(y(u) - z_d)\psi(v-u)\,dx\,dt \\ &\qquad + \nu \int_Q |u|^{\,-2} u(v-u)\,dx\,dt \geqslant 0 \;\; \forall v \in \mathscr{U}_{\mathrm{ad}}. \end{aligned}\right\} \tag{14.10}$$

We shall transform (14.10) by introducing the adjoint state $p(u)$ defined as the solution in $L^{q'}(Q)$, $1/q + 1/q' = 1$, of[77]

$$\left.\begin{aligned} &-\frac{\partial p(u)}{\partial t} + A^*(t)p(u) = |y(u) - z_d|^{q-2}(y(u) - z_d) \quad \text{in } Q, \\ &p(u) = 0 \quad \text{on } \Sigma, \\ &p(x,T;u) = 0 \quad \text{in } \Omega. \end{aligned}\right\} \tag{14.11}$$

Then (we have done what needed to be done!) the first term in (14.10) equals

$$\int_Q \left(-\frac{\partial}{\partial t} p(u) + A^*(t)p(u) \right) \psi(v-u)\,dx\,dt = \int_Q p(u)\,(v-u)\,dx\,dt$$

and (14.10) is equivalent to

$$\int_Q (p(u) + \nu|u|^{q-2} u)\,(v-u)\,dx\,dt \geqslant 0 \;\; \forall v \in \mathscr{U}_{\mathrm{ad}}. \tag{14.12}$$

Hence we have

Theorem 14.1. *We assume that the state is given by* (14.1) *and the cost function is given by* (14.5). *Then the unique optimal control is characterized by* (14.1) (*where* $v = u$), (14.11) *and* (14.12).

---

[77] Note that $|y(u) - z_d|^{q-2}(y(u) - z_d) \in L^{q'}(Q)$; problem (14.11) admits a unique solution such that

$$\frac{\partial p(u)}{\partial x_i}, \quad \frac{\partial p(u)}{\partial t}, \quad \frac{\partial^2}{\partial x_i \partial x_j} p(u) \in L^{q'}(Q).$$

Example 14.1. Take $\mathscr{U}_{\mathrm{ad}}=\{v|v\in\mathscr{U}=L^q(Q),\ v\geqslant 0$ almost everywhere in $Q\}$. Then (14.12) is equivalent to

$$\left.\begin{aligned} &p(u)+\nu|u|^{q-2}u\geqslant 0, \quad u\geqslant 0,\\ &(p(u)+\nu|u|^{q-2}u)u=0 \quad \text{almost everywhere in } Q.\end{aligned}\right\} \tag{14.13}$$

We are again confronted with a *unilateral problem.*

14.3. Remarks on Decoupling

In Theorem 14.1 let us now consider the case where there are no constraints: $\mathscr{U}_{\mathrm{ad}}=\mathscr{U}$.

Then the optimal control is determined by the following system of equations

$$\left.\begin{aligned} &\left.\begin{aligned} &\frac{\partial y}{\partial t}+Ay+\frac{1}{\nu^{q'-1}}|p|^{q'-2}p=f\\ &-\frac{\partial p}{\partial t}+A^*p=|y-z_d|^{q-2}(y-z_d)\end{aligned}\right\}\ \text{in } Q,\\ &\left.\begin{aligned} &y(x,0)=y_0(x)\\ &p(x,T)=0\end{aligned}\right\}\ \text{in } \Omega,\\ &y=0, \quad p=0 \ \text{ on } \Sigma,\end{aligned}\right\} \tag{14.14}$$

and

$$u=-\frac{1}{\nu^{q'-1}}|p|^{q'-2}p. \tag{14.15}$$

(In fact (14.12) reduces to $p(u)+\nu|u|^{q-2}u=0$).

To decouple the system of equations (14.14) we argue as in Section 4. We introduce (compare with Lemma 4.1) the solution $\{\varphi,\psi\}$ of the system

$$\left.\begin{aligned} &\frac{\partial\varphi}{\partial t}+A\varphi+\frac{1}{\nu^{q'-2}}|\psi|^{q'-2}\psi=f \quad \text{in } \Omega\times]s,T[,\\ &-\frac{\partial\psi}{\partial t}+A^*\psi=|y-z_d|^{q-2}(y-z_d) \quad \text{in } \Omega\times]s,T[,\\ &\varphi(x,s)=h(x), \quad \psi(x,T)=0 \quad \text{in } \Omega,\\ &\varphi=0, \quad \psi=0 \quad \text{in } \Gamma\times]s,T[.\end{aligned}\right\} \tag{14.16}$$

This (non-linear) system of equations admits a unique solution since it corresponds to an optimal control problem analogous to the one considered in the previous section, with the exception that it is considered over $\Omega \times]s, T[$ instead of $\Omega \times]0, T[$ and with h in place of y_0.

We then define the mapping

$$h \to \psi(x,s) \quad \text{of } L^q(\Omega) \to L^{q'}(\Omega), \tag{14.17}$$

which is a non-affine mapping. We may write

$$\psi(0,s) = \Phi(h,s)$$

and we arrive at the identity

$$p(s) = \Phi(y(s),s), \qquad 0 < s < T, \tag{14.18}$$

where

$$\eta, t \to \Phi(\eta,t) \quad \text{is a function from } L^q(\Omega) \times [0,T] \to L^{q'}(\Omega). \tag{14.19}$$

We may then obtain an identity for Φ: we shall obtain a functional-partial differential equation (as in Section 4.8 and 5.5). Substituting (14.18) in the second equation of (14.14) (and performing a *formal* calculation) we get:

$$\frac{\partial \Phi}{\partial \eta}(y(t),t) \cdot \frac{\partial y}{\partial t} - \frac{\partial \Phi}{\partial t}(y(t),t) + A^* \Phi(y(t),t) = |y(t) - z_d|^{q-2}(y(t) - z_d)$$

and utilizing the first equation of (14.14) we deduce

$$-\frac{\partial \Phi}{\partial \eta}(y(t),t)\left(-Ay - \frac{1}{\nu^{q'-1}}|\Phi|^{q'-2}\Phi + f\right) - \frac{\partial \Phi}{\partial t}(y(t),t)$$
$$+ A^* \Phi(y(t),t) = |y(t) - z_d|^{q-2}(y(t) - z_d).$$

Since we are concerned with an identity in $y(t)$, we deduce that

$$\left.\begin{aligned} &-\frac{\partial \Phi}{\partial t}(\eta,t) + \frac{\partial \Phi}{\partial \eta}(\eta,t) \cdot A\eta + A^* \Phi(\eta,t) + \frac{1}{\nu^{q'-1}} \frac{\partial \Phi}{\partial \eta}(\eta,t) \cdot |\Phi|^{q'-2}\Phi \\ &-\frac{\partial \Phi}{\partial \eta}(\eta,t) \cdot f = |\eta - z_d|^{q-2}(\eta - z_d) \quad \forall \eta, \end{aligned}\right\} \tag{14.20}$$

to which we must add

$$\Phi(\eta,T) = 0 \quad \forall \eta, \tag{14.21}$$

and

$$\Phi(\eta,t) = 0 \quad \text{on } \Sigma. \tag{14.22}$$

This (complicated) set of equations replaces the Riccati integro-differential equation of Section 4 and 5.

15. Existence Results for Optimal Controls

15.1. Orientation

The present section is the analogue of Section 7, Chapter 2 for systems described by evolution equations.

We shall consider systems whose state is given by non-linear evolution equations. We shall be only considering particular examples. More general results which are available are indicated in the notes. First order necessary conditions of optimality are presented in the next section.

15.2. Non-linear Problem with Distributed Control (I)

We consider (as in Section 3) a second order[78] parabolic operator $\partial/\partial t + A(t)$ and we assume that the state is given by

$$\frac{\partial}{\partial t} y(v) + A(t) y(v) + v y(v) = f \quad \text{in } Q, \quad f \in L^2(\Omega), \tag{15.1}$$

with (for example)

$$y(v)|_{\Sigma} = 0, \tag{15.2}$$

and the initial condition

$$y(x,0;v) = y_0(x), \quad x \in \Omega, \quad y_0 \in L^2(\Omega). \tag{15.3}$$

We assume in (15.1) that

$$v \in \mathscr{U} = L^{\infty}(Q), \tag{15.4}$$

which allows us to conclude that the problem (15.1), (15.2), (15.3) admits a unique solution satisfying

$$y(v) \in L^2(0,T; H_0^1(\Omega)). \tag{15.5}$$

Remark 15.1. Clearly the mapping $v \to y(v)$ is *not* linear.

Remark 15.2. The existence and uniqueness of $y(v)$ may be proved under more general conditions on v. We may take v to be an element of $L^{\rho}(Q)$, with appropriate ρ—cf. in particular Ladyzenskaya-Solonnikov-Uralts'eva [1].

Cost Function. To fix ideas, let us take the cost function to be

$$J(v) = \int_{\Omega} [y(x,T;v) - z_d(x)]^2 \, dx + \nu \|v\|_{L^{\infty}(Q)}, \quad \nu > 0. \tag{15.6}$$

---

[78] The fact that the operator is of second order is not essential in any way to the arguments in Sections 15.2, 15.3, 15.4.

Remark 15.3. From (15.1) and (15.5), it follows that

$$\frac{\partial}{\partial t} y(v) \in L^2(0,T; H^1(\Omega)),$$

allowing us to conclude that

$$y(0,T;v) \quad \text{makes sense and belongs to } L^2(\Omega).$$

Remark 15.4. In (15.6), if $\mathcal{U}_{\text{ad}}$ (cf. below) is bounded in $L^\infty(Q)$ we may take $\nu=0$.

We now consider the set of admissible controls:

$$\begin{aligned}\mathcal{U}_{\text{ad}} = {} & \text{convex set in } L^\infty(Q) \text{ which is closed in the weak} \\ & \text{topology of the dual of } L^1(Q),\end{aligned} \tag{15.7}$$

and we seek,

$$\operatorname{Inf} J(v), \quad v \in \mathcal{U}_{\text{ad}}. \tag{15.8}$$

Theorem 15.1. *We assume that the operator $\partial/\partial t + A$ is a second order parabolic operator satisfying* (1.28) *and that the set of admissible controls is given by* (15.7). *Then if the cost function is given by* (15.6), *there exists an optimal control in* $\mathcal{U}_{\text{ad}}$.

Proof. Let v_n be a minimising sequence: $J(v_n) \to \inf J(v)$, $v_n \in \mathcal{U}_{\text{ad}}$. Set $y_n = y(v_n)$. By virtue of the term $\nu \|v\|_{L^\infty(Q)}$ in $J(v)$, v_n is bounded in $L^\infty(Q)$. Then

$$y_n \text{ is bounded in } L^2(0,T; H_0^1(\Omega)). \tag{15.9}$$

To see this, we note that from (15.1) we may deduce

$$\frac{1}{2}\frac{d}{dt}|y_n(t)|^2 + a(t; y_n(t), y_n(t)) + \int_\Omega v_n y_n(x,t)^2 \, dx = \int_\Omega f y_n \, dx,$$

whence (after integrating from 0 to t)[79], by virtue of the fact that v_n is bounded in $L^\infty(Q)$,

$$|y_n(t)|^2 + \alpha \int_0^t \|y_n(\sigma)\|^2 \, d\sigma \leqslant C_1 \int_0^t |y_n(\sigma)|^2 \, d\sigma + C_2,$$

whence it follows (by applying Gronwall's lemma) that y_n is bounded in

$$L^\infty(0,T; L^2(\Omega))$$

and thus we have (15.9).

---

[79] $|\ |$ (resp. $\|\ \|$) denotes the norm in $L^2(\Omega)$ (resp. $H_0^1(\Omega)$).

But from (15.1), we have,

$$\frac{\partial y_n}{\partial t} = f - A(t)\,y_n - v_n y_n$$

and the right hand side is bounded in $L^2(0,T;H^{-1}(\Omega))$. Hence

$$\frac{\partial y_n}{\partial t} \text{ is bounded in } L^2(0,T;H^{-1}(\Omega)). \tag{15.10}$$

But since the injection map of $H_0^1(\Omega)\to L^2(\Omega)$ is *compact* (cf. the proof of Theorem 7.1, Chapter 2) and using a "*compactness lemma*" (cf. Lions [3], Proposition 4.2, Chapter 4)[80] it follows from (15.9), (15.10) that we may extract a subsequence, again denoted by y_n, such that

$$\left.\begin{aligned} &y_n\to y \quad \textit{weakly} \text{ in } L^2(0,T;H_0^1(\Omega)),\\ &\frac{\partial y_n}{\partial t}\to\frac{\partial y}{\partial t} \quad \textit{weakly} \text{ in } L^2(0,T;H^1(\Omega)),\\ &y_n\to y \quad \textit{strongly} \text{ in } L^2(Q). \end{aligned}\right\} \tag{15.11}$$

We may assume that

$$v_n\to v \quad \text{"weak star" in } L^\infty(Q), \quad \text{and } v\in\mathscr{U}_{\mathrm{ad}}. \tag{15.12}$$

But then (cf. proof of Theorem 7.1, Chapter 2) we have

$$y_n v_n \to y v \quad \text{in } \mathscr{D}'(Q) \quad \text{for example,}$$

and consequently y satisfies (15.1), (15.2) and (15.3) (since $y_n(x,0)\to y(x,0)$ weakly in $L^2(\Omega)$). Hence $y=y(v)$ and $\liminf J(v_n)\geqslant J(v)$ and therefore v is an optimal control. □

15.3. Non-linear Problem with Distributed Control. Singular Perturbation

Let (compare with Remark 7.6, Chapter 2) ξ_0, ξ_1 be given in $L^\infty(Q)$ with

$$0<\beta\leqslant\xi_0(x,t)\leqslant\xi_1(x,t) \text{ a.e. in } Q, \tag{15.13}$$

and let

$$\mathscr{U}_{\mathrm{ad}}=\{v\,|\,\xi_0(x,t)\leqslant v(x,t)\leqslant\xi_1(x,t) \quad \text{a.e. in } Q\}. \tag{15.14}$$

---

[80] See also Lions [13], Chapter 1.

For $v \in \mathscr{U}_{ad}$, we set

$$A(v)\cdot\psi = -\sum_{i,j=1}^{n} \frac{\partial}{\partial x_i}\left(a_{ij}(x,t)\,v(x,t)\frac{\partial\psi}{\partial x_j}\right), \tag{15.15}$$

with the functions a_{ij} satisfying (1.28).

We then define the state of the system by

$$\frac{\partial}{\partial t}y(v) + A(v)\,y(v) = f, \quad f \in L^2(Q), \tag{15.16}$$

$$\left.\begin{aligned} &y(v)|_{\Sigma} = 0, \\ &y(x,0;v) = y_0(x), \quad x \in \Omega. \end{aligned}\right\} \tag{15.17}$$

Problem (15.16), (15.17) admits a unique solution. In fact, we apply the results of Section 1 with $a(t;\varphi,\psi)$ given by

$$a_v(t;\varphi,\psi) = \sum_{i,j=1}^{n} \int_{\Omega} v(x,t)\,a_{ij}(x,t)\frac{\partial\varphi}{\partial x_j}\frac{\partial\psi}{\partial x_i}\,dx, \quad \varphi,\ \psi \in H^1(\Omega). \tag{15.18}$$

By virtue of (15.13), we have:

$$a_v(t;\varphi,\varphi) \geqslant \alpha_\beta \int_{\Omega} |\operatorname{grad}\varphi|^2\,dx \geqslant C\|\varphi\|^2_{H_0^1(\Omega)}, \quad \varphi \in H_0^1(\Omega),$$

(and $a_v(t;\varphi,\varphi) + \lambda|\varphi|^2 \geqslant C\|\varphi\|^2_{H^1(\Omega)}$, $\lambda > 0$ arbitrary, $\forall\varphi \in H^1(\Omega)$).

Let us now consider, for example, the cost function given as in (15.6) but with $v = 0$ ($\mathscr{U}_{ad}$ is bounded):

$$J(v) = \int_{\Omega} [y(x,T;v) - z_d(x)]^2\,dx. \tag{15.19}$$

As in Remark 7.6, Chapter 2, we are not able to decide whether an optimal control exists. The difficulty is that as $v_n \to v$ weak star in $L^\infty(Q)$ and $y_n \to y$ as in (15.11), it is not true in general that $A(v_n)\,y_n \to A(v)\,y$, and hence we are not able to conclude that there exists a $w \in \mathscr{U}_{ad}$ such that $y = y(w)$.

To avoid—but not to overcome—this difficulty, there appears to be three possibilities (similar to that considered in (i), (ii), (iii), Chapter 2, Section 7.2). As in Chapter 2, Section 7.3, we shall present details of the 3rd possibility by utilising a *singular perturbation*[81].

We thus introduce (as in (7.23), Chapter 2)

$$\left.\begin{aligned} &b(\varphi,\psi) = \text{continuous bilinear form on } H_0^2(\Omega), \text{ such that} \\ &b(\psi,\psi) \geqslant \beta\|\psi\|^2_{H_0^2(\Omega)}, \quad \beta > 0, \quad \forall\psi \in H_0^2(\Omega). \end{aligned}\right. \tag{15.20}$$

[81] See other possibilities in Ekeland's forthcoming thesis.

Let B be the corresponding operator

$$b(\varphi,\psi)=(B\varphi,\psi).$$

For $\varepsilon>0$, if we set

$$a_{\varepsilon,v}(t;\varphi,\psi)=\varepsilon b(\varphi,\psi)+a_v(t;\varphi,\psi), \qquad \varphi,\psi\in H_0^2(\Omega), \tag{15.21}$$

it may be seen that

$$a_{\varepsilon,v}(t;\psi,\psi)\geqslant\varepsilon\|\psi\|^2_{H_0^2(\Omega)}+c\|\psi\|^2_{H_0^1(\Omega)}, \qquad c>0. \tag{15.22}$$

Then we introduce the operator $\partial/\partial t+(\varepsilon B+A(v))$ (4th order parabolic) which is termed the *singular perturbation of the operator* $\partial/\partial t+A(v)$ and we define the state $y_\varepsilon(v)$ by[82]

$$\left.\begin{aligned}&\frac{\partial}{\partial t}y_\varepsilon(v)+(\varepsilon B+A(v))y_\varepsilon(v)=f \quad \text{in } Q,\\ &y_\varepsilon(v)\in L^2(0,T;H_0^2(\Omega)),\\ &\qquad\qquad y_\varepsilon(x,0;v)=y_0(x), \qquad x\in\Omega.\end{aligned}\right\} \tag{15.23}$$

We then have the following results which are analogous to Theorems 7.2 and 7.3 of Chapter 2:

Theorem 15.2. *As* $\varepsilon\to 0$, *we have:*

$$y_\varepsilon(v)\to y(v) \quad \text{in } L^2(0,T;H_0^1(\Omega)), \tag{15.24}$$

$$\left.\begin{aligned}&\sqrt{\varepsilon}\,y_\varepsilon(v)\to 0 \quad \text{in } L^2(0,T;H_0^2(\Omega)),\\ &\frac{\partial}{\partial t}y_\varepsilon(v)\to\frac{\partial}{\partial t}y(v) \quad \text{in } L^2(0,T;H^{-2}(\Omega)).\end{aligned}\right\} \tag{15.25}$$

Theorem 15.3. *For* $\varepsilon>0$ *fixed, there exists an optimal control* u_ε *in* $\mathscr{U}_{\text{ad}}$, *that is, there exists* u_ε *such that*

$$\int_\Omega[y_\varepsilon(x,T;u_\varepsilon)-z_d(x)]^2\,dx\leqslant\int_\Omega[y_\varepsilon(x,T,v)-z_d)]^2\,dx \quad \forall v\in\mathscr{U}_{\text{ad}}. \tag{15.26}$$

Proof of Theorem 15.2 (cf. D. Huet [1]). Utilising (15.22), we first verify that

$$y_\varepsilon(v)\ (\text{resp. } \sqrt{\varepsilon}\,y_\varepsilon(v)) \quad \text{is bounded in } L^2(0,T;H_0^1(\Omega))\ (\text{resp. } L^2(0,T;H_0^2(\Omega)).$$

Then utilising (15.23) in the form

$$\frac{\partial}{\partial t}y_\varepsilon(v)=f-(\varepsilon B+A(v))\,y_\varepsilon(v),$$

[82] According to (15.22), problem (15.23) admits a unique solution.

we deduce the analogues of (15.24), (15.25) where the convergence is weak convergence. We now pass to *strong* convergence by considering the expression

$$X_\varepsilon = \tfrac{1}{2}|y_\varepsilon(T;v) - y(T,v)|^2 + \varepsilon \int_0^T b_\varepsilon(y_\varepsilon(v), y_\varepsilon(v))\,dt$$
$$+ \int_0^T a_v(t; y_\varepsilon(v) - y(v), y_\varepsilon(v) - y(v))\,dt$$

for which we verify convergence towards zero.

Proof of Theorem 15.3. We suppress the index "ε". We consider a minimising sequence v_n and we set $y_n = y(v_n)$. Since ε is fixed and $\varepsilon > 0$, from (15.22) we deduce that

$$y_n \text{ ranges in a bounded subset of } L^2(0,T;H_0^2(\Omega)), \tag{15.27}$$

and then utilising equation (15.23) we deduce that

$$\frac{\partial y_n}{\partial t} \text{ ranges in a bounded subset of } L^2(0,T;H^{-2}(\Omega)). \tag{15.28}$$

Since the injection map of $H_0^2(\Omega) \to H_0^1(\Omega)$ is compact (cf. for example Lions-Magenes [1], Chapter 1) and utilising (as in the proof of Theorem 15.1) the *compactness lemma* (Lions [3], Proposition 4.2, Chapter 4), we deduce from (15.27), (15.28) that we may extract a sequence—again denoted by y_n—such that

$$\left.\begin{aligned} y_n &\to y \quad \textit{weakly in } L^2(0,T;H^2(\Omega)),\\ \frac{\partial y_n}{\partial t} &\to \frac{\partial y}{\partial t} \quad \textit{weakly in } L^2(0,T;H^{-2}(\Omega)),\\ y_n &\to y \quad \textit{strongly in } L^2(0,T;H_0^1(\Omega)). \end{aligned}\right\} \tag{15.29}$$

We may also assume, by extracting a subsequence, that

$$v_n \to v \quad \textit{weak-star in } L^\infty(Q), \qquad v \in \mathcal{U}_{\text{ad}},$$

and we may *now* pass to the limit, since

$$a_{ij} v_n \frac{\partial y_n}{\partial x_j} \to a_{ij} v \frac{\partial y}{\partial x_j} \quad \text{in (for example) } \mathcal{D}'(\Omega),$$

allowing us to conclude $y = y(v)$ and the proof is completed as in Theorem 15.1. □

Remark 15.5 (Analogous to Remark 7.7, Chapter 2). It is not known if

$$\inf_{v \in \mathcal{U}_{\text{ad}}} \int_\Omega (y_\varepsilon(x,T;v) - z_d)^2\,dx \to \inf_{v \in \mathcal{U}_{\text{ad}}} \int_\Omega (y(x,T;v) - z_d)^2\,dx \quad \text{as } \varepsilon \to 0.$$

Remark 15.6. Generally speaking, if the control appears in the coefficients containing derivatives of order $\leqslant 2m$, we shall add a *singular perturbation* εB, B being of order $2m+2$.

15.4. Non-linear Problem. Boundary Control

We consider the operator $A=A(x,t,\partial/\partial x)$ as in Section 15.2; the state $y(v)$ is given by

$$\frac{\partial}{\partial t}y(v)+Ay(v)=f \quad \text{in } Q, \quad f\in L^2(Q), \tag{15.30}$$

with

$$\frac{\partial y(v)}{\partial \nu_A}+v\beta(y(v))=0 \quad \text{on } \Sigma, \quad v\in L^\infty(\Gamma), \tag{15.31}$$

where

$$\beta(y)=|y|^{\rho-2}y, \quad \rho\geqslant 2, \tag{15.32}$$

and with

$$y(x,0;v)=y_0(x), \quad y_0\in L^2(\Omega). \tag{15.33}$$

We assume that $v\in\mathscr{U}_{\text{ad}}$, where

$$\left.\begin{aligned}&\mathscr{U}_{\text{ad}}=\{v|0<\beta\leqslant\xi_0(x,t)\leqslant v(x,t)\leqslant\xi_1(x,t) \text{ a. e. on } \Sigma,\\ &\xi_0,\xi_1\in L^\infty(\Sigma)\}.\end{aligned}\right\} \tag{15.34}$$

(15.30) ... (15.33) is solved in the following sense: we set

$$V=\{\psi|\psi\in H^1(\Omega), \psi|_\Gamma\in L^\rho(\Gamma)\}, \tag{15.35}$$

the space being endowed with the norm (7.34) Chapter 2 which is a reflexive Banach space; for $\varphi,\psi\in V$, we set

$$a_v(t;\varphi,\psi)=\sum_{i,j=1}^n\int_\Omega a_{ij}(x,t)\frac{\partial\varphi}{\partial x_j}\frac{\partial\psi}{\partial x_i}dx+\int_\Gamma v\beta(\varphi)\psi\,d\Gamma. \tag{15.36}$$

Then

Theorem 15.4. *There exists one and only one function* $y(v)$ *such that*

$$y(v)\in L^2(0,T;H^1(\Omega)), \tag{15.37}$$

$$y(v)|_\Sigma\in L^\rho(\Sigma), \tag{15.38}$$

$$\frac{d}{dt}(y(v),\psi)+a_v(t;y(t;v),\psi)=(f(t),\psi) \quad \forall\psi\in V, \tag{15.39}$$

$$y(0;v)=y_0. \tag{15.40}$$

Remark 15.7. From (15.37) we get (15.30) and formally (15.31). Then

$$\frac{\partial}{\partial t} y(v) \in L^2(0,T; H^{-1}(\Omega))$$

and (15.40) makes sense. Therefore it is reasonable to define $y(v)$ as the solution of (15.30) ... (15.33) by means of Theorem 15.4.

Proof of Theorem 15.4. This is a consequence of general results on monotone evolution equations (cf. Browder [2], Brezis [1], Lions [7]; cf. also I. M. Visik [2])[83]. It then suffices to note that $a_v(t;\varphi,\psi)$ has properties analogous to that of (7.38), (7.39), (7.40) of Chapter 2. □

Cost Function. We again consider $J(v)$ given by (15.19).
We then have

Theorem 15.5. *If* $\mathscr{U}_{\mathrm{ad}}$ *is given by* (15.34), *then there exists an optimal control for* $J(v)$ *given by* (15.19).

Proof. Let v_n be a minimising sequence. Set $y_n = y(v_n)$. By virtue of (15.34), we have

$$a_v(t;\psi,\psi) \geqslant \alpha \int_\Omega (\operatorname{grad}\psi)^2\,dx + \beta \int_\Gamma |\psi|^p\,d\Gamma.$$

Hence

$$a_v(t;\psi,\psi) \geqslant c_1 \|\psi\|^2_{H^1(\Omega)}, \qquad c_1 > 0,$$

and

$$a_v(t;\psi,\psi) \geqslant c_2 \int_\Gamma |\psi|^p\,d\Gamma, \qquad c_2 > 0,$$

c_i's independent of v. It follows that

$$y_n \text{ ranges in a bounded subset of } L^2(0,T;H^1(\Omega)), \tag{15.41}$$

$$y_n|_\Sigma \text{ ranges in a bounded subset of } L^p(\Sigma), \tag{15.42}$$

and utilising the fact that $\partial y_n/\partial t = f - A y_n$,

$$\frac{\partial y_n}{\partial t} \text{ ranges over a bounded subset of } L^2(0,T;H^{-1}(\Omega)). \tag{15.43}$$

But the injection of $H^1(\Omega) \to H^{\frac{1}{2}+\varepsilon}(\Omega)$ ($\varepsilon > 0$ fixed but arbitrary, $\varepsilon < \frac{1}{2}$) is compact. Then utilising the *compactness lemma* of Proposition 4.2, Chapter 4, Lions [3], we deduce from (15.41), (15.43) that we may extract a sequence $\{y_n, v_n\}$—again denoted by $\{y_n, v_n\}$, such that

---

[83] See Lions [13], Chapter 2.

$$\left.\begin{aligned} &y_n \to y \quad \textit{weakly in } L^2(0,T;H^1(\Omega)),\\ &\frac{\partial y_n}{\partial t} \to \frac{\partial y}{\partial t} \quad \textit{weakly in } L^2(0,T;H^{-1}(\Omega)),\\ &y_n \to y \quad \textit{strongly in } L^2(0,T;H^{\frac{1}{2}+\varepsilon}(\Omega)), \end{aligned}\right\} \tag{15.44}$$

and $v_n \to v$ weak-star in $L^\infty(\Sigma)$, $v \in \mathscr{U}_{ad}$.

But since the mapping $\psi \to \psi|_\Sigma$ is a continuous mapping from $H^{\frac{1}{2}+\varepsilon}(\Omega) \to L^2(\Gamma)$, we have,

$$y_n \to y \quad \textit{strongly in } L^2(\Sigma) \tag{15.45}$$

and hence we may assume (by extracting a new sequence) that

$$y_n \to y \quad \text{a.e. on } \Sigma, \tag{15.46}$$

and from (15.42) that

$$y_n \to y \quad \textit{weakly in } L^\rho(\Sigma). \tag{15.47}$$

Then $\beta(y_n) \to \beta(y)$ in $L^{\rho-\eta}(\Sigma)$, $\eta > 0$ and consequently $y = y(v)$. The proof is now completed in the same manner as before. □

15.5. Utilization of Convexity and the Maximum Principle for Second Order Parabolic Equations

Notation and Hypotheses. We consider the operator $A = A(x,t,\partial/\partial x)$ as in Section 15.2 [84]. We are given:

$$b_0, b_1 \in L^\infty(Q), \tag{15.48}$$

$$\mathscr{U}_{ad} = \{v \mid v \in L^\infty(Q),\ \xi_0 \leqslant v(x,t) \leqslant \xi_1,\ \xi_i \text{ given in } \mathbb{R}\}, \tag{15.49}$$

$$\left.\begin{aligned} &\text{the function } f(x,t;\lambda) \text{ which is continuous for } x \in \bar{\Omega},\\ &t \in [0,T], \quad \xi_0 \leqslant \lambda \leqslant \xi_1, \text{ the function } \lambda \to f(x,t;\lambda),\\ &\lambda \in [\xi_0, \xi_1] \text{ being } \textit{convex}. \end{aligned}\right\} \tag{15.50}$$

For $v \in \mathscr{U}_{ad}$, we denote by $f(x,t;v)$ the function $x,t \to f(x,t;v(x,t))$ and we define the state $y(v)$ by

$$\left.\begin{aligned} &\frac{\partial}{\partial t} y(v) + A\,y(v) + (b_0 v + b_1) y = f(x,t;v),\\ &y(v)|_\Sigma = 0,\\ &y(x,0;v) = y_0(x), \quad x \in \Omega, \quad y_0 \text{ given in } L^2(\Omega). \end{aligned}\right\} \tag{15.51}$$

[84] This time the fact that A is of second order plays an essential role.

Cost Function. Let g be given in $L^2(\Omega)$, $g \geqslant 0$ (a.e.); we then take as cost function,

$$J(v) = \int_{\Omega} y(x, T; v) g(x) \, dx. \tag{15.52}$$

We shall prove

Theorem 15.6. *Under the hypothesis of* (15.48), (15.49), (15.50) *and with the cost function being given by* (15.52), *there exists an optimal control* $u \in \mathscr{U}_{\mathrm{ad}}$.

For the proof of this theorem we require a classical lemma of Calculus of Variations (expressing the semi-continuity of a functional).

Lemma 15.1. *Let* $\mathcal{O}$ *be an open set in* $\mathbb{R}^N$, K *a closed, bounded, convex subset of* $\mathbb{R}^P$ *and* $\xi, k \to \Phi(\xi, k)$ *be a continuous function from* $\bar{\mathcal{O}} \times K \to \mathbb{R}$ *such that*

$$k \to \Phi(\xi, k) \quad \text{is convex on } K, \qquad \forall \xi \in \mathcal{O}. \tag{15.53}$$

Let u_n *be a sequence of functions with*

$$\left.\begin{aligned} & u_n \in L^\infty(\mathcal{O}; \mathbb{R}^P) (= (L^\infty(\mathcal{O}))^P), \\ & u_n(\xi) \in K \quad \text{a.e.}, \\ & u_n \to u \quad \text{weak star in } L^\infty(\mathcal{O}; \mathbb{R}^P). \end{aligned}\right\} \tag{15.54}$$

Then

$$\liminf \int_{\mathcal{O}} \Phi(\xi, u_n(\xi)) \, d\xi \geqslant \int_{\mathcal{O}} \Phi(\xi, u(\xi)) \, d\xi. \tag{15.55}$$

Proof. 1). Set $\psi_n(\xi) = \Phi(\xi, u_n(\xi))$; ψ_n ranges in a bounded subset of $L^\infty(\mathcal{O})$. We may extract a subsequence, again denoted by ψ_n, such that $\psi_n \to \psi_*$ weak star in $L^\infty(\mathcal{O})$. Then

$$\liminf \int_{\mathcal{O}} \Phi(\xi, u_n(\xi)) \, d\xi = \int_{\mathcal{O}} \psi_*(\xi) \, d\xi. \tag{15.56}$$

Let $\mathcal{O}_E$ = set of Lebesgue points for the functions u and ψ_*. We know that

$$\text{Measure } (\mathcal{O} - \mathcal{O}_E) = 0. \tag{15.57}$$

If we take $\xi_0 \in \mathcal{O}_E$, then

$$\left.\begin{aligned} & \Phi(\xi_0, u(\xi_0)) = \lim_{|\Delta| \to 0} \frac{1}{|\Delta|} \int_{\Delta} \Phi(\xi, u(\xi)) \, d\xi, \\ & \Delta = \text{cube} \subset \mathcal{O} \quad \text{with centre } \xi_0, \quad |\Delta| = \text{measure of } \Delta. \end{aligned}\right\} \tag{15.58}$$

We shall prove that

$$\Phi(\xi_0, u(\xi_0)) \leqslant \psi_*(\xi_0), \tag{15.59}$$

which then is true (from (15.57)) a.e. on $\mathcal{O}$, from which we deduce that

$$\int_{\mathcal{O}} \Phi(\xi, u(\xi))\, d\xi \leqslant \int_{\mathcal{O}} \psi_*(\xi)\, d\xi,$$

which when combined with (15.56) proves (15.55).

2. To prove (15.59), we use the convexity property (15.33), which implies the existence of an affine function

$$k \to L_{\xi_0}(k) \quad \text{of } K \to \mathbb{R}$$

such that

$$\left.\begin{aligned} &\Phi(\xi_0, k) \geqslant L_{\xi_0}(k) \ \ \forall k \in K, \\ &\Phi(\xi_0, u(\xi_0)) = L_{\xi_0}(u(\xi_0)). \end{aligned}\right\} \tag{15.60}$$

Then

$$\Phi(\xi_0, u(\xi_0)) = \lim_{|\Delta| \to 0} \frac{1}{|\Delta|} \int_{\Delta} L_{\xi_0}(u(\xi))\, d\xi. \tag{15.61}$$

But

$$\begin{aligned} &\Phi(\xi_0, u_n(\xi)) \leqslant \Phi(\xi, u_n(\xi)) + 0(|\Delta|) \ \ \forall \xi \in \Delta, \quad \forall n, \\ &(0(|\Delta|) \to 0 \quad \text{if } |\Delta| \to 0), \end{aligned}$$

and then using (15.60), we get

$$L_{\xi_0}(u_n(\xi)) \leqslant \Phi(\xi, u_n(\xi)) + 0(|\Delta|),$$

whence

$$\frac{1}{|\Delta|} \int_{\Delta} L_{\xi_0}(u_n(\xi))\, d\xi \leqslant \frac{1}{|\Delta|} \int_{\Delta} \Phi(\xi, u_n(\xi))\, d\xi + 0(|\Delta|). \tag{15.62}$$

But since L_{ξ_0} is affine, $L_{\xi_0}(u_n) \to L_{\xi_0}(u)$ weak star in $L^\infty(\mathcal{O})$, and hence $\frac{1}{|\Delta|} \int_{\Delta} L_{\xi_0}(u_n(\xi))\, d\xi \to \frac{1}{|\Delta|} \int_{\Delta} L_{\xi_0}(u(\xi))\, d\xi$. Since $\Phi(\xi, u_n) = \psi_n \to \psi_*$ weak star in $L^\infty(\mathcal{O})$, we deduce from (15.60) that

$$\frac{1}{|\Delta|} \int_{\Delta} L_{\xi_0}(u(\xi))\, d\xi \leqslant \frac{1}{|\Delta|} \int_{\Delta} \psi_*(\xi)\, d\xi + 0(|\Delta|).$$

If we now let $|\Delta|\to 0$, and utilise (15.60) and the fact that ξ_0 is a Lebesgue point of ψ_*, we deduce (15.59). □

Proof of Theorem 15.6. Let v_n be a minimising sequence,

$$J(v_n)\to j=\inf_{v\in\mathscr{U}_{\text{ad}}} J(v).$$

As in the proof of Theorem 15.1, we may verify that $y_n=y(v_n)$ ranges in a bounded subset of $L^2(0,T;H_0^1(\Omega))$ (note that $f(x,t;v_n)$ is bounded, in particular in $L^\infty(Q)$). Utilising (15.51) we deduce that

$$\frac{\partial}{\partial t}y_n \quad \text{ranges in a bounded subset of } L^2(0,T;H^{-1}(\Omega)),$$

and (as in the proof of Theorem 15.1) we may extract a sequence, again denoted by $\{y_n,v_n\}$, such that

$$\left.\begin{aligned} &y_n\to y \quad \textit{weakly} \text{ in } L^2(0,T;H_0^1(\Omega))\\ &\frac{\partial y_n}{\partial t}\to\frac{\partial y}{\partial t} \quad \textit{weakly} \text{ in } L^2(0,T;H^{-1}(\Omega)),\\ &y_n\to y \quad \textit{strongly} \text{ in } L^2(Q),\end{aligned}\right\} \tag{15.63}$$

$$v_n\to v \quad \textit{weak star} \text{ in } L^\infty(Q), \qquad v\in\mathscr{U}_{\text{ad}}, \tag{15.64}$$

and we may also assume that

$$f(x,t;v_n)\to f_* \quad \textit{weak star} \text{ in } L^\infty(Q). \tag{15.65}$$

Then $b_0 v_n y_n\to b_0 v y$ in $\mathscr{D}'(Q)$ for example, and consequently,

$$\left.\begin{aligned} &\frac{\partial y}{\partial t}+Ay+(b_0v+b_1)y=f_*,\\ &y\in L^2(0,T;H_0^1(\Omega)), \qquad y(x,0)=y_0(x).\end{aligned}\right\} \tag{15.66}$$

We now utilise Lemma 15.1 under the following conditions:

$$\begin{aligned} &\mathcal{O}=Q, \qquad \xi=(x,t)\in\mathcal{O},\\ &\Phi(\xi,k)=h(\xi)f(x,t;k), \qquad h\in\mathscr{D}(Q), \qquad h\geqslant 0,\\ &k\in K=[\xi_0,\xi_1],\\ &u_n=v_n.\end{aligned}$$

Hence we deduce from (15.55) that

$$\liminf\int_Q hf(x,t;v_n)\,dx\,dt\geqslant\int_Q hf(x,t;v)\,dx\,dt$$

which when combined with (15.65) proves that

$$\int_Q h f_* \, dx\,dt \geqslant \int_Q h f(x,t;v)\, dx\,dt \quad \forall h \in \mathscr{D}(Q), \qquad h \geqslant 0$$

and consequently

$$f_*(x,t) \geqslant f(x,t;v) \quad \text{a.e. in } Q. \tag{15.67}$$

But from (15.66), (15.67) and the fact that A is a *second order elliptic operator*, it follows (according to the maximum principle for second order parabolic equations—cf. for example Ladyzenskaya-Solonnikov-Ouralts'eva [1]) that

$$y_*(x,t) \geqslant y(x,t;v) \quad \text{a.e. in } Q. \tag{15.68}$$

Hence $y_*(x,T) \geqslant y(x,T;v)$ a.e. and consequently

$$\int_\Omega y_*(x,T) g(x)\, dx \geqslant J(v).$$

But from (15.63), $J(v_n) \to \int_\Omega y_*(x,T) g(x)\, dx$. Therefore $j \geqslant J(v)$ and hence $j = J(v)$ and v is optimal. □

15.6. Control of Systems Governed by Evolution Inequalities

As in Chapter 2, Section 7.5, we may consider systems whose state is given by the solution of an evolution inequality, for example by

$$\left.\begin{aligned}
\frac{\partial}{\partial t} y(v) + A y(v) = f & \quad \text{in } Q,\\
y(x,0;v) = 0 & \quad \text{in } \Omega,\\
y(v) \geqslant 0 & \quad \text{a.e. on } \Sigma,\\
\frac{\partial y}{\partial \nu_A}(v) \geqslant v & \quad \text{a.e. on } \Sigma, \qquad v \in L^2(\Sigma),\\
\left(\frac{\partial y}{\partial \nu_A}(v) - v\right) y(v) = 0 & \quad \text{a.e. on } \Sigma.
\end{aligned}\right\} \tag{15.69}$$

(For the solution of (15.69), cf. Lions-Stampacchia [1], Brezis [1]).

As in Chapter 2, Section 7.5, we may give existence theorems for optimal controls. But as yet this direction of research has been very little explored[85].

[85] Due to the very large number of problems in mechanics which have to be formulated in terms of variational inequalities (cf. Duvaut-Lions [1], [2]), it seems likely that these problems will be relevant in applications.

16. First Order Necessary Conditions

16.1. Statement of the Theorem

We adopt a notation which is similar to Chapter 2, Section 8. We assume that

$$\left.\begin{array}{l}\mathscr{U}_{\mathrm{ad}}=\{v \mid x,t\to v(x,t) \text{ is a measurable function of } Q\to\mathbb{R}^M,\\ \quad v(x,t)\in K \text{ a.e., } K=\text{closed, convex subset of } \mathbb{R}^M\}.\end{array}\right\} \tag{16.1}$$

Let $A=A(x,t,\partial/\partial x)$ be given and assume that (1.28) holds. Let

$$\left.\begin{array}{l}b(x,t;\lambda),\ f(x,t;\lambda) \text{ be continuous, bounded functions of}\\ Q\times K\to\mathbb{R};\end{array}\right\} \tag{16.2}$$

if $v\in\mathscr{U}_{\mathrm{ad}}$, $b(x,t;v)$ denotes the function

$$x,t\to b(x,t;v(x,t)).$$

The notation is similar for $f(x,t;v)$. We shall often write $b(v)$, $f(v)$. We set

$$\left.\begin{array}{l}A(v)\psi=A\psi+b(x,t;v)\psi=A\psi+b(v)\psi,\\ \Lambda(v)=\dfrac{\partial}{\partial t}+A(v).\end{array}\right\} \tag{16.3}$$

The state $y(v)$ of the system is given by

$$\frac{\partial}{\partial t}y(v)+Ay(v)+b(v)y(v)=f(v) \quad \text{in } Q, \tag{16.4}$$

$$y(v)|_{\Sigma}=0, \tag{16.5}$$

$$y(x,0;v)=y_0(x), \qquad x\in\Omega, \quad y_0 \text{ given in } L^\infty(\Omega). \tag{16.6}$$

The cost function is given by

$$\left.\begin{array}{l}J(v)=\displaystyle\int_\Omega y(x,T;v)g(x)\,dx,\\ g \text{ given in } L^\infty(\Omega), \qquad g\geqslant 0.\end{array}\right\} \tag{16.7}$$

We define the adjoint state $p(v)$ by

$$\left.\begin{array}{r}-\dfrac{\partial}{\partial t}p(v)+A^*p(v)+b(v)p(v)=0 \quad \text{in } Q,\\ p(v)=0 \quad \text{on } \Sigma,\\ p(x,T;v)=g(x), \qquad x\in\Omega.\end{array}\right\} \tag{16.8}$$

We shall prove

Theorem 16.1. *We assume that A is given with the conditions* (1.28) *being satisfied and that* (16.1), (16.2), (16.7) *hold. We assume that an optimal control u exists. Then, we have, almost everywhere in Q*

$$\begin{aligned} &p(x,t;u(x,t))[f(x,t;u(x,t))-b(x,t;u(x,t))\,y(x,t;u(x,t))] \\ &\quad = \inf_{k\in K} p(x,t;k)[f(x,t;k)-b(x,t;k)\,y(x,t;k)]. \end{aligned} \tag{16.9}$$

16.2. Proof of Theorem 16.1

The outline of the proof of this theorem is the same as that of the proof of Theorem 8.1, Chapter 2.

16.2.1. "Algebraic" Transformation

By definition, if u is optimal

$$\int_\Omega [y(x,T;v)-y(x,T;u)]\,g(x)\,dx \geqslant 0 \quad \forall v\in\mathcal{U}_{\text{ad}}. \tag{16.10}$$

Multiplying the first equation in (16.8) (with $v=u$) by $y(v)-y(u)$ and applying Green's Formula, we get:

$$\begin{aligned} &\int_\Omega [y(x,T;v)-y(x,T;u)]\,g(x)\,dx \\ &= \int_Q p(u)\Lambda(u)(y(v)-y(u))\,dx\,dt \\ &= \int_\Omega p(u)\big(\Lambda(v)\,y(v)-\Lambda(u)\,y(u)-(\Lambda(v)-\Lambda(u))\,y(u) \\ &\qquad\qquad -(\Lambda(v)-\Lambda(u))(y(v)-y(u))\big)\,dx\,dt \\ &= \int_Q p(u)[f(v)-f(u)-(b(v)-b(u))\,y(u)-(b(v)-b(u))(y(v)-y(u))]\,dx\,dt \end{aligned}$$

and hence (16.10) is equivalent to

$$\left.\begin{aligned} &\int_Q p(u)[f(v)-(b(v)-b(u))\,y(u) \\ &\qquad -(b(v)-b(u))(y(v)-y(u))]\,dx\,dt \geqslant 0 \quad \forall v\in\mathcal{U}_{\text{ad}}. \end{aligned}\right\} \tag{16.11}$$

16.2.2. Utilization of (16.11)

Same comments as in subsection 8.2.2, Chapter 2. Let

$\{x_0, t_0\} \in Q$,

$\mathcal{O}_j = G_j \times I_j$, cube with centre $\{x_0, t_0\}$, $\mathcal{O}_j \subset Q$,

G_j = cube in $\mathbb{R}^n$ with centre x_0, I_j interval in $\mathbb{R}$ with mid-point t_0,

$|\mathcal{O}_j|, |G_j| \ldots$ measure of $\mathcal{O}_j, G_j$,

$v_j = (1-\chi_j)u + \chi_j k, \quad k \in K, \quad \chi_j$ = characteristic function of $\mathcal{O}_j$.

Set

$$b_j = \frac{1}{|\mathcal{O}_j|} \int_\Omega p(u)(f(v_j) - f(u))\, dx\, dt,$$

$$c_j = \frac{1}{|\mathcal{O}_j|} \int_Q p(u)(b(v_j) - b(u))\, y(u)\, dx\, dt,$$

$$d_j = \frac{1}{|\mathcal{O}_j|} \int_\Omega p(u)(b(v_j) - b(u))(y(v_j) - y(u))\, dx\, dt.$$

From (16.11) where we take $v = v_j (\in \mathcal{U}_{ad})$ we deduce that

$$b_j - c_j - d_j \geqslant 0.$$

Choose $\{x_0, t_0\}$ to be an element in the set of Lebesgue points of functions $p(u), f(u), b(u), y(u)$. Then if we assume for the moment that

$$d_j \to 0 \quad \text{as } j \to \infty (|\mathcal{O}_j| \to 0, |I_j| \to 0), \tag{16.12}$$

we may deduce that

$$\begin{aligned} &p(x_0, t_0; u(x_0, t_0))[f(x_0, t_0; k) - f(x_0, t_0; u(x_0, t_0))] \\ &- p(x_0, t_0; u(x_0, t_0))[b(x_0, t_0; k) - b(x_0, t_0; u(x_0, t_0))] \\ &\times y(x_0, t_0; u(x_0; t_0)) \geqslant 0 \end{aligned} \tag{16.13}$$

and this is true almost everywhere on Q, whence (16.9).

16.2.3. Proof of (16.12)

We first note that (cf. Ladyzenskaya-Solonnikov-Ouralts'eva [1], Theorem 7.1, Chapter 3)

$$p(u) \in L^\infty(Q).$$

Therefore

$$|d_j| \leqslant \frac{c}{|\mathcal{O}_j|} \int_{\mathcal{O}_j} |y(v_j) - y(u)| \, dx$$

whence

$$|d_j| \leqslant \frac{c}{|\mathcal{O}_j|^{\frac{1}{2}}} \left(\int_{\mathcal{O}_j} |y(v_j) - y(u)|^2 \, dx \right)^{\frac{1}{2}}. \tag{16.14}$$

Let us set

$$\psi_j = y(v_j) - y(u). \tag{16.15}$$

We have

$$\left.\begin{aligned} A(u)\psi_j = f(v_j) - f(u) - (b(v_j) - b(u))\, y(v_j) = g_j \quad \text{in } Q, \\ \psi_j = 0 \quad \text{on } \Sigma, \\ \psi_j(\mathrm{x}, 0) = 0. \end{aligned}\right\} \tag{16.16}$$

From this we deduce that

$$\|\psi_j\|_{L^2(0,T,H^1(\Omega))} \leqslant C \|g_j\|_{L^2(Q)}. \tag{16.17}$$

But, according to Ladyzenskaya-Solonnikov-Ouralts'eva, *loc. cit.*, $y(v_j)$ ranges in a bounded subset of $L^\infty(Q)$ and consequently

$$\|g_j\|_{L^2(Q)} \leqslant C |\mathcal{O}_j|^{\frac{1}{2}}. \tag{16.18}$$

From Sobolev's Theorem, there exists an $r > 2$ such that $H^1(\Omega) \subset L^r(\Omega)$ and from (16.17), (16.18) we deduce that

$$\|\psi_j\|_{L^2(0,T,L^r(\Omega))} \leqslant C |\mathcal{O}_j|^{\frac{1}{2}}. \tag{16.19}$$

But from Holder's inequality

$$\|\psi_j\|_{L^2(\mathcal{O}_j)} \leqslant |G_j|^{\frac{1}{2}} \|\psi_j\|_{L^2(0,T,L^r(\Omega))}.$$

Combining the above with (16.19) and substituting in (16.14)

$$|d_j| \leqslant C |G_j|^{\frac{1}{2}}$$

whence the result follows. □

16.3. Remarks

Remark 16.1. In the preceeding, we may replace hypotheses (16.2) by less restrictive hypotheses of the type that the functions in question belong to appropriate L^r spaces (cf. Ladyzenskaya-Solonnikov-Ouralts'eva [1]). On the other hand, the fact that the operator A is of *second order* plays an essential role (at least in the proof we have presented).

Remark 16.2. If we conclude from (16.8) that

$$p(x,t;v)>0 \quad \text{a.e.}$$

(for example if $b\equiv 0$), then we may deduce from (16.9) that

$$\begin{aligned} &f(x,t;u(x,t))-b(x,t;u(x,t))\,y(x,t;u(x,t)) \\ &=\inf_{k\in K}\left[f(x,t;k)-b(x,t;k)\,y(x,t;u(x,t))\right]. \end{aligned} \tag{16.20}$$

Then $y(x,t;u(x,t))=y$ satisfies

$$\frac{\partial y}{\partial t}+Ay-\inf_{k\in K}\left[f(x,t;k)-b(x,t;k)\,y(x,t)\right]=0,$$

$$y=0 \quad \text{on } \Sigma,$$

$$y(x,0)=y_0(x).$$

17. Time Optimal Control

17.1. Problem Statement

We consider the system whose state is given by

$$\frac{d}{dt}y(t;v)+A(t)\,y(t;v)=f+Bv, \text{ [86]} \tag{17.1}$$

$$y(0;v)=y_0. \tag{17.2}$$

Let $\mathscr{U}_{ad}$ be a given closed, convex subset of $\mathscr{U}$ and let y_1 be a given element in H.

We assume *(controllability)*

$$\left.\begin{array}{l} \text{there exists a } v\in\mathscr{U}_{ad} \text{ such that} \\ y(\tau;v)=y_1 \quad \text{for an appropriate } \tau. \text{ [87]} \end{array}\right\} \tag{17.3}$$

The *optimal time* is defined by

$$\tau_0=\inf\tau, \quad \tau \text{ such that (17.3) holds.} \tag{17.4}$$

[86] Written symbolically. The control may be boundary control.

[87] Once for all, we assume that $\tau\in[0,T]$ and $\tau\leqslant T$. Since T may be arbitrary, this does not impose any restriction.

The problems which we shall study are:

(i) *existence of an optimal control*, that is, existence of $u \in \mathscr{U}_{ad}$ such that

$$y(\tau_0; u) = y_1; \tag{17.5}$$

(ii) *properties* of the optimal control, if it exists.

These problems are treated in Sections 17.2, 17.3.

Remark 17.1. The same problem may clearly be posed for systems governed by *nonlinear* equations.

17.2. Existence Theorem

Theorem 17.1. *We assume that* (1.1), (1.2), (17.3) *and* (2.1) *hold and that* $\mathscr{U}_{ad}$ *is bounded. Then there exists an optimal control, that is* $u \in \mathscr{U}_{ad}$, *such that* (17.5) *is satisfied.*

Proof. Let τ_n be such that

$$y(\tau_n; v_n) = y_1, \qquad v_n \in \mathscr{U}_{ad}, \tag{17.6}$$

$$\tau_n \to \tau_0. \tag{17.7}$$

Set $y_n = y(v_n)$. Since $\mathscr{U}_{ad}$ is bounded, we may verify that

$$y_n \text{ (resp. } y_n') \quad \text{ranges in a bounded set in } L^2(0,T;V) \quad (\text{resp. } L^2(0,T;V')). \tag{17.8}$$

We may then extract a subsequence, again denoted by $\{v_n, y_n\}$, such that

$$\left.\begin{aligned} v_n \to v \quad & \textit{weakly} \text{ in } \mathscr{U}, \quad v \in \mathscr{U}_{ad}, \\ y_n \to y \quad (\text{resp. } y_n' \to y') \quad & \textit{weakly} \text{ in } L^2(0,T;V) \quad (\text{resp. } L^2(0,T;V')). \end{aligned}\right\} \tag{17.9}$$

We deduce from the equality $y_n' + A y_n = f + B v_n$ that

$$y' + A y = f + B v$$

and $y(0) = y_0$ and hence

$$y = y(v). \tag{17.10}$$

But

$$y(\tau_n; v_n) - y(\tau_0; v) = y(\tau_n; v_n) - y(\tau_0; v_n) + y(\tau_0; v_n) - y(\tau_0; v). \tag{17.11}$$

Now from (17.9) $y(\tau_0; v_n) \to y(\tau_0; v)$ weakly in H and

$$\|y(\tau_n; v_n) - y(\tau_0; v_n\|_{V'} = \left\| \int_{\tau_0}^{\tau_n} y'(t; v_n)\, dt \right\|_{V'}$$

$$\leqslant (\tau_n - \tau_0)^{\frac{1}{2}} \left(\int_{\tau_0}^{\tau_n} \|y'(t; v_n)\|_{V'}^2\, dt \right)^{\frac{1}{2}} \leqslant C(\tau_n - \tau_0)^{\frac{1}{2}};$$

hence (17.11) shows that

$$y(\tau_n; v_n) - y(\tau_0; v) \to 0 \quad \text{weakly in } V'.$$

But from (17.6) it follows that $y(\tau_0; v) = y_1$. □

Remark 17.2. For a far more general existence theorem relative to a system governed by a *non-linear* parabolic operator, cf. Lions [2]—cf. also Chapter 4, Section 10.3.

Example 17.1. Take $\mathscr{U} = L^2(\Sigma)$. Let the state $y(v)$ be given by (cf. Section 3.2)

$$\left.\begin{aligned} &\frac{\partial}{\partial t} y(v) + A\, y(v) = f, \\ &\frac{\partial y}{\partial \nu_A}(v) = v, \\ &y(x, 0; v) = y_0(x). \end{aligned}\right\} \tag{17.12}$$

Theorem 17.1 may be applied to this example. Hence the theorem covers the case of boundary control.

Remark 17.3. Theorem 17.1 may be modified without any difficulty to cover the case of systems with Dirichlet boundary conditions (cf. Section 9) and where the control is exercised through the boundary.

17.3. Bang-Bang Theorem

We consider the same data and hypotheses as in Section 17.2 with

$$A \text{ independent of } t, \tag{17.13}$$

$$\mathscr{U}_{\text{ad}} = \{v \mid |v(t)| \leqslant 1 \text{ a.e.}\}. \tag{17.14}$$

Then (cf. for example Section 1.6)—A is the infinitesimal generator of a *semi-group* $G(t)$ in H.

Remark 17.4. What follows is equally applicable to the case where H is a reflexive Banach space. We only have to make the hypothesis that A is the infinitesimal generator of a semi-group in H.

We shall now prove the following result due to Fattorini [1]:

Theorem 17.2. *We assume that* (1.1), (1.2), (17.13), (17.14) *and* (17.3) *hold. Let* u *be an optimal control, that is, an element of* $\mathcal{U}_{ad}$ *satisfying* (17.5) *(we know from Theorem 17.1 that such elements exist). Then*

$$|u(t)|=1 \quad \text{a.e. on }]0,\tau_0[. \tag{17.15}$$

We shall establish a number of lemmas before giving a proof of the theorem. Let us at once isolate

Corollary 17.1. *Under the hypotheses of Theorem 17.2, there exists an optimal control which is unique.*

Proof. Let $u_1, u_2 \in \mathcal{U}_{ad}$, $y(\tau_0; u_i) = y_1$, $i=1,2$. Then

$$y(\tau_0;(1-\Theta)u_1+\Theta u_2)=y_1, \qquad 0<\Theta<1 \quad \text{and} \quad u=(1-\Theta)u_1+u_2$$

does not satisfy (17.5) unless $u_1(t)=u_2(t)$ a.e. □

Notation.

$$\left.\begin{array}{l} K_s=\{h \mid h=G*v(s),^{88} \quad v\in L^\infty(0,s;H)\}, \\ K_s(e)=\{h \mid h=G*v(s),\ v\in L^\infty(0,s;H) \text{ with support in } e\cap[0,s], \\ \quad e=\text{measurable subset of } [0,T]\}. \end{array}\right\} \tag{17.16}$$

Lemma 17.1. *With the notation of* (17.16), *we have:*

$$K_s = K_{s'} \quad \forall s,s'. \tag{17.17}$$

We then set

$$K = K_s \quad \forall s. \tag{17.18}$$

We have

$$G(s)H \subset K. \tag{17.19}$$

Proof. 1. Let $\tau > s$. Then we may easily verify

$$\left.\begin{array}{l} G*v(s) = G*\tilde{v}(\tau), \\ \text{where} \\ \tilde{v}(t) = \begin{cases} 0 & \text{if } 0<t<\tau-s, \\ v(t-(\tau-s)) & \text{if } t>\tau-s. \end{cases} \end{array}\right\} \tag{17.20}$$

---

[88] $G*v(s)=\int_0^s G(s-t)v(t)\,dt$.

Consequently

$$K_s \subset K_\tau. \tag{17.21}$$

2. Let $\tau < s$. We may again verify that

$$\left.\begin{aligned} &G*v(\tau) = G*w(s), \\ &\text{where} \\ &w(t) = v(t+\tau-s) + \frac{1}{s} G(t) \int_0^{\tau-s} G(\tau-s-t)v(t)\,dt, \end{aligned}\right\} \tag{17.22}$$

which proves the reverse inclusion of (17.21), whence (17.17) (and (17.18)).

3. We check that

$$G(s)h = G*\left(\frac{1}{s} Gh\right)(s), \quad \text{whence (17.19).} \quad \square \tag{17.23}$$

The following lemma is fundamental.

Lemma 17.2. *For almost all* $t \in e$ *we have*

$$K_t(e) = K. \tag{17.24}$$

Proof. 1. Clearly $K_t(e) \subset K_t$ and hence it suffices to prove that for almost all $t \in e$, $K \subset K_t(e)$. Let $h \in K_t$. Then $h \in K_s$ with s arbitrarily small—and therefore for any $t_1 < t$, we may represent h in the form (cf. (17.20))

$$h = \int_{t_1}^{t} G(t-\sigma)v(\sigma)\,d\sigma, \qquad v \in L^\infty(0,T;H). \tag{17.25}$$

Now suppose that we can find a sequence t_n such that

$$\left.\begin{aligned} &t_n < t_{n+1} < \cdots t, \qquad t_n \to t, \\ &\operatorname{mes}([t_n, t_{n+1}] \cap e) \geq \rho(t_{n+1} - t_n), \qquad \rho > 0, \\ &\frac{t_{n+1} - t_n}{t_{n+2} - t_{n+1}} \leq c. \end{aligned}\right\} \tag{17.26}$$

We shall see in part 2. of the proof that

for almost all $t \in e$, we may find a sequence t_n such that (17.26) holds. (17.27)

In (17.25) let us choose as t_1 the first element of the sequence $\{t_n\}$. From (17.25) we deduce that we may write

$$h=\sum_{n=1}^{\infty} G(t-t_{n+1}) \int_{t_n}^{t_{n+1}} G(t_{n+1}-\sigma)\, v(\sigma)\, d\sigma$$

$$=\sum_{n=1}^{\infty} \int_{e\cap t_n+1,\, t_{n+2}} \frac{1}{\operatorname{mes}(e\cap[t_{n+1},t_{n+2}])} G(t-\sigma_1)$$

$$\times G(\sigma_1-t_{n+1})d\sigma_1 \int_{t_n}^{t_{n+1}} G(t_{n+1}-\sigma_2)v(\sigma_2)d\sigma_2 = \int_0^t G(t-\sigma)w(\sigma)d\sigma,$$

where

$$w(\sigma)=\begin{cases} \dfrac{1}{\operatorname{mes}(e\cap[t_{n+1},t_{n+2}])} G(\sigma-t_{n+1}) \displaystyle\int_{t_n}^{t_{n+1}} G(t_{n+1}-\sigma_2)v(\sigma_2)d\sigma_2 \\ \text{in the set } e\cap[t_{n+1},t_{n+2}], \qquad n=1,2,\ldots; \\ 0 \text{ otherwise.} \end{cases}$$

From (17.26), we have

$$|w(\sigma)| \leqslant \frac{1}{\rho(t_{n+2}-t_{n+1})} C(t_{n+1}-t_n) \leqslant \text{constant}.$$

Hence

$$h=G*w(t), \quad w\in L^{\infty}(0,T;H), \quad w \text{ has support in } e, \text{ and } h\in K_t(e).$$

The lemma is proved with the exception of

2. *Proof of (17.27)*. This is a result in measure theory. We first define

$$e_m=\left\{\sigma \,|\, \sigma\in e, \operatorname{mes}\left(e\cap\left[\sigma-\frac{1}{k},\sigma\right]\right)\geqslant\frac{1}{2k},\ k\geqslant m\right\}$$

and

d_m = set of points of density of e_m (cf. Riesz-Nagy [1], Chapter 1, Section 6). It is known that $\operatorname{mes}\left(e-\bigcap_{m\geqslant 1} d_m\right)=0$ and hence it suffices to prove (7.27) for $t\in d_m$.

But then we may construct $\{t_n\}$, $t_n\in d_m$ $\forall n$, $t_{n+1}=t_n+s_n<t$, $s_n>0$, $s_n/s_{n+1}\leqslant e$ (in a manner such that the last property of (17.26) holds) and since $t_n\in d_m\subset e_m$ $\forall n$ there exists a $\rho>0$ such that

$$\text{measure } ([t_n,t_{n+1}]\cap e)\geqslant\rho(t_{n+1}-t_n). \quad \square$$

Lemma 17.3. *Let $u \in \mathscr{U}_{ad}$ be optimal with respect to $\{y_0, y_1\}$, that is, $y(\tau; u) = y_1$ with τ minimum. Then for any $s < \tau$, u is optimal with respect to $\{y_0, y(s; u)\}$.*

In other words, if $w \in \mathscr{U}_{ad}$ satisfies

$$y(\sigma; w) = y(s; u) \quad \textit{for some appropriate } \sigma, \tag{17.28}$$

we necessarily have $\sigma \geqslant s$.

Proof. Assume that (17.28) holds with $\sigma < s$. Then define v by

$$v(t) = \begin{cases} w(t) & \text{in }]0, \sigma[, \\ u(t+s-\sigma) & \text{in }]\sigma, T-(s-\sigma)[. \end{cases}$$

Then from (17.28):

$$y(\sigma; v) = y(\sigma; w) = y(s; u)$$

and

$$\frac{d}{dt} y(t; v) + A y(t; v) = u(t+s-\sigma) \quad \text{for } t \geqslant \sigma.$$

Hence

$$y(t-(s-\sigma); v) = y(t; u), \qquad t \geqslant s,$$

and hence

$$y(\tau-(s-\sigma); v) = y(\tau; u) = y.$$

But since $\tau-(s-\sigma) < \tau$, this contradicts the hypothesis that u is optimal with respect to $\{y_0, y_1\}$. □

Lemma 17.4. *Let $v \in \mathscr{U}_{ad}$ be such that*

$$|v(t)| \leqslant 1-\varepsilon \quad \text{almost everywhere, } \varepsilon > 0, \tag{17.29}$$

$$y(\tau; v) = y_1 \quad \text{for some appropriate } \tau. \tag{17.30}$$

Then $\tau > \tau_0$.

Proof. To prove this, we will verify that there exists a $s < \tau$ and $w \in \mathscr{U}_{ad}$ such that

$$y(s; w) = y(\tau; v) \tag{17.31}$$

(which proves that τ is not optimal). For this, we note that (17.31) may be written as

$$\int_0^\tau G(\tau-\sigma) v(\sigma) d\sigma + G(\tau) y_0 - G(s) y_0 = \int_0^s G(s-\sigma) w(\sigma) d\sigma. \tag{17.32}$$

But it may be easily verified that the left hand side of (17.32) may be written as

$$\int_0^s G(s-\sigma) \tilde{v}(\sigma) d\sigma,$$

with

$$\tilde{v}(\sigma)=v(\sigma+\tau-s)+\frac{1}{s}G(\sigma)\int_0^{\tau-s}G(\tau-s-\sigma_1)v(\sigma_1)d\sigma_1$$

$$+\frac{1}{s}\left[G(\sigma+\tau-s)y_0-G(\sigma)y_0\right]$$

and for $\tau-s$ sufficiently small, we have by virtue of (17.29) $|\tilde{v}(\sigma)|\leqslant 1$. Therefore we may take $w=\tilde{v}$. □

Proof of Theorem 17.2. Assume that (17.15) did not hold. Then there would exists $e\subset[0,\tau_0]$, measure $(e)>0$, such that

$$|u(t)|\leqslant 1-\varepsilon, \quad \varepsilon>0, \quad t\in e. \tag{17.33}$$

It follows from the fundamental Lemma 17.2, that we can find an s such that $K_s(e)=K=K_s$. In other words, there exists a $\bar{g}\in L^\infty(0,T;H)$, with support in e, such that

$$\int_0^s G(s-\sigma)\bar{g}(\sigma)d\sigma=\int_0^s G(s-\sigma)u(\sigma)d\sigma. \tag{17.34}$$

Let us introduce the control

$$v=\begin{cases}(1-\delta)u+\delta\bar{g} & \text{in } (0,\tau_0), \quad \delta>0 \text{ almost everywhere,}\\ 0 & \text{for } t>\tau_0.\end{cases} \tag{17.35}$$

We shall choose δ in a manner such that

$$|v(t)|\leqslant 1-\varepsilon_1 \quad \text{almost everywhere,} \quad \varepsilon_1>0 \quad \text{appropriate.} \tag{17.36}$$

This is possible. To see this, we note that on e, we have from (17.33):

$$|v(t)|\leqslant(1-\delta)(1-\varepsilon)+\delta|\bar{g}(t)|\leqslant 1-\varepsilon_2$$

for δ sufficiently small, and outside e,

$$|v(t)|=(1-\delta)|u(t)|\leqslant(1-\delta) \quad (\text{since } u\in\mathscr{U}_{\text{ad}}).$$

But

$$y(s;v)=(1-\delta)\int_0^s G(s-\sigma)u(\sigma)d\sigma+\delta\int_0^s G(s-\sigma)\bar{g}(\sigma)d\sigma+G(s)y_0$$

and using (17.34) we have

$$y(s;v)=y(s;u). \tag{17.37}$$

But then from Lemma 17.4 (since we have (17.36)) u is not optimal with respect to $\{y_0, y(s;u)\}$ and hence from Lemma 17.3 u is not optimal with respect to $\{y_0, y_1\}$ contradicting the hypothesis. □

Remark 17.5. It is possible to proceed further using much simpler arguments when the semi-group G is a *group*—in other words when *we can reverse time* (cf. Chapter 4, Section 1). Indeed we have the following result:

Theorem 17.3. *We assume that* $-A$ *is the infinitesimal generator of a group* $G(t)$. *We assume that*

$$\mathscr{U}_{ad}=\{v\,|\,v(t)\in H_{ad}\subset H,\ H_{ad}=\text{neighborhood of } 0 \text{ in } H\}. \tag{17.38}$$

We further assume that there exists a $v\in\mathscr{U}_{ad}$ *such that* (17.3) *holds and that there exists an optimal control* u *(that is,* $y(\tau_0;u)=y_1$, $\tau_0=$ *optimal time defined in* (17.4)).

Then

$$u(t)\in\partial H_{ad} \quad (\text{boundary of } H_{ad}), \text{ almost everywhere.} \tag{17.39}$$

Proof. If (17.39) does not hold, there exists an $e\subset[0,\tau_0]$ such that

$$\left.\begin{aligned}&\operatorname{mes}(e)>0,\\&\text{distance }(u(t),\partial H_{ad})\geqslant c_1>0 \text{ almost everywhere for } t\in e.\end{aligned}\right\} \tag{17.40}$$

Let χ_e be the characteristic function of e. Then

$$y(\tau_0;u)=y_1=G(\tau_0)\,y_0+\int_0^{\tau_0} G(\tau_0-\sigma)u(\sigma)\,d\sigma$$

may be written as—by virtue of the fact that G is a *group:*

$$y(\tau_0;u)=G(\tau_1)\,y_0+\int_0^{\tau_1} G(\tau_1-\sigma)\tilde{u}(\sigma)\,d\sigma, \tag{17.41}$$

with

$$\left.\begin{aligned}&\tilde{u}(\sigma)=u(\sigma)+\frac{1}{\operatorname{mes}(e)}\chi_e(\sigma)\,G(\sigma-\tau_1)\,[y(\tau_0;u)-y(\tau_1;u)],\\&\tau_1<\tau_2.\end{aligned}\right\} \tag{17.42}$$

We may choose τ_1 sufficiently near to τ_0 such that

$$|u(\sigma)-\tilde{u}(\sigma)|\leqslant\frac{c_1}{2}\quad\text{for } \sigma\in e$$

and therefore from (17.40), $u\in\mathscr{U}_{ad}$. Then, from (17.41)

$$y(\tau_1;\tilde{u})=y(\tau_0;u)=y_1$$

which contradicts the fact τ_0 is optimal. ☐

Theorem 17.4. *Assume that the hypotheses of Theorem* 17.3 *hold. We further assume that* H_{ad} *is convex. Then there exists an* $h_0\in H$ *such that*

$$(h_0, y(\tau_0;v)-y(\tau_0;u))\geqslant 0\quad\forall v\in\mathscr{U}_{ad}. \tag{17.43}$$

Proof 1. Define the set—which is clearly convex:

$$K=\{h\,|\,h\in H,\ h=y(\tau_0;v)-y(\tau_0;0),\ v\in\mathscr{U}_{\mathrm{ad}}\}. \tag{17.44}$$

Let us verify that

$$0 \text{ is in the } \textit{interior} \text{ of } K, \tag{17.45}$$

$$y_1-y(\tau_0;0)\ \text{(which belongs to } K\text{) is not in the interior of } K. \tag{17.46}$$

To prove (17.45), let $h\in H$ be arbitrary. We have to prove that $\varepsilon h\in K$ for $|\varepsilon|$ sufficiently small. Now

$$h=\int_0^{\tau_0} G(\tau_0-\sigma)\,G(\sigma-\tau_0)\,h\,d\sigma,$$

and therefore

$$\varepsilon h=y(\tau_0;\varepsilon G(\sigma-\tau_0)h)-y(\tau_0;0)\in K \quad \text{for } |\varepsilon| \text{ sufficiently small.}$$

We prove (17.46) by contradiction. If $y_1-y(\tau_0;0)$ were in the interior of K, for an appropriate $\varepsilon>0$ we would have

$$y_1-y(\tau_0;0)+\varepsilon(y_1-G(\tau_0)\,y_0)\in K$$

and there would exist a $v\in\mathscr{U}_{\mathrm{ad}}$ such that

$$y_1-y(\tau_0;0)+\varepsilon(y_1-G(\tau_0)\,y_0)=y(\tau_0;v)-y(\tau_0;0),$$

whence

$$y_1=y(\tau_0;0)+\int_0^{\tau_0} G(\tau_0-\sigma)\frac{1}{1+\varepsilon}\,v(\sigma)\,d\sigma=y(\tau_0;w)$$

and therefore $w(\sigma)\in$ interior of $\mathscr{U}_{\mathrm{ad}}$ contradicting (17.39).

2. The result now follows as a consequence of (17.45), (17.46) and of Corollary 6, Chapter 5, Section 9 of Dunford-Schwartz [1]. □

Remark 17.6. (17.43) may be interpreted as follows: let us introduce the *adjoint state* by

$$-\frac{dp}{dt}+A^*p=0 \quad \text{in }]0,\tau_0[, \tag{17.47}$$

$$p(\tau_0)=h_0. \tag{17.48}$$

Then (17.43) is equivalent to

$$\int_0^{\tau_0} (p(t),v(t)-u(t))\,dt\geqslant 0 \quad \forall v\in\mathscr{U}_{\mathrm{ad}}, \tag{17.49}$$

—whence we may pass to local conditions in t (Chapter 2, Theorem 2.1):

$$(p(t), h-u(t)) \geqslant 0 \quad \text{almost everywhere in }]0,\tau_0[, \ \forall h \in H_{ad}. \tag{17.50}$$

Example. If H_{ad} = unit ball in H, we would have

$$u(t) = -\frac{p(t)}{|p(t)|}. \tag{17.51}$$

18. Miscellaneous

18.1. Equations with Delay

18.1.1. Definition of the State

Let us first consider evolution equations with time delay with the control absent. A typical problem is the following[89]: let $w>0$ be given; we seek $y(t)$ satisfying

$$y \in L^2(0,T;V), \quad y' \in L^2(0,T;V')\text{—that is, } y \in W(0,T), \tag{18.1}$$

$$\left.\begin{aligned} y'(t)+A(t)y(t)+y(t-\omega)=f_1(t), \quad t>\omega, \\ y(t)=g(t) \quad \text{in }]0,\omega[, \end{aligned}\right\} \tag{18.2}$$

with f_1 given in $L^2(\omega,T;V')$ and g given in $W(0,\omega)$.[90]

Let us formulate this problem in a more convenient function space setting. We introduce the operator $M \in \mathscr{L}(L^2(0,T;V); L^2(0,T;V))$ (for example) by

$$My(t) = \begin{cases} y(t-\omega) & \text{if } t>\omega, \\ 0 & \text{if } t<\omega. \end{cases} \tag{18.3}$$

Let us define f and y_0 by

$$f(t) = \begin{cases} f_1(t) & \text{if } t>\omega, \\ g'+Ag & \text{if } t<\omega, \end{cases} \quad \text{hence } f \in L^2(0,T;V'), \tag{18.4}$$

and

$$y_0 = g(0) \quad (\text{hence } y_0 \in H). \tag{18.5}$$

---

[89] We restrict ourselves to simple examples. For far more general results in the absence of control, cf. Artola [1], Baiocchi [1]. For the case of finite-dimensional systems the book of Halanay [1] and the bibliography of this book may be consulted.

[90] This may be generalized: cf. Artola [1], Baiocchi [1].

We may replace (assuming (1.1) and (1.2) hold) (18.2) by,

$$\left.\begin{aligned} y'+Ay+My=f \quad \text{in }]0,T[,\\ y(0)=y_0. \end{aligned}\right\} \tag{18.6}$$

Indeed, in $]\omega, T[$, (18.6) reduces to the first equation of (18.2) and in $]0,\omega[$, (18.6) becomes

$$y'+Ay=g'+Ag, \qquad y(0)=g(0)$$

and from the uniqueness in Theorem 1.2, we have $y=g$ in $]0,\omega[$.

The extension of the problem of evolution with delay (18.6) to more general situations is immediate. Let

$$\left.\begin{aligned} &t\to\omega(t) \quad \text{be a bounded measurable function which}\\ &\text{is positive in } [0,T]. \end{aligned}\right\} \tag{18.7}$$

For $y\in W(0,T)$, we define

$$My(t)=\begin{cases} y(t-\omega(t)) & \text{if } t-\omega(t)\geqslant 0,\\ 0 & \text{if } t-\omega(t)<0. \end{cases} \tag{18.8}$$

We then look for a $y\in W(0,T)$ which is a solution of (18.6), M being given by (18.8).

It may be shown (Artola [1], Baiocchi [1], page 303) that *if* (1.1), (1.2) *as well as* (18.7) *hold and if* M *is given by* (18.8), *then problem* (18.6) *admits a unique solution.*

18.1.2. Control Problem

Consider now $B\in\mathscr{L}(\mathscr{U};\mathscr{H})$ (the notation being standard) and the system whose state $y(v)$ is given by

$$\left.\begin{aligned} y'(v)+A(t)y(v)+My(v)=f+Bv,\\ y(0;v)=y_0. \end{aligned}\right\} \tag{18.9}$$

Assume that the cost function is given by

$$J(v)=\int_0^T |y(t;v)-z_d|^2\,dt+(Nv,v)_{\mathscr{U}}. \tag{18.10}$$

Let $\mathscr{U}_{\text{ad}}$ be a closed, convex subset of $\mathscr{U}$ and let $u\in\mathscr{U}_{\text{ad}}$ be the optimal control. We introduce the adjoint state in the following manner: we note that

$$M\in\mathscr{L}(C^0([0,T];H);L^2(0,T;H)),$$

and hence

$$M \in \mathscr{L}(W(0,T); L^2(0,T;H)). \tag{18.11}$$

We then introduce the adjoint M^* such that

$$M^* \in \mathscr{L}(L^2(0,T;H); W(0,T)') \tag{18.12}$$

and thereby define the adjoint state $p(u)$ by

$$\left.\begin{aligned} -\frac{dp(u)}{dt} + A^*(t)p(u) + M^*p(u) = y(u) - z_d, \\ p(T;u) = 0. \end{aligned}\right\} \tag{18.13}$$

Remark 18.1. To be more accurate, if we assume, for example, that $w(t)$ is monotonic and regular such that M^* is defined by a formula analogous to M, then from Baiocchi [1], problem (18.3) admits a unique solution. Otherwise, the isomorphism obtained in Baiocchi, loc. cit. has to be transformed.

The optimal control u is characterized by

$$\int_0^T (y(t;u) - z_d, y(t;v) - y(t;u))\,dt + (Nu, v-u)_{\mathscr{U}} \geqslant 0 \quad \forall v \in \mathscr{U}_{\mathrm{ad}}.$$

Utilizing (18.13), we get,

$$(\Lambda_{\mathscr{U}}^{-1} B^* p(u) + Nu, v-u)_{\mathscr{U}} \geqslant 0 \quad \forall v \in \mathscr{U}_{\mathrm{ad}}. \tag{18.14}$$

Hence the optimal control u is determined by the simultaneous solution of (18.9) (with $v=u$), (18.13) and (18.14).

Example 18.1. For the case where $w(t)=w$, we obtain,

$$\left.\begin{aligned} y'(u) + A(t)y(u) + y(t-\omega;u) = f, \quad & t>\omega, \\ y'(u) + A(t)y(u) = f \quad & \text{in }]0,\omega[, \\ y(0;u) = y_0, & \\ -p'(u) + A^*(t)p(u) + p(t+\omega;u) = y(t;u) - z_d, \quad & 0<t<T-\omega, \\ -p'(u) + A^*(t)p(u) = y(t;u) - z_d, \quad & T-\omega<t<T, \\ p(T;u) = 0, & \end{aligned}\right\} \tag{18.15}$$

to which we must adjoin (18.14).

18.2. Spaces which are not Normable

In the general theory of systems (cf. A. V. Balakrishnan [3]) it is quite natural to consider controls belonging to spaces *which cannot be normed.* We shall present an (obvious) example of such a situation.

We define

$$\left.\begin{aligned}&\mathscr{D}'_+(V')=\mathscr{L}(\mathscr{D}_-;V'),\\&\quad\mathscr{D}_-=\text{space of functions from } C^\infty \text{ to } R_t,\\&\qquad\text{with support bounded on the right}^{91};\end{aligned}\right\}\tag{18.16}$$

$\mathscr{D}'_+(V')$ is the space of distributions with values in V' and with support bounded on the left (that is, if $f\in\mathscr{D}'_+(V')$, then $f=0$ for $t<t_F$).

Take

$$\mathscr{U}=\mathscr{D}'_+(V'),\tag{18.17}$$

and

$$\mathscr{U}_{\text{ad}}=\text{closed subset of } \mathscr{D}'_+(V').\tag{18.18}$$

If we assume that the function $t\to a(t;\varphi,\psi)$ is defined on $\mathbb{R}$ and is infinitely differentiable, then (Lions [1]) there exists a unique $y(v)$ in $\mathscr{D}'_+(V')$ such that

$$\frac{d}{dt}y(v)+A(t)\,y(v)=f+v.\tag{18.19}$$

Let us define the *affine* cost function

$$J(v)=(g,y(v))=\int_{-\infty}^{+\infty}(g(t),y(t;v))\,dt,\tag{18.20}$$

where $g\in\mathscr{D}(V)$ (that is, g is infinitely differentiable with values in V and with support bounded on the right)[92]. If u is an optimal control, assumed to exist, then if

$$J(u)\leqslant J(v)\quad\forall v\in\mathscr{U}_{\text{ad}},$$

we introduce the adjoint state $p(u)$ by

$$\left.\begin{aligned}&-\frac{d}{dt}p(u)+A^*p(u)=g,\\&p(u)\in\mathscr{D}_-(V),\end{aligned}\right\}\tag{18.21}$$

and "$J(u)\leqslant J(v)$" $\forall v\in\mathscr{U}_{\text{ad}}$ is equivalent to

$$\int_{-\infty}^{+\infty}p(t;u)(v-u)\,dt\geqslant 0\quad\forall v\in\mathscr{U}_{\text{ad}}.\tag{18.22}$$

---

[91] Endowed with the Schwartz topology; cf. L. Schwartz [2]; "$\varphi\in\mathscr{D}_-$" signifies that φ is C^∞ and $\varphi(t)=0$ for $t\geqslant t_\varphi$.

[92] See L. Schwartz [3].

Notes

The results of section 1 have been recapitulated here for the convenience of the reader. The existence proof follows the method of Faedo-Galerkin. Other proofs may be found in Lions [3] (energy method) and in Lions-Magenes [1], Chapter 3 (elliptic regularization).

Also consult A. Friedman [1], Ladyzenskaya-Solonnikov-Uralts'eva [1].

Here we have not studied evolution inequalities[93] (see bibliographical notes of Chapter 1).

Section 1.6 presents material on semi-groups which is absolutely indispensable for the study of what follows in the chapter. For this theory consult the books of Hille-Phillips [1], T. Kato [1], K. Yosida [1]. Clearly semi-group methods are related to methods of Laplace Transform in t for which we refer the reader to Garnir [1], Lions [4].

Section 1 is only concerned with *linear* problems; cf. sections 15 and 16 for nonlinear problems.

Section 2 is an extension to systems described by partial differential evolution equations of results which are well known for systems described by ordinary differential equations (cf. R. E. Kalman [1], [2]); in this case the set of equations given in Theorem 2.1 are generally referred to as "two-point boundary value problems". The novel phenomenon exhibited here is that there are additional lateral boundary conditions (lateral inequalities).

In sections 4 and 5 we present an extension of classical synthesis (or feedback) in the case of finite dimensions (cf. Athans-Falb [1], Bucy [1], Bucy-Joseph [1], Lee-Markus [1]) to the infinite dimensional case with unbounded operators.

An extension to infinite dimensions, but for bounded operators, was presented in Falb-Kleinman [1], P. Faurre [1]. Here we encounter serious difficulties to justify the formal calculations of section 4.3. This is due to the nonlinear character (quadratic nonlinearity of "Riccati type") of the equations. The method we adopt to define $P(t)$ (and $r(t)$) is given here for the first time in this framework; cf. other examples in Balakrishnan-Lions [1], Bensoussan [3]. A direct study of the nonlinear parabolic integro-differential equation characterizing the kernel $P(x,\xi,t)$ (cf. Section 5)[94] is made in Da Prato [1], Temam [1]. For a

[93] Problems of control for systems governed by inequalities or evolution inequalities are still largely open. See section 15.6 for some brief remarks.

[94] By "Direct Study" we mean an approach not using "variational theory". There are some technical differences in the hypotheses needed in the various approaches.

detailed numerical analysis of this type of equations, we refer to Nédelec [1].

There are other examples where the Calculus of Variations is useful to solve nonlinear partial differential equations. In this connection the work of Fleming [6] should be mentioned. There Fleming, following an idea of Hopf [1] and Conway-Hopf [1] considers a nonlinear hyperbolic equation as the Hamilton-Jacobi equation of a stochastic calculus of variations problem.

In finite dimensions, the case where $J(v)$ is not coercive has been considered by P. Faurre [2].

Clearly y and p may be expressed in terms of the "fundamental solution" of the parabolic evolution operator $\frac{\partial}{\partial t} + A(t)$ and in particular when $A(t)=A$ the infinitesimal generator of a semi-group, y and p may be expressed in terms of the semi-group. Additional results along these lines may be found in the work of A. V. Balakrishnan [1], [2], and E. I. Axelband [1]. All classical methods may equally well be used: expansion in eigen-functions, Laplace and Mellin Transforms, etc. cf. Y. Sakawa [1], [2].

Section 6 is also an extension to infinite dimensions of results given by Kalman [1], [2] (as $T \to +\infty$). As in sections 4 and 5, the methods used to define P_∞ and r_∞ are different from those used in the finite dimensional case. In the scalar case both in linear and nonlinear situations, a more precise study has been made by Kalman-Bucy [1].

The Riccati integro-differential equations encountered in sections 4, 5, 6 have previously been introduced (in a less general setting) by a number of authors (we refer the reader to H. Erzberger-M. Kim [1], S. G. Tzafestas and J. M. Nightingale [1][95], P. K. C. Wang [1] and the bibliography of the last mentioned paper). However the methods used by Erzberger-Kim and Wang lead to the following difficulty (which has not been overcome in this study): one looks a priori for an operator P and a function r such that the identity $p=Py+r$ holds; this requires a direct study of the Riccati integro-differential equation. On the other hand, in this study P and r are directly defined and it is then shown that the identity $p=Py+r$ holds and we finally show that P and r satisfy equations which naturally arise. In this manner existence of a global solution of $P(t)$ and of the kernel $P(x,\xi,t)$ (by applying Schwartz's kernel theorem) is assured.

The Riccati equation is connected to the extremal field theory of Calculus of Variations (see Hestenes [1], invariant embedding method of Bellman, cf. Bellman-Kalaba-Wing [1], the sweep method of Gelfand,

[95] I wish to thank G. de Calan for giving me this reference.

cf. Gelfand-Fomin [1], section 31). Analogous methods are often used in astrophysics, notably (invariance method of V. A. Ambarzumian [1], S. Chandrasekhar [1], Riccáti transformation, cf. for example Hummer-Rybicki [1]).

The results of section 4.5 demonstrate the convergence of the Faedo-Galerkin approximation technique (the method is classical for evolution equations without control; cf. S. Faedo [1], J. W. Green [1], Lions [3]).

For systems governed by ordinary differential equations, the results of Section 4.7 are classical; cf. R. E. Kalman [1], [2]. Section 4.8 is (as emphasized in the text) formal. We have utilized dynamic programming (Bellman [1], cf. also N. Moiseyev [1]); analogous results were presented in Butkovskii [1], Wang [1]. The results of Section 4.8 can without doubt be justified, but it is certainly difficult to use the Hamilton-Jacobi equation (4.98) which is a partial differential-functional differential equation. Equations of this type have appeared in a different context in Donsker-Lions [1] and in a somewhat similar context in the work of W. H. Fleming [1], [2] and of R. Mortensen [1], [2]. These equations may be solved by utilizing integration in function spaces as a basic tool (cf. Donsker [1])[96,97]. The backward uniqueness property of Section 7.2 has been studied in Lions-Malgrange [1], by adapting a method due to Carleman (cf. Treves [1], Chapter 2, Section 2.2). A different method which depends on convexity has been used by S. Agmon-L. Nirenberg [1], [2]. Cf. also S. G. Krein [1], [2], S. Zaidman [1].

The results of Sections 8 and 9 are presented in a more general framework in Lions-Magenes [1], Chapter 6 (where amongst other things the case of a parabolic operator of order $2m$ with general boundary conditions is studied; however unilateral problems and decoupling have not been studied) and in Lions [1], note 3.

A proof of property (9.17) has not been given in the text. We refer the reader to Lions-Magenes [1], Chapter 4. A different proof, which is more elementary but has content which is less general may be given by using Rellich's formula for the elliptic case (see J. Necas [1]) as a point of departure and then using Laplace Transform in t.

For examples, the reader may consult the book of Butkovskii [1] which we have already mentioned and the papers of W. L. Brogan [1], D. M. Wiberg [1] (these authors do not use the method of transposition to pose boundary value problems with non-zero boundary conditions—but they introduce directly distributions carried by the boundary of Q; this procedure is complicated when the dimension of the space is >1).

[96] In the present context if one starts with a representation of the state as an integral in function space this is natural.

[97] In particular, using the work of E. Nelson [1] we arrive at similar problems for systems governed by the Schroedinger equation.

The results of section 10.3 are due to H. O. Fattorini [5] (who proves controllability directly, without using duality). Other results for abstract parabolic operators and for parabolic operators of arbitrary order in 1 dimension (and hyperbolic operators) may be found in the work of Fattorini [5]. For numerical problems related to Section 10.2, cf. R. Lattes-J. L. Lions [1]. In this book the method of quasi-reversibility is introduced. It would be interesting to extend this method to cover the case of "super-uniqueness" of Section 10.3.

A study of controllability via initial conditions which is far more advanced than that presented in Section 10.3 may be found in Fattorini [2], [3], [4].

See also W. Miranker [1].

The results of Section 11.4 are due to A. Bensoussan [1] where a numerical study of this type of problem may be found (cf. also a different method in Lattes-Lions [1]).

We may also explicitly write the state of the system by using the fundamental solution of the operator $\frac{\partial}{\partial t} + A(t)$. The problem of controllability considered in Section 11.3 (for example) may then be posed in terms of operators. This route has been followed by R. Conti [1], [2], S. K. Mitter [1]—one of their objectives being the extension to infinite dimensions of a criterion of Antosiewicz [1] relative to finite dimensions.

The system (for example)

$$
(*)\qquad
\begin{aligned}
\frac{\partial y(v)}{\partial t} + A\,y(v) &= 0 \quad \text{in } Q,\\
y(v)|_{\Sigma} &= v \quad \text{on } \Sigma,\\
y(x,0;v) &= y_0(x) \quad \text{in } Q,
\end{aligned}
$$

with observation $C\,y(v)$ is said to be observable if

$$v=0, \qquad C\,y(0)=0 \;\Rightarrow\; y_0=0.$$

(cf. the work of Kalman already cited and L. Markus [1]). For example, if $C\,y(v) = \frac{\partial}{\partial v_A}\,y(v)$ then the system (*) is observable from the uniqueness of the Cauchy problem. When the system is not observable, we may, theoretically, make it observable by passing to the quotient space (but by passing to a sub-space which is generally infinite-dimensional and which may be very difficult to characterize explicitly); cf. A. V. Balakrishnan [3].

The result of Theorem 12.1 is a variant of classical results due to Mandelbrojt [1], Fenchel [1]. Functions which satisfy (12.6) have been systematically utilized by J. J. Moreau [1] where Theorem 12.1 (with a different statement) may be found. The presentation given in the text is by far not the most general; consult the work of Minty, Moreau (cf. bibliography of Moreau [1]) and Rockafellar [1], [22], [3] as well as the course notes of J. Cea [1]. For Section 12.2 see the work of Russell [2]. A result more general than that of Theorem 12.1 is given in Brezis [3]. Other results and applications may be found in A. Bensoussan [3].

Theorem 12.1 is clearly also applicable to elliptic situations of Chapters 1 and 2, as well as to hyperbolic problems considered in the next chapter.

For other applications of duality the reader may consult J. D. Pearson [1].

The results of Section 13.1 are one of the many possible adaptations of the classical theory of Lagrange multipliers to infinite dimensions. There we have followed an exposition due to J. P. Aubin [1]. The fundamental work is due to F. John [1] and Kuhn-Tucker [1].

Additional results (partially heuristic) related to Section 14.3 are given in Lions [12], where the non-linear feedback law is expanded in a Taylor series, the coefficients of which satisfy a sequence of integro-partial differential equations.

Results more general than that presented in Sections 15.2 and 15.4 may be found in Lions [2] (for a different aspect see Lions [8]). Theorem 15.6 is due to W. Fleming [5].

A variant of Theorem 16.1 may be found in the paper of Fleming already cited (other results may also be found there). For the regularity of the solution of parabolic equations (which plays a fundamental role—this is really the only difficulty of the problem) consult also Oleinik-Ilin-Kalashnikov [1]. As in the elliptic case (see notes of Chapter 2) we should be wary of "axiomatic" proofs of theorems of the type 16.1 (and of the type "Pontryagin's maximum principle) where the only difficulty of the problem is taken to be part of the hypotheses. Theorem 17.1 has been given in a less general framework in Yu. Egorov [1]; cf. also Vallee [1].

Theorem 17.2 is due to H. O. Fattorini [1]. The work of A. Friedman [1] on this subject contains an error (T cannot be kept fixed in Lemma 2, attributed to Fattorini by mistake). The work of Yu. V. Egorov [2] contains a detailed study of this problem, but we were unable to follow all points of the proof given by the author. The results are probably all correct[98]. For other aspects of this question see A. V. Balakrishnan [1], P. L. Falb [1].

---

[98] Mrs. O. A. Oleinik has brought to our attention the comments of Yu. V. Egorov in *Referativni Journal* regarding the work of Friedman [1].

Theorem 17.2 is a (partial) extension of finite-dimensional "bang-bang" results: Bellman-Glicksberg-Gross [1], Halkin [1], La Salle [1], Markus [2].

Theorem 17.3 is due to Friedman [1] and Fattorini [6]. We have followed the method of Fattorini (in Fattorini [6] a more general result for $A(t)$ depending on t can be found).

We may consider controls belonging to more general spaces than those considered here. This could be done by immersion of the control space into Sobolev spaces of *Hilbertian* structure having "negative exponents". A study of a case where the control is an element of a space of Gevrey type may be found in Lions-Magenes [2]. Section 18.2 presents an example (trivial) where the control is a distribution. Problems where the control and observation contain noise have not been considered. For a systematic study of this problem and control problems for systems described by stochastic partial differential equations, cf. A. Bensoussan [2], [3] and H. J. Kushner [1]. For the finite dimensional case, cf. Kalman-Bucy [1].

We also have not initiated a study of problems for systems governed by partial differential equations where there are two different controls, antagonistic (game theory) or not. Certain results of Varaiya [1] can most probably be adapted to this case.

We would also like to bring to the attention of the reader the work of Mischenko [1], Psenichny [1], [2] on differential games. It would be interesting to extend these results (as far as possible) to systems governed by partial differential equations.

The case where there are two non-antagnostic controls lead to problems where the *state of the system* itself is defined by one of (the very many) *unilateral problems* considered in this book. Such situations lead to the consideration of the important problem of *optimal control of systems governed by unilateral problems* (problem touched upon in Chapter 2, Section 7.5). A systematic study of this important problem remains to be done.

Finally, we have not considered control problems for systems governed by multi-valued differential equations with unbounded operators as coefficients (cf. for the "bounded" base, C. Castaing [1]).

Unilateral problems encountered in this chapter bear a certain similarity to problems in dynamic plasticity (cf. Cristescu [1]) (see also the comments in Chapter 1).

CHAPTER IV

Control of Systems Governed by Hyperbolic Equations or by Equations which are well Posed in the Petrowsky Sense

1. Second Order Evolution Equations

1.1. Notation and Hypotheses

We use the same notation as in Chapter 3, Section 1. Therefore we consider Hilbert spaces V, H with

$$V \subset H, \quad V \text{ dense in } H, \quad V \text{ separable.} \tag{1.1}$$

Identifying H with its dual and denoting V' to be the dual of V, we have

$$V \subset H \subset V'. \tag{1.2}$$

We consider a family of operators $A(t) \in \mathscr{L}(V; V')$. We set

$$(A(t)\varphi, \psi) = a(t; \varphi; \psi) \quad \forall \varphi, \psi \in V, \tag{1.3}$$

and we shall assume that

$$\left.\begin{array}{l}\forall \varphi, \psi \in V, \text{ the function } t \to a(t; \varphi, \psi) \text{ is continuously}\\ \text{differentiable in } [0, T];\end{array}\right\} \tag{1.4}$$

$$a(t; \varphi, \psi) = a(t; \psi, \varphi) \quad \forall \varphi, \psi \in V; \tag{1.5}$$

$$\left.\begin{array}{l}\text{there exists a } \lambda \in \mathbb{R} \text{ such that}^1\\ a(t; \varphi, \varphi) + \lambda|\varphi|^2 \geqslant \alpha\|\varphi\|^2 \quad \forall \varphi \in V, \quad \alpha > 0, \quad t \in [0, T].\end{array}\right\} \tag{1.6}$$

Remark 1.1. The preceeding hypotheses are not the most general which ensure the validity of what follows. But the *methods* used in the sequel are general. For other considerations (relative to the present section) consult Lions [3].

---

[1] $|\varphi| = \|\varphi\|_H, \quad \|\varphi\| = \|\varphi\|_V.$

1.2. Problem Statement. An Existence and Uniqueness Result

We consider the following evolution problem:

$$\left.\begin{array}{l}\text{find } y \text{ satisfying,}\\ y\in L^2(0,T;V), \quad \dfrac{dy}{dt}\in L^2(0,T;H),\end{array}\right\} \tag{1.7}$$

$$\frac{d^2y}{dt^2}+A(t)y=f \quad \text{in }]0,T[, \tag{1.8}$$

with

$$f \text{ given in } L^2(0,T;H), \tag{1.9}$$

and with the initial conditions:

$$y(0)=y_0, \quad y_0 \text{ given in } V, \tag{1.10}$$

$$\frac{dy}{dt}(0)=y_1, \quad y_1 \text{ given in } H. \tag{1.11}$$

Remark 1.2. From (1.8) it follows that

$$\frac{d^2y}{dt^2}=f-A(t)y\in L^2(0,T;V')$$

whence it follows that, in particular, (cf. Chapter 3, Theorem 1.1) $\dfrac{dy}{dt}$ is (a.e. equal to a function) continuous from $[0,T]\to V'$ and from (1.7) that y is, in particular, continuous from $[0,T]\to H$. Hence (1.10), (1.11) make sense.

We shall prove in the sequel

Theorem 1.1. *Under the hypotheses of* (1.4), (1.5), (1.6), *problem* (1.7)...(1.11) *admits a unique solution. The mapping*

$$\{f,y_0,y_1\}\to\left\{y,\frac{dy}{dt}\right\} \tag{1.12}$$

is a linear continuous map of $L^2(0,T;H)\times V\times H\to L^2(0,T;V)\times L^2(0,T;H)$.

Remark 1.3. We have an even stronger result (cf. Lions-Magenes [1], Chapter 3):

$$\left.\begin{array}{ll} y & \text{is continuous from } [0,T]\to V,\\ \dfrac{dy}{dt} & \text{is continuous from } [0,T]\to H.\end{array}\right\} \tag{1.13}$$

Remark 1.4. An *essential difference* between the present problem and the evolution problem treated in Chapter 3, Section 1, is that now we can *reverse the direction of time* whereas this was impossible in the parabolic case. In other words, if we consider problem (1.7), (1.8) (with (1.9)) together with

$$y(T)=\bar{y}_0 \quad \text{given in } V, \tag{1.14}$$

$$\frac{dy}{dt}(T)=\bar{y}_1 \quad \text{given in } H, \tag{1.15}$$

(in place of (1.10), (1.11)), then as in Theorem 1.1, it still admits a unique solution which depends continuously on the data.

1.3. Proof of Uniqueness

Let y satisfy conditions (1.7), ..., (1.11) with $f=0$, $y_0=0$, $y_1=0$. Let $s\in]0,T[$. We set

$$\psi(t)=\begin{cases}-\int_t^s y(\sigma)d\sigma, & t<s,\\ 0, & t>s.\end{cases}$$

We have,

$$\int_0^T \left(\frac{d^2y}{dt^2}+A(t)y,\psi(t)\right)dt=0$$

and integrating by parts, which is permissible, we get[2]

$$\int_0^s [a(t;\psi',\psi)-(y',y)]dt=0. \tag{1.16}$$

We set

$$\frac{d}{dt}a(t;\varphi,\psi)=a'(t;\varphi,\psi) \quad \forall\,\varphi,\psi\in V. \tag{1.17}$$

Then (1.16) gives us

$$\int_0^s \frac{d}{dt}[a(t;\psi,\psi)-|y(t)|^2]dt-\int_0^s a(t;\psi,\psi)dt=0,$$

[2] In general $z'=dz/dt$.

hence

$$a(0;\psi(0),\psi(0))+|y(s)|^2=\int_0^s a'(t;\psi,\psi)\,dt$$

and hence (utilising (1.6))

$$\|\psi(0)\|^2+|y(s)|^2\leqslant c_1\left(\int_0^s \|\psi\|^2\,dt+|\psi(0)|^2\right). \tag{1.18}$$

But if we set

$$w(t)=\int_0^t y(\sigma)\,d\sigma,$$

(1.18) may be written as

$$\|w(s)\|^2+|y(s)|^2\leqslant c_1\left(\int_0^s \|w(t)-w(s)\|^2\,dt+|w(s)|^2\right)$$

whence

$$(1-2c_1 s)\,\|w(s)\|^2+|y(s)|^2\leqslant c_2\int_0^s (\|w(t)\|^2+|y(t)|^2)\,dt. \tag{1.19}$$

If we now choose s_0 with $1-2c_1 s_0=\frac{1}{2}$ for example, then (1.19) gives

$$\|w(s)\|^2+|y(s)|^2\leqslant c_3\int_0^s (\|w(t)\|^2+|y(t)|^2)\,dt \quad \text{for } 0\leqslant s\leqslant s_0;$$

and therefore $y=0$ in $[0,s_0]$. But the length s_0 being independent of the choice of the origin, we deduce that $y=0$ in $[s_0,2s_0]$, ... whence uniqueness. □

1.4. Proof of Existence

(Compare with Chapter 3, Section 1.4).

As in Chapter 3, (1.14), let $w_1,\dots,w_m,\dots$ be a basis in V. We consider

$$\left.\begin{aligned} y_{0m}&=\sum_{i=1}^m \xi_{im}^0 w_i, \quad y_{0m}\to y_0 \text{ in } V \text{ as } m\to\infty,\\ y_{1m}&=\sum_{i=1}^m \xi_{im}^1 w_i, \quad y_{1m}\to y_1 \text{ in } H \text{ as } m\to\infty. \end{aligned}\right\} \tag{1.20}$$

We define $y_m(t)$, an "approximate solution" to the problem by,

$$\left.\begin{aligned}&\frac{d^2}{dt^2}(y_m(t), w_j)+a(t; y_m(t), w_j)=(f(t), w_j), \quad 1\leqslant j\leqslant m,\\ &y_m(t) = \sum_{i=1}^{m} g_{im}(t) w_i,\\ &y_m(0)=y_{0m},\\ &y'_m(0)=y_{1m}.\end{aligned}\right\} \tag{1.21}$$

We may verify without any difficulty that this system of m linear differential equations admits a unique solution. To obtain *a priori* estimates of $y_m(t)$, we multiply the first equation in (1.21) by $g'_{jm}(t)$ and sum over j. We obtain

$$(y''_m(t), y'_m(t))+a(t; y_m(t), y'_m(t))=(f(t), y'_m(t));$$

whence

$$\frac{d}{dt}[|y'_m(t)|^2+a(t; y_m(t), y_m(t))]-a'(t; y_m(t), y_m(t))=2(f(t), y'_m(t)).$$

We deduce

$$|y'_m(t)|^2+a(t; y_m(t), y_m(t))=|y_{1m}|^2+a(0; y_{0m}, y_{0m}) + \int_0^t a'(\sigma; y_m(\sigma), y_m(\sigma))\,d\sigma+2\int_0^t (f(\sigma), y'_m(\sigma))\,d\sigma$$

whence

$$\left.\begin{aligned}&|y'_m(t)|^2+\|y_m(t)\|^2\leqslant c(\|y_{0m}\|^2+|y_{1m}|^2)+c\,|y_m(t)|^2\\ &\qquad +c\int_0^t (\|y_m(\sigma)\|^2+|f(\sigma)|\,|y'_m(\sigma)|)\,d\sigma.\end{aligned}\right\} \tag{1.22}$$

But

$$|y_m(t)|\leqslant |y_{0m}|+\int_0^t |y'_m(\sigma)|\,d\sigma$$

and if we set

$$|y_m(t)|^2+\|y_m(t)\|^2=Y_m(t): \tag{1.23}$$

$$Y_m(t)\leqslant c\Big[\|y_{0m}\|^2+|y_{1m}|^2+\int_0^t |f(\sigma)|^2\,d\sigma\Big]+c\int_0^t Y_m(\sigma)\,d\sigma. \tag{1.24}$$

From Gronwall's lemma, we deduce that

$$y_m(t)\leqslant C=\text{constant independent of } m, \quad t\in[0,T]. \tag{1.25}$$

It follows from (1.25) that

$$\left.\begin{array}{l}\text{as } m\to\infty,\\ \quad y_m \text{ ranges over a bounded subset of } L^\infty(0,T;V),\\ \dfrac{dy_m}{dt} \text{ ranges over a bounded subset of } L^\infty(0,T;H).\end{array}\right\}\tag{1.26}$$

Therefore we may extract a sequence y_μ such that

$$\left.\begin{array}{ll} y_\mu\to z & \text{in } L^\infty(0,T;V)\\ & \quad\text{(in the weak topology of the dual of } L^1(0,T;V)),\\ \dfrac{dy}{dt}\to\dfrac{dz}{dt} & \text{in } L^\infty(0,T;H)\\ & \quad\text{(in the weak topology of the dual of } L^1(0,T;H)).\end{array}\right\}\tag{1.27}$$

Then, in particular, $y_\mu(0)\to z(0)$ and since from (1.20)

$$y_\mu(0)=y_{0\mu}\to y_0 \quad \text{in } V,$$

we have

$$z(0)=y_0. \tag{1.28}$$

Let $\varphi\in\mathscr{C}^1[0,T]$, $\varphi(T)=0$, $\varphi_j(t)=\varphi(t)w_j$. Let us multiply the first equation in (1.21) by $\varphi(t)$ (we take $m=\mu>i$). We get,

$$\int_0^T [(-y'_\mu,\varphi'_j)+a(t;y_\mu,\varphi_j)]\,dt=\int_0^T (f,\varphi_j)\,dt+(y_{1\mu},\varphi_j(0))$$

and by virtue of (1.27) we may pass to the limit in μ, giving us

$$\int_0^T [-(z',w_j)\varphi'+a(t;z,w_j)\varphi]\,dt=\int_0^T (f,w_j)\varphi\,dt+(y_1,w_j)\varphi(0). \tag{1.29}$$

Taking $\varphi\in\mathscr{D}(]0,T[)$, we deduce that

$$\frac{d^2}{dt^2}(z,w_j)+a(t;z,w_j)=(f,w_j)$$

and this for $\forall j$, hence

$$\frac{d^2z}{dt^2}+A(t)z=f.$$

We then deduce that

$$(z'(0),w_j)\varphi(0)=(y_1,w_j)\varphi(0)\ \ \forall w_j,$$

hence

$$z'(0)=y_1,$$

which proves that y is a solution to the problem. □

Remark 1.5. In fact we have the stronger result:

$$y\in L^\infty(0,T;V), \qquad \frac{dy}{dt}\in L^\infty(0,T;H).$$

Indeed, as we have indicated (without proof) in Remark 1.3, we may push this result a little further.

Remark 1.6. We also have the following result:

$$\left.\begin{aligned} &y_m\to y \quad \textit{strongly} \text{ in } L^2(0,T;V),\\ &\frac{dy_m}{dt}\to\frac{dy}{dt} \quad \textit{weakly} \text{ in } L^2(0,T;H).\end{aligned}\right\} \tag{1.30}$$

Indeed, we already have weak convergence. We then consider

$$X_m=\int_0^T[a(t;y_m-y,y_m-y)+|y_m'(t)-y'(t)|^2]\,dt$$

We verify that $X_m\to 0$ whence we deduce (1.30).

1.5. Examples (I)

We use the same notation as in Section 1, Chapter 3. We consider functions a_{ij} satisfying

$$\left.\begin{aligned} &a_{ij}\in\mathscr{C}^1([0,T];L^\infty(\Omega)),\\ &a_{ij}=a_{ji}\ \forall\, i,j,\\ &\sum_{i,j=1}^n a_{ij}(x,t)\xi_i\xi_j\geqslant\alpha(\xi_1^2+\cdots+\xi_n^2), \qquad \alpha>0, \qquad \xi_i\in\mathbb{R}.\end{aligned}\right\} \tag{1.31}$$

Let V be a sub-space of $H^1(\Omega)$, with

$$H_0^1(\Omega)\subset V\subset H^1(\Omega) \quad \text{(with the possibility of equality).} \tag{1.32}$$

We take $H=L^2(\Omega)$.

For $\varphi,\psi\in V$, we set

$$a(t;\varphi,\psi)=\sum_{i,j=1}^n\int_\Omega a_{ij}(x,t)\frac{\partial\varphi}{\partial x_j}\frac{\partial\psi}{\partial x_i}\,dx. \tag{1.33}$$

We are then in a position to apply Theorem 1.1. Concrete results depend on the choice of V. Let us give three examples.

Example 1.1. $V=H_0^1(\Omega)$.

Then Theorem 1.1 implies the existence and uniqueness of y satisfying

$$y, \frac{\partial y}{\partial x_i}, \frac{\partial y}{\partial t} \in L^2(Q), \qquad Q=\Omega\times]0,T[, \tag{1.34}$$

$$\frac{\partial^2 y}{\partial t^2} + A\left(x,t,\frac{\partial}{\partial x}\right) y = f \quad \text{in } Q, \tag{1.35}$$

where

$$A\psi = A\left(x,t,\frac{\partial}{\partial x}\right)\psi = A(t)\psi = -\sum_{i,j=1}^{n} \frac{\partial}{\partial x_i}\left(a_{ij}(x,t)\frac{\partial\psi}{\partial x_j}\right), \tag{1.36}$$

with

$$y=0 \quad \text{on } \Sigma=\Gamma\times]0,T[, \tag{1.37}$$

and

$$\left.\begin{aligned} y(x,0) &= y_0(x), \qquad y_0\in H_0^1(\Omega). \\ \frac{\partial y}{\partial t}(x,0) &= y_1(x), \qquad y_1\in L^2(\Omega). \end{aligned}\right\} \tag{1.38}$$

Example 1.2. $V=H^1(\Omega)$.

In this case, Theorem 1.1 implies the existence and uniqueness of y satisfying (1.34), (1.35) and

$$\frac{\partial y}{\partial \nu_A} = 0 \quad \text{on } \Sigma \; [3] \tag{1.39}$$

and (1.38) (with $y_0\in H^1(\Omega)$).

Example 1.3. We now choose $\Gamma_0\subset\Gamma$ (with positive measure) and

$$V=\{\psi\,|\,\psi\in H^1(\Omega),\ \psi=0 \text{ on } \Gamma_0\}. \tag{1.40}$$

Then Theorem 1.1 implies the existence and uniqueness of y satisfying (1.34), (1.35) and

$$y=0 \quad \text{on } \Gamma_0\times]0,T[,$$

$$\frac{\partial y}{\partial \nu_A} = 0 \quad \text{on } \Gamma_1\times]0,T[, \qquad \Gamma_1=\Gamma-\Gamma_0,$$

with (1.38) (where $y_0\in V$).

---

[3] Formal verification as in Chapter 3, Example 1.2; for the justification cf. Lions-Magenes [1], Chapter 5.

Clearly by taking $\Gamma_0=\Gamma$ (resp. $\Gamma_0=\varphi$) we get back Example 1.1 (resp. Example 1.2).

Remark 1.7. The operator $\partial^2/\partial t^2+A$ is a *second order hyperbolic operator.*

The problems considered in Examples 1.1, 1.2, 1.3 are mixed problems in the sense of Hadamard. Cf. Hadamard [1], O. A. Ladyzenskaya [1].

1.6. Examples (II)

Let us take the "second order evolution" analogue of the problem considered in Chapter 2, Section 2.7. We consider a function $a(x,t)$ such that

$$a\in\mathscr{C}^1([0,T];\ L^\infty(\Omega)). \tag{1.41}$$

We introduce

$$V=\{\psi\,|\,\psi,\ \Delta\psi\in L^2(\Omega)\},\qquad H=L^2(\Omega), \tag{1.42}$$

and

$$a(t;\varphi,\psi)=\int_\Omega a(x,t)\,\Delta\varphi\,\Delta\psi\,dx\quad\forall\,\varphi,\psi\in V. \tag{1.43}$$

We may then apply Theorem 1.1. We thus obtain the existence and uniqueness of y satisfying

$$y,\ \ \frac{\partial y}{\partial t},\ \ \Delta y\in L^2(Q), \tag{1.44}$$

$$\frac{\partial^2 y}{\partial t^2}+\Delta(a\,\Delta y)=f\quad\text{in } Q, \tag{1.45}$$

$$\Delta y=0,\qquad \frac{\partial\Delta y}{\partial n}=0\quad\text{on } \Sigma=\Gamma\times]0,T[, \tag{1.46}$$

$$\left.\begin{aligned}&y(x,0)=y_0(x),\qquad \frac{\partial y}{\partial t}(x,0)=y_1(x),\qquad x\in\Omega,\\ &y_0\in V,\qquad y_1\in L^2(\Omega).\end{aligned}\right\} \tag{1.47}$$

Remark 1.8. The operator

$$\psi\to\frac{\partial^2\psi}{\partial t^2}+\Delta(a\,\Delta\psi)$$

is an operator which is well-posed in the sense of Petrowsky—cf. Gelfand-Silov [1], Volume 3.

1.7. Orientation

We note that in all the examples considered in the previous section the *boundary conditions on Σ were homogeneous*, that is, $B_j y=0$, $j=1$, or $j=1,2$ where $B=$boundary differential operator of order $0,1\ldots$. In order to be able to consider *boundary control problems*, this situation clearly has to be generalised.

We shall first consider problems where the control is distributed over Q and then where the control is distributed over Ω. After that we shall consider boundary control problems.

2. Control Problems

2.1. Notation. Immediate Properties

As far as possible, we adopt a notation similar to that of Chapter 3, Section 2. We are given B with

$$B\in\mathscr{L}(\mathscr{U}; L^2(0,T;H)). \tag{2.1}$$

For f, y_0, y_1 given, with $f\in L^2(0,T;H)$, $y_0\in V$, $y_1\in H$, we denote by $y(v)$ the *state* of the system which is a solution of

$$\frac{d^2}{dt^2}y(v)+A(t)y(v)=f+Bv, \tag{2.2}$$

$$y(0;v)=y_0, \quad \frac{dy}{dt}(0;v)=y_1, \tag{2.3}$$

$$y(v)\in L^2(0,T;V), \quad \frac{dy}{dt}\in L^2(0,T;H)^4 \tag{2.4}$$

The *observation* may be introduced as in Chapter 3, Section 2.1. If we define $W(0,T)$ as the space spanned by y with $v=0$ and as $\{f,y_0,y_1\}$ spans $L^2(0,T;H)\times V\times H$, then we take $z(v)$ and $J(v)$ as in Chapter 3, Section 2.1.

We take N as in Chapter 3, (2.6) and $J(v)$ as in Chapter 3, (2.7). Let $\mathscr{U}_{\mathrm{ad}}$ (set of admissible controls) = closed, convex subset of $\mathscr{U}$. There exists an optimal control u characterised by (2.14), Chapter 3.

We shall state our results more explicitly by choosing the observation in a less abstract fashion.

[4] The difference from Chapter 3 is that upto now we have taken $f\in L^2(0,T;H)$ whereas in Chapter 3 we could take $f\in L^2(0,T;V')$ from the start.

We may consider the following cases:

$$\text{we take } \mathscr{C} \in \mathscr{L}(L^2(0,T;V);\mathscr{H}) \quad \text{and we observe} \quad z(v) = \mathscr{C}(y(v)); \tag{2.5}$$

$$\text{we take } \mathscr{C} \in \mathscr{L}(L^2(0,T;H);\mathscr{H}) \quad \text{and we observe} \quad z(v) = \mathscr{C}(y'(v)); \tag{2.6}$$

$$\text{we take } D_0 \in \mathscr{L}(V;\mathscr{H}) \quad \text{and we observe} \quad z(v) = D_0(y(T;v)); \tag{2.7}$$

$$\text{we take } D \in \mathscr{L}(H;\mathscr{H}) \quad \text{and we observe} \quad z(v) = D(y'(T;v)). \tag{2.8}$$

Clearly we may consider the four cases simultaneously (provided we introduce four Hilbert Spaces $\mathscr{H}_1, \ldots, \mathscr{H}_4$). It is however preferable to consider the four cases separately since this allows us to isolate the difficulties.

2.2. Case (2.5)

The cost function is

$$J(v) = \|C(y(v)) - z_d\|_{\mathscr{H}}^2 + (Nv, v)_{\mathscr{U}}. \tag{2.9}$$

The optimal control is characterised by

$$(Cy(u) - z_d, Cy(v) - Cy(u))_{\mathscr{H}} + (Nu, v-u)_{\mathscr{U}} \geqslant 0 \quad \forall v \in \mathscr{U}_{\text{ad}}. \tag{2.10}$$

But $C^* \in \mathscr{L}(\mathscr{H}'; L^2(0,T;V'))$ and if $\Lambda_{\mathscr{H}} = \Lambda$ is the canonical isomorphism of $\mathscr{H}$ onto $\mathscr{H}'$, (2.10) is equivalent to

$$\int_0^T (C^* \Lambda(Cy(u) - z_d), y(v) - y(u))\, dt + (Nu, v-u)_{\mathscr{U}} \geqslant 0 \quad \forall v \in \mathscr{U}_{\text{ad}}. \tag{2.11}$$

In order to proceed further we have to resolve the following difficulty: we formally introduce the *adjoint state* $p(u)$ by

$$\left.\begin{aligned} &\frac{d^2}{dt^2} p(u) + A(t) p(u) = C^* \Lambda(Cy(u) - z_d), \\ &p(T;u) = 0, \\ &\frac{d}{dt} p(T;u) = 0. \end{aligned}\right\} \tag{2.12}$$

But since

$$C^* \Lambda(Cy(u) - z_d) \in L^2(0,T;V')$$

we cannot apply the results of Section 1 [5]. Let us therefore impose a stronger hypothesis on C:

$$C\in\mathscr{L}(L^2(0,T;H);\mathscr{H}). \tag{2.13}$$

Then

$$C^*\Lambda(Cy(u)-z_d)\in L^2(0,T;H)$$

and from Theorem 1.1, (2.12) admits a unique solution which satisfies

$$p(u)\in L^2(0,T;V),\quad \frac{d}{dt}p(u)\in L^2(0,T;H). \tag{2.14}$$

We shall now transform (2.11) as follows: we scalar multiply (by considering the scalar product in H) both sides of (2.12) by $y(v)-y(u)$ which gives us

$$\int_0^T\left(\frac{d^2}{dt^2}p(u)+A(t)p(u),y(v)-y(u)\right)dt=(Cy(u)-z_d,Cy(v)-Cy(u))_{\mathscr{H}}. \tag{2.15}$$

We now apply *Green's formula* to the left hand side of (2.15).

We note that if $\varphi\in L^2(0,T;V)$, $\varphi'\in L^2(0,T;H)$, $\varphi''\in L^2(0,T;V')$ and if ψ has the same properties, then

$$\int_0^T(\varphi'',\psi)dt=(\varphi'(T),\psi(T))-(\varphi'(0),\psi(0))-(\varphi(T),\psi'(T))$$
$$+(\varphi(0),\psi'(0))+\int_0^T(\varphi,\psi'')dt.$$

From this we deduce that

$$\int_0^T\left(p(u),\left(\frac{d^2}{dt^2}+A(t)\right)y(v)-\left(\frac{d^2}{dt^2}+A(t)\right)y(u)\right)dt$$
$$=(Cy(u)-z_d,Cy(v)-Cy(u))_{\mathscr{H}},$$

that is,

$$=\int_0^T(p(u),B(v-u))dt=(B^*p(u),v-u)$$

(where the last bracket denotes the scalar product between $\mathscr{U}'$ and $\mathscr{U}$)

$=(\Lambda_{\mathscr{U}}^{-1}B^*p(u),v-u)_{\mathscr{U}}$, and (2.11) may be written as

$$(\Lambda_{\mathscr{U}}^{-1}B^*p(u)+Nu,v-u)_{\mathscr{U}}\geqslant 0\quad\forall v\in\mathscr{U}_{\text{ad}}. \tag{2.16}$$

Remark 2.1. Condition (2.16) is similar to (2.18), Chapter 3.

Remark 2.2. The preceeding calculations show that

$$\tfrac{1}{2}J'(u)=B^*p(u)+\Lambda_{\mathscr{U}}Nu. \tag{2.17}$$

---

[5] By generalising Section 1 we can overcome this difficulty. This will be done later; cf. also Lions-Magenes [1], Chapter 3 and the work of Baiocchi [1].

Let us summarise the results we have obtained in (compare with Theorem 2.1, Chapter 3):

Theorem 2.1. *Let us assume that* (1.4), (1.5), (1.6) *hold and that the cost function is given by* (2.9) *with* C *satisfying* (2.13). *The optimal control is then characterised by the following system of equations and inequalities:*

$$\left.\begin{aligned}&\frac{d^2}{dt^2}y(u)+A(t)y(u)=f+Bu,\\&y(0;u)=y_0,\qquad y'(0;u)=y_1,\end{aligned}\right\}\tag{2.18}$$

$$\left.\begin{aligned}&\frac{d^2}{dt^2}p(u)+A(t)p(u)=C^*\Lambda(Cy(u)-z_d),\\&p(T;u)=0,\qquad p'(T;u)=0,\end{aligned}\right\}\tag{2.19}$$

$$\left.\begin{aligned}&(\Lambda_{\mathscr{U}}^{-1}B^*p(u)+Nu,v-u)_{\mathscr{U}}\geqslant 0\quad\forall v\in\mathscr{U}_{\text{ad}},\\&u\in\mathscr{U}_{\text{ad}},\end{aligned}\right\}\tag{2.20}$$

with

$$\left.\begin{aligned}&y(u),p(u)\in L^2(0,T;V),\\&y'(u),p'(u)\in L^2(0,T;H).\end{aligned}\right\}\tag{2.21}$$

Remark 2.3. If $\mathscr{U}_{\text{ad}}=\mathscr{U}$, (2.20) reduces to

$$\Lambda_{\mathscr{U}}^{-1}B^*p(u)+Nu=0.\tag{2.22}$$

We may then eliminate u which leads to the following system of equations:

$$\left.\begin{aligned}&\frac{d^2y}{dt^2}+Ay+BN^{-1}\Lambda_{\mathscr{U}}^{-1}B^*p=f,\\&\frac{d^2p}{dt^2}+Ap-C^*\Lambda Cy=-C^*\Lambda z_d,\\&y(0)=y_0,\qquad y'(0)=y_1,\\&p(T)=0,\qquad p'(T)=0.\end{aligned}\right\}\tag{2.23}$$

We shall study this system of equations in detail in Section 4.

2.3. Case (2.6)

The cost function is

$$J(v)=\|C(y'(v))-z_d\|_{\mathscr{H}}^2+(Nv,v)_{\mathscr{U}}.\tag{2.24}$$

The optimal control u is characterised by

$$(C(y'(u))-z_d,C(y'(v))-C(y'(u)))_{\mathscr{H}}+(Nu,v-u)_{\mathscr{U}}\geqslant 0\quad\forall v\in\mathscr{U}_{\text{ad}}.\tag{2.25}$$

But

$$C^* \in \mathscr{L}(\mathscr{H}'; L^2(0,T;H)),$$

which implies that (2.25) is equivalent to

$$\int_0^T (C^* \Lambda(Cy'(u) - z_d, y'(v) - y'(u)))dt + (Nu, v-u)_{\mathscr{U}} \geqslant 0 \quad \forall v \in \mathscr{U}_{\text{ad}}. \tag{2.26}$$

As in Section 2.2, our objective now is to transform (2.26) into a more explicit condition. We introduce the *adjoint state* $p(u)$ as the solution of

$$p''(u) - Ap(u) + \int_t^T A'(\sigma)p(\sigma;u)d\sigma = C^* \Lambda(Cy'(u) - z_d), \tag{2.27}$$

$$\left.\begin{aligned} p(T;u)=0,\\ p'(T;u)=0;\end{aligned}\right\} \tag{2.28}$$

in (2.27), $A'(\sigma)$ is defined by

$$a'(\sigma;\varphi,\psi) = (A'(\sigma)\varphi,\psi) \quad \forall \varphi,\psi \in V.$$

Problem (2.27), (2.28) *admits a unique solution—at least if we make an additional hypothesis*[6]*:*

$$A'(\sigma) \in \mathscr{L}(V;H). \tag{2.29}$$

Indeed, reversing the direction of time and changing notation, problem (2.27), (2.28) is equivalent to solving

$$\left.\begin{aligned} y'' + Ay + \int_0^t A'(\sigma)y(\sigma)d\sigma = f,\\ y(0)=0, \qquad y'(0)=0.\end{aligned}\right\} \tag{2.30}$$

We introduce (as in Section 1.4) an approximate solution y_m by

$$\left.\begin{aligned} (y_m'', w_j) + (A(t)y_m, w_j) + \int_0^t (A'(\sigma)y_m(\sigma), w_j)d\sigma = (f(t), w_j), \quad 1 \leqslant j \leqslant m,\\ y_m(0)=0, \qquad y_m'(0)=0.\end{aligned}\right\} \tag{2.31}$$

From this we deduce

$$\frac{d}{dt}[|y_m'(t)|^2 + a(t; y_m(t), y_m(t))] - a'(t; y_m(t), y_m(t))$$
$$+2\int_0^t (A'(\sigma)y_m(\sigma), y_m'(t))d\sigma = 2(f(t), y_m'(t)),$$

---

6 We do not consider how the problem can be solved when this additional hypothesis is not in effect.

whence

$$|y'_m(t)|^2 + \|y_m(t)\|^2 \leqslant C \int_0^t (|y'_m(\sigma)|^2 + \|y_m(\sigma)\|^2) d\sigma$$
$$+ C \int_0^t \left(\int_0^\sigma \|y_m(\sigma_1)\| d\sigma_1 \right) |y'_m(\sigma)| d\sigma$$

and the remainder of the argument proceeds as in Section 1.4. Uniqueness is proved as in Section 1.3.

We are now in a position to transform (2.26). Let us first carry out a formal calculation which we shall justify later. Scalar multiply (2.27) by $y'(v)-y'(u)$—and here we are faced with the first difficulty which has to be overcome; the scalar products have no meaning (for the moment). Setting

$$q(t) = \int_t^T A'(\sigma) p(\sigma) d\sigma,$$

we obtain

$$\begin{aligned} &\int_0^T (p''(u) + A p(u), y'(v) - y'(u)) dt + \int_0^T (q, y'(v) - y'(u)) dt \\ &\qquad = (C y'(u) - z_d, C y'(v) - C y(u))_{\mathscr{H}}. \end{aligned} \tag{2.32}$$

Integrating by parts, the first term of (2.32) becomes

$$\begin{aligned} &- \int_0^T (p'(u), y''(v) - y''(u)) dt + \int_0^T a(t; p(u), y'(v) - y'(u)) dt \\ &+ \int_0^T (A'(t) p(u), y(v) - y(u)) dt \\ &= - \int_0^T (p'(u), B(v-u)) dt + \int_0^T [a(t; p'(u), y(v) - y(u)) \\ &\quad + a(t; p(u), y'(v) - y'(u)) + a'(t; p(u), y(v) - y(u))] dt \\ &= - \int_0^T (p'(u), B(v-u)) dt + \int_0^T \frac{d}{dt} a(t; p(u), y(v) - y(u)) dt \\ &= - \int_0^T (p'(u), B(v-u)) dt = (-B^* p'(u), v-u) \\ &= (-\Lambda_{\mathscr{U}}^{-1} B^* p'(u), v-u)_{\mathscr{U}} \end{aligned}$$

and (2.26) becomes

$$(-\Lambda_{\mathscr{U}}^{-1} B^* p'(u) + N u, v-u)_{\mathscr{U}} \geqslant 0 \quad \forall v \in \mathscr{U}_{\mathrm{ad}}. \tag{2.33}$$

It now remains to justify the formal calculations. We shall reduce the problems to equations on $\mathbb{R}_t$ and then regularise. We first introduce ψ as the solution of (we extend $A(t)$ on the whole of $\mathbb{R}$, where the extended $A(t)$ has properties similar to $A(t)$ on $(0,T)$)

$$\left.\begin{aligned} \psi'' + A(t) &= \begin{cases} B(v-u) & \text{on } (0,T), \\ 0 & \text{for } t>T, \end{cases} \\ \psi(0) &= 0, \quad \psi'(0)=0, \end{aligned}\right\} \tag{2.34}$$

and we extend ψ by 0 for $t<0$. We then introduce π as the solution of

$$\left.\begin{aligned} \pi'' + A(t)\pi + \int_t^T A'(\sigma)\pi(\sigma)d\sigma &= \begin{cases} C^* \Lambda(C y'(u) - z_d) & \text{on } (0,T), \\ 0 & \text{for } t<0, \end{cases} \\ \pi(T) = 0, \quad \pi'(T) &= 0 \end{aligned}\right\} \tag{2.35}$$

which we extend by 0 for $t>T$.

We have:

$$\psi = y(v) - y(u) \quad \text{on } (0,T), \qquad \pi = p(u) \quad \text{on } (0,T)$$

and

$$\left.\begin{aligned} &(C y'(u) - z_d, C y'(v) - C y'(u))_{\mathscr{H}} \\ &= \int_{-\infty}^{+\infty} \left(\pi'' + A\pi + \int_t^T A'(\sigma)\pi(\sigma)d\sigma, \psi'\right) dt. \end{aligned}\right\} \tag{2.36}$$

Let $\rho_n(t) = \rho_n$ be a regularising sequence on $\mathbb{R}_t$. Then the second term of (2.36) becomes[7]

$$\lim_{n\to\infty} \int_{-\infty}^{+\infty} \left((\pi'' + A\pi) * \rho_n + \left(\int_t^T A'(\sigma)\pi(\sigma)d\sigma\right) * \rho_n, \psi' * \rho_n\right) dt = \lim_{n\to\infty} X_n, \tag{2.37}$$

and we may integrate by parts in X_n. We get

$$\begin{aligned} X_n &= \int_{-\infty}^{+\infty} [(-\pi' * \rho_n, \psi'' * \rho_n) + ((A\pi) * \rho_n, \psi' * \rho_n) + ((A'\pi) * \rho_n, \psi * \rho_n)] dt \\ &= \int_{-\infty}^{+\infty} (-\pi' * \rho_n, B(v-u) * \rho_n) dt \\ &\quad + \int_{-\infty}^{+\infty} [(\pi' * \rho_n, (A\psi) * \rho_n) + ((A\pi) * \rho_n, \psi' * \rho_n) + ((A'\pi) * \rho_n, \psi * \rho_n)] dt. \end{aligned}$$

---

[7] In general

$$g * \rho_n = \int_{-\infty}^{+\infty} g(t-\sigma)\rho_n(\sigma)d\sigma.$$

If we can show that the last integral $\to 0$, then the result will be proved.

This may be written as

$$Y_n = \int_{-\infty}^{+\infty} [(\pi' * \rho_n, A(\psi * \rho_n)) + (A(\pi * \rho_n), \psi' * \rho_n) + (A'(\pi * \rho_n), \psi * \rho_n)] dt$$
$$+ Y_n^1 + Y_n^2 + Y_n^3,$$

where

$$Y_n^1 = \int_{-\infty}^{+\infty} (\pi' * \rho_n, (A\psi) * \rho_n - A(\psi * \rho_n)) dt,$$

$$Y_n^2 = \int_{-\infty}^{+\infty} ((A\pi) * \rho_n - A(\pi * \rho_n), \psi' * \rho_n) dt,$$

$$Y_n^3 = \int_{-\infty}^{+\infty} ((A'\pi) * \rho_n - A'(\pi * \rho_n), \psi * \rho_n) dt.$$

The first integral Y_n equals

$$\int_{-\infty}^{+\infty} \frac{d}{dt} a(t; \pi * \rho_n, \psi * \rho_n) dt = 0,$$

and hence

$$Y_n = Y_n^1 + Y_n^2 + Y_n^3.$$

Now from the vector version of Friedrichs's lemma (cf. for example Lions [3], Lemma 7.2, Chapter 4) we have

$$\frac{d}{dt} [(A\psi) * \rho_n - A(\psi * \rho_n)] \to 0 \quad \text{in } L^2(\mathbb{R}_t; V'),$$

and hence

$$Y_n^1 = -\int_{-\infty}^{+\infty} \left(\pi * \rho_n, \frac{d}{dt} [(A\psi) * \rho_n - A(\psi * \rho_n)] \right) dt \to 0.$$

In a similar manner $Y_n^2 \to 0$ and $Y_n^3 \to 0$. This justifies (2.33).

We have thus proved

Theorem 2.2. *We assume that* (1.4), (1.5), (1.6) *hold and that the cost function is given by* (2.24) *with C satisfying* (2.6). *We assume that* (2.29) *holds. Then the optimal control u is determined by the following system of equations and inequalities:*

$$\left.\begin{aligned}&\frac{d^2}{dt^2}y(u)+A(t)y(u)=f+Bu,\\&\qquad y(0;u)=y_0,\quad y'(0;u)=y_1,\end{aligned}\right\}\tag{2.38}$$

$$\left.\begin{aligned}&\frac{d^2}{dt^2}p(u)+A(t)p(u)+\int_t^T A'(\sigma)p(\sigma;u)d\sigma=C*\Lambda(y'(u)-z_d),\\&\qquad p(T;u)=0,\quad p'(T;u)=0,\end{aligned}\right\}\tag{2.39}$$

$$(-\Lambda_{\mathscr{U}}^{-1}B*p'(u)+Nu,v-u)_{\mathscr{U}}\geqslant 0\quad\forall v\in\mathscr{U}_{\mathrm{ad}}.\tag{2.40}$$

Remark 2.4. In the case of no control constraints, $\mathscr{U}_{\mathrm{ad}}=\mathscr{U}$, (2.40) reduces to

$$-\Lambda_{\mathscr{U}}^{-1}B*p'(u)+Nu=0.\tag{2.41}$$

Eliminating u we obtain the system of equations

$$\left.\begin{aligned}&\frac{d^2}{dt^2}y+A(t)y-BN^{-1}\Lambda_{\mathscr{U}}^{-1}B*p'=f,\\&\frac{d^2}{dt^2}p+A(t)p+\int_t^T A'(\sigma)p(\sigma)d\sigma=C^*\Lambda(Cy'-z_d),\\&y(0)=y_0,\quad y'(0)=y_1,\quad p(T)=0,\quad p'(T)=0.\end{aligned}\right\}\tag{2.42}$$

Remark 2.5. It is worth mentioning again that we *know* (from the way we have obtained them) that the system of equations and inequalities (2.38), (2.39), (2.40) admits a unique solution. A direct proof of this result (without any reference to the control problem) is not at all evident.

2.4. Case (2.7)

The cost function is

$$J(v)=\|D_0(y(T;v))-z_d\|^2+(Nv,v).\tag{2.43}$$

The optimal control u is characterised by

$$(D_0(y(T;u))-z_d,D_0(y(T;v)-y(T;u)))_{\mathscr{H}}+(Nu,v-u)_{\mathscr{U}}\geqslant 0\quad\forall v\in\mathscr{U}_{\mathrm{ad}},\tag{2.44}$$

that is, by

$$(D_0^*\Lambda(D_0y(T;u)-z_d),y(T;v)-y(T;u))+(Nu,v-u)_{\mathscr{U}}\geqslant 0\quad\forall v\in\mathscr{U}_{\mathrm{ad}},\tag{2.45}$$

where in (2.45) the first bracket denotes the scalar product between V' and V.

The *adjoint state* is formally defined by

$$\frac{d^2}{dt^2} p(u) + A(t) p(u) = 0 \quad \text{in }]0, T[, \tag{2.46}$$

$$p(T; u) = 0, \tag{2.47}$$

$$p'(T; u) = D_0^* \Lambda(D_0 y(T; u) - z_d). \tag{2.48}$$

But in order to apply Theorem 1.1 $p'(T; u)$ must be given in H, [8] which leads to the hypothesis

$$D_0 \in \mathscr{L}(H; \mathscr{H}). \tag{2.49}$$

Then problem (2.46), (2.47), (2.48) admits a unique solution.

Let us scalar multiply (2.46) by $y(t; v) - y(t; u)$ and apply Green's formula, which is valid. This gives us

$$(D_0^* \Lambda(D_0 y(T; u) - z_d), y(T; v) - y(T; u))$$
$$+ \int_0^T \left(p(u), \left(\frac{d^2}{dt^2} + A \right) (y(v) - y(u)) \right) dt = 0,$$

and (2.45) gives us

$$(-B^* p(u), v - u) + (Nu, v - u)_{\mathscr{U}} \geqslant 0 \quad \forall v \in \mathscr{U}_{\text{ad}}, \tag{2.50}$$

which may also be written as

$$(-\Lambda_{\mathscr{U}}^{-1} B^* p(u) + Nu, v - u)_{\mathscr{U}} \geqslant 0 \quad \forall v \in \forall_{\text{ad}}. \tag{2.51}$$

We have thus proved

Theorem 2.3. *Let us assume that* (1.4), (1.5), (1.6) *hold and that the cost function is given by* (2.43) *with* D_0 *satisfying* (2.49). *Then the optimal control* u *is determined by the simultaneous solution of the system*

$$\left.\begin{aligned} &\frac{d^2}{dt^2} y(u) + A(t) y(u) = f + Bu, \\ &\qquad y(0; u) = y_0, \quad y'(0; u) = y_1, \end{aligned}\right\} \tag{2.52}$$

$$\left.\begin{aligned} &\frac{d^2}{dt^2} p(u) + A(t) p(u) = 0, \\ &\qquad p(T; u) = 0, \quad p'(T; u) = D_0^* \Lambda(D_0 y(T; u) - z_d), \end{aligned}\right\} \tag{2.53}$$

and

$$(-\Lambda_{\mathscr{U}}^{-1} B^* p(u) + Nu, v - u)_{\mathscr{U}} \geqslant 0 \quad \forall v \in \mathscr{U}_{\text{ad}}. \tag{2.51}$$

---

[8] Otherwise we have to make use of solutions which are weaker; cf. Section 3 following.

Remark 2.6. In the case where there are no control constraints, that is, $\mathscr{U}_{\text{ad}}=\mathscr{U}$, (2.51) reduces to

$$-\Lambda_{\mathscr{U}}^{-1} B^* p(u)+Nu=0, \tag{2.54}$$

and we may eliminate u in (2.51), (2.52), (2.53).

2.5. Case (2.8)

The cost function is

$$J(v)= \|D(y'(T;v))-z_d\|_{\mathscr{H}}^2+(Nv,v)_{\mathscr{U}}. \tag{2.55}$$

The optimal control u is characterised by

$$(Dy'(T;u)-z_d, D(y'(T;v)-y'(T;u)))_{\mathscr{H}}+(Nu,v-u)_{\mathscr{H}}\geqslant 0 \quad \forall v\in\mathscr{U}_{\text{ad}}, \tag{2.56}$$

or by

$$(D^*\Lambda(Dy'(T;u)-z_d), y'(T;v)-y'(T;u))+(Nu,v-u)_{\mathscr{U}}\geqslant 0 \quad \forall v\in\mathscr{U}_{\text{ad}}. \tag{2.57}$$

The *adjoint state* $p(u)$ is *formally* defined by

$$\left.\begin{aligned} &\frac{d^2}{dt^2}p(u)+A(t)p(u)=0 \quad \text{in }]0,T[, \\ &p(T;u)=D^*\Lambda(Dy'(T;u)-z_d), \\ &p'(T;u)=0. \end{aligned}\right\} \tag{2.58}$$

But then $p(T;u)$ is given in H and it is imperative that we generalise the state of affairs in Section 1. This will be done in the following section and by the same token the present problem will also be solved.

3. Transposition and Applications to Control

3.1. Transposition of Theorem 1.1.

Assume that the hypotheses of Theorem 1.1 hold. It follows that for f given in $L^2(0,T;H)$, there exists a unique φ satisfying

$$\left.\begin{aligned} \varphi''+A(t)=f, \\ \varphi(T)=0, \quad \varphi'(T)=0, \end{aligned}\right\} \tag{3.1}$$

with $\varphi\in L^2(0,T;V)$, $\varphi'\in L^2(0,T;H)$.

Let X denote the space spanned by φ as f ranges over $L^2(0,T;H)$. Endowed with the norm $\|\varphi\|_X=\|f\|_{L^2(0,T;H)}$, this is a Hilbert space and the operator

$$\varphi \to \varphi'' + A\varphi$$

is an isomorphism of X onto $L^2(0, T; H)$.

By *transposition*, we deduce the following:

$$\left.\begin{array}{l}\text{let } \varphi \to L(\varphi) \text{ be a continuous linear form on } X\text{; there} \\ \text{exists a unique } y \in L^2(0,T;H) \text{ such that} \\ \int_0^T (y, \varphi'' + A\varphi)\,dt = L(\varphi) \quad \forall \varphi \in X.\end{array}\right\} \tag{3.2}$$

Amongst other things, the mapping $L \to y$ is continuous from $X' \to L^2(0,T;H)$.

But, from Remark 1.3, if we set

$$L(\varphi) = \int_0^T (f, \varphi)\,dt + (y_1, \varphi(0)) - (y_0, \varphi'(0)) \tag{3.3}$$

where

$$f \in L^1(0,T;V'), \tag{3.4}$$

$$y_0 \in H, \tag{3.5}$$

$$y_1 \in V', \tag{3.6}$$

we define a continuous linear form on X. Hence we may apply (3.2) together with (3.3)…(3.6). Let us state the result:

Theorem 3.1. *We assume that the hypotheses of Theorem* 1.1 *hold. Let* f, y_0, y_1 *be given as in* (3.4), (3.5), (3.6) *respectively. There exists a* $y \in L^2(0,T;H)$ *which is unique such that*

$$\int_0^T (y, \varphi'' + A\varphi)\,dt = \int_0^T (f, \varphi)\,dt + (y_1, \varphi(0)) - (y_0, \varphi'(0)) \quad \forall \varphi \in X. \tag{3.7}$$

The mapping $\{f, y_0, y_1\} \to y$ *is continuous from*

$$L^1(0,T;V) \times H \times V' \to L^2(0,T;H).$$

Interpretation of (3.7). We restrict ourselves to a formal interpretation of (3.7). For the justification, cf. Lions-Magenes [1], Chapter 3.

In (3.7) let us first take a $\varphi(t)$ with compact support in $]0,T[$. We deduce that

$$\frac{d^2y}{dt^2} + A(t)y = f \quad \text{in }]0,T[. \tag{3.8}$$

If we "scalar multiply" (3.8) by φ and integrate by parts, we obtain,

$$\int_0^T (f,\varphi)\,dt = -(y'(0),\varphi(0)) + (y(0),\varphi'(0)) + \int_0^T (y,\varphi''+A\varphi)\,dt,$$

which comparing with (3.7) gives us the initial conditions

$$y(0)=y_0, \quad y'(0)=y_1. \tag{3.9}$$

3.2. Application (I)

Let us again consider the problem of Section 2.5. A precise meaning can now be attached to (2.58). We apply Theorem 3.1 by reversing the direction of time. Hence $p(u)$ is defined as the unique element of $L^2(0,T;H)$ such that

$$\left.\begin{aligned} &\int_0^T (p(u),\varphi''+A(t)\varphi)\,dt = (Dy'(T;u)-z_d, D\varphi'(T))_{\mathscr{H}} \quad \forall\varphi \text{ such that} \\ &\quad \varphi''+A(t)\varphi\in L^2(0,T;H), \quad \varphi(0)=0, \quad \varphi'(0)=0. \end{aligned}\right\} \tag{3.10}$$

Therefore, in (3.10) we can take

$$\varphi = y(v)-y(u)$$

giving us

$$(Dy'(T;u)-z_d, D(y'(T,v)-y'(T;u))) = \int_0^T (p(u),B(v-u))\,dt.$$

Thus (2.57) is equivalent to

$$(B^* p(u), v-u) + (Nu, v-u)_{\mathscr{U}} \geqslant 0 \quad \forall v\in\mathscr{U}_{\mathrm{ad}}, \tag{3.11}$$

or again

$$(\Lambda_{\mathscr{U}}^{-1} B^* p(u) + Nu, v-u)_{\mathscr{U}} \geqslant 0 \quad \forall v\in\mathscr{U}_{\mathrm{ad}}. \tag{3.12}$$

Hence

Theorem 3.2. *Let us assume that* (1.4), (1.5), (1.6) *hold and let the cost function be given by* (2.55). *Then the optimal control* u *is determined by the solution of*

$$\left.\begin{aligned} &\frac{d^2}{dt^2} y(u) + A(t) y(u) = f + Bu, \\ &y(0;u)=y_0, \quad y'(0;u)=y_1, \end{aligned}\right\} \tag{3.13}$$

to which (3.10) *and* (3.12) *are adjoined.*

3.3. Application (II)

The study begun in Section 2.4 can now be completed. The cost function is given by (2.43) with $D_0 \in \mathscr{L}(V; \mathscr{H})$. Then in (2.48),

$$D_0^* \Lambda(D_0 y(T; u) - z_d) \in V'$$

and we now apply Theorem 3.1 to formulate and solve (2.46), (2.47), (2.48): we define $p(u)$ as the unique element of $L^2(0, T; H)$ which satisfies

$$\left.\begin{array}{l} \int_0^T (p(u), \varphi'' + A(t)\varphi)\, dt = -(D_0 y(T; u) - z_d, D_0 \varphi(T))_{\mathscr{H}} \\ \forall \varphi \quad \text{such that} \\ \varphi'' + A(t)\varphi \in L^2(0, T; H), \qquad \varphi(0) = 0, \qquad \varphi'(0) = 0. \end{array}\right\} \tag{3.14}$$

In (3.14), let us take

$$\varphi = y(v) - y(u).$$

We then deduce that (2.44) is equivalent to

$$(-\Lambda_{\mathscr{U}}^{-1} B^* p(u) + N u, v - u)_{\mathscr{U}} \geqslant 0 \quad \forall v \in \mathscr{U}_{\text{ad}}. \tag{3.15}$$

Hence

Theorem 3.3. *Let us assume that* (1.4), (1.5), (1.6) *hold. The cost function is given by* (2.43). *Let us assume that* $D_0 \in \mathscr{L}(V; \mathscr{H})$. *The optimal control* u *is determined by the solution of the system of equations and inequalities* (3.13), (3.14), (3.15).

3.4. Application (III)

We may also complete the study undertaken in Section 2.2. We assume that $C \in \mathscr{L}(L^2(0, T; V); \mathscr{H})$. We then solve (2.12) by applying Theorem 3.1. Hence $p(u)$ is defined as the unique element of $L^2(0, T; H)$ which satisfies

$$\left.\begin{array}{l} \int_0^T (p(u), \varphi'' + A(t)\varphi)\, dt = (C y(u) - z_d, C\varphi)_{\mathscr{H}} \quad \forall \varphi \quad \text{such that} \\ \varphi'' + A(t)\varphi \in L^2(0, T; H), \qquad \varphi(0) = 0, \qquad \varphi'(0) = 0. \end{array}\right\} \tag{3.16}$$

Taking $\varphi = y(v) - y(u)$, we obtain

Theorem 3.4. *Let us assume that* (1.4), (1.5), (1.6) *hold. Let the cost function be given by* (2.9). *Let us assume that* $C \in \mathscr{L}(L^2(0, T; V); \mathscr{H})$.

Then the unique optimal control u is determined by the solution of the system of equations (3.13), (3.16) *and the inequality*

$$(\Lambda_{\mathcal{U}}^{-1} B^* p(u) + N u, v - u)_{\mathcal{U}} \geqslant 0 \quad \forall v \in \mathcal{U}_{\text{ad}}. \tag{3.17}$$

Remark 3.1. Let us assume that

$$\mathcal{U} = L^2(0, T; E), \tag{3.18}$$

$$\left.\begin{aligned} &\mathcal{U}_{\text{ad}} = \{v \mid v(t) \in E_{\text{ad}} \text{ a.e., } v \in \mathcal{U}\}, \\ &E_{\text{ad}} = \text{closed, convex subset of } E. \end{aligned}\right\} \tag{3.19}$$

Then applying Theorem 2.1, Chapter 2, we see that condition (3.17)[9] is equivalent to

$$\big((\Lambda_{\mathcal{U}}^{-1} B^* p(u) + N u)(t), e - u(t)\big)_E \geqslant 0 \quad \text{a.e. on } (0, T), \quad \forall e \in E_{\text{ad}}. \tag{3.20}$$

4. Examples

4.1. Examples of Hyperbolic Problems. Distributed Control Distributed Observation

We take A to be defined as in Section 1.5.

Let us take C = identity considered as an element of

$$\mathcal{L}(L^2(0, T; H); L^2(0, T; H)).$$

Let us assume that

$$\mathcal{U} = L^2(Q), \qquad B = \text{identity}. \tag{4.1}$$

If $V = H_0^1(\Omega)$,[10] the state is given by

$$\left.\begin{aligned} \frac{\partial^2 y(v)}{\partial t^2} + A y(v) &= f + v, \\ y(v) &= 0 \quad \text{on } \Sigma, \\ y(x,0; v) = y_0(x), \quad \frac{\partial y}{\partial t}(x,0; v) &= y_1(x) \quad \text{in } \Omega \end{aligned}\right\} \tag{4.2}$$

---

[9] The present remark applies to all the different cases.

[10] If $V = H^1(\Omega)$, then $\partial y(v)/\partial v_A = 0$ on Σ. Cf. Section 1.5.

and the cost function by

$$J(v)=\int_{\Omega}(y(v)-z_d)^2\,dx\,dt+(Nv,v)_{\mathscr{U}}. \tag{4.3}$$

Theorem 2.1 shows that the optimal control u is determined by (4.2) (where $v=u$) and by

$$\left.\begin{aligned} &\frac{\partial^2}{\partial t^2}p(u)+A(t)p(u)=y(u)-z_d \quad \text{in } Q,\\ &p(u)=0 \quad \text{on } \Sigma,\\ &p(x,T;u)=0,\quad \frac{\partial p}{\partial t}(x,T;u)=0 \quad \text{on } \Omega, \end{aligned}\right\} \tag{4.4}$$

and

$$\int_{Q}(p(u)+Nu)(v-u)\,dx\,dt\geqslant 0 \quad \forall v\in\mathscr{U}_{\mathrm{ad}}. \tag{4.5}$$

Example 4.1. Let us take

$$\mathscr{U}_{\mathrm{ad}}=\{v\,|\,v\geqslant 0 \text{ a.e. in } Q\}. \tag{4.6}$$

Then the optimal control is determined by the solution of the unilateral problem

$$\left.\begin{aligned} &\frac{\partial^2 p}{\partial t^2}+Ap=y-z_d \quad \text{in } Q,\\ &\frac{\partial^2 y}{\partial t^2}+Ay-f\geqslant 0 \quad \text{in } Q,\\ &p+N\left(\frac{\partial^2 y}{\partial t^2}+Ay-f\right)\geqslant 0 \quad \text{in } Q,\\ &\left(\frac{\partial^2 y}{\partial t^2}+Ay-f\right)\left[p+N\left(\frac{\partial^2 y}{\partial t^2}+Ay-f\right)\right]=0 \quad \text{in } Q,\\ &y=0 \quad \text{on } \Sigma, \qquad p=0 \quad \text{on } \Sigma,\\ &y(x,0)=y_0(x),\quad \frac{\partial y}{\partial t}(x,0)=y_1(x) \quad \text{on } \Omega,\\ &p(x,T)=0,\quad \frac{\partial p}{\partial t}(x,0)=0 \quad \text{on } \Omega. \end{aligned}\right\} \tag{4.7}$$

Then

$$u=\frac{\partial^2 y}{\partial t^2}+Ay-f.$$

Remark 4.1. As in Chapter 3, Example 3.1 we may carry (4.7) a little further. We find that if $(\partial^2/\partial t^2 + A) z_d = f$ a.e. in Q, and if $N = \nu$ (identity), $\nu > 0$, then

$$u = -\frac{1}{\nu} \inf(0, p) \quad \text{a.e. in } Q, \tag{4.8}$$

and the optimal control is determined by the solution of the non-linear system of equations

$$\left.\begin{aligned} &\frac{\partial^2 y}{\partial t^2} + A y + \frac{1}{\nu} \inf(0,p) = f \quad \text{in } Q, \\ &\frac{\partial^2 p}{\partial t^2} + A p = y - z_d \quad \text{in } Q, \\ &y|_\Sigma = 0, \quad p|_\Sigma = 0, \\ &y(x,0) = y_0(x), \quad \frac{\partial y}{\partial t}(x,0) = 0, \quad p(x,T) = 0, \quad \frac{\partial p}{\partial t}(x,T) = 0 \quad \text{on } \Omega. \end{aligned}\right\} \tag{4.9}$$

Let us note that $p \in H^1(Q)$ (in particular) allowing us to conclude from (4.8) that

$$u \in H^1(Q); \tag{4.10}$$

this is a result on the *regularity of the optimal control.*

Remark 4.2. If we take $C =$ identity considered as an element of $\mathscr{L}(L^2(0,T;V); L^2(0,T;V))$, with $V = H_0^1(\Omega)$ for example, the cost function becomes

$$J(v) = \int_Q \left[(y(v) - z_d)^2 + \sum_{i=1}^n \left(\frac{\partial}{\partial x_i} (y(v) - z_d) \right)^2 \right] dx + (Nv, v)_{\mathscr{U}}. \tag{4.11}$$

Then (compare with Chapter 3, (3.13)) the *adjoint state* is defined by

$$\left.\begin{aligned} &\frac{\partial^2}{\partial t^2} p(u) + A p(u) = (-\Delta_x + I)(y(u) - z_d) \quad \text{in } Q, \\ &p(u) = 0 \quad \text{on } \Sigma, \\ &p(x,T;u) = 0, \quad \frac{\partial p}{\partial t}(x,T;u) = 0 \quad \text{in } \Omega. \end{aligned}\right\} \tag{4.12}$$

But now $(-\Delta_x + I)(y(u) - z_d) \in L^2(0,T;V')$ and we have to apply the results of Section 3 (here Theorem 3.4).

4.2. Examples of Hyperbolic Systems. Distributed Control Observation of the Final State

Let us assume that the state $y(v)$ is given by (4.2) and the observation by

$$y(x,T;v) \quad \text{in } L^2(\Omega). \tag{4.13}$$

Hence

$$C\,y(v)=D_0(y(T;v)), \qquad D_0=\text{identity from } L^2(\Omega)\to L^2(\Omega).$$

Let the cost function be given by

$$J(v)=\int_\Omega [y(x,T;v)-z_d(x)]^2\,dx+(N v,v)_{L^2(Q)}. \tag{4.14}$$

We are now in a situation where we may apply Theorem 2.3; we then obtain the optimal control u as the simultaneous solution of (4.2) (with $v=u$),

$$\left.\begin{aligned} &\frac{\partial^2}{\partial t^2}p(u)+A\,p(u)=0 \quad \text{in } Q,\\ &p(u)=0 \quad \text{on } \Sigma,\\ &p(x,T;u)=0, \qquad x\in\Omega,\\ &\frac{\partial p}{\partial t}(x,T;u)=y(x,T;u)-z_d(x), \qquad x\in\Omega, \end{aligned}\right\} \tag{4.15}$$

and

$$(-p(u)+Nu,v-u)_Q\geqslant 0 \;\;\forall v\in\mathcal{U}_{ad}. \tag{4.16}$$

Example 4.2. Let us again consider $\mathcal{U}_{ad}$ given by (4.6). Then (4.16) is equivalent to

$$\left.\begin{aligned} u\geqslant 0 &\quad \text{a.e. in } Q,\\ -p(u)+Nu\geqslant 0 &\quad \text{a.e. in } Q,\\ u(-p(u)+Nu)=0 &\quad \text{a.e. in } Q. \end{aligned}\right\} \tag{4.17}$$

Remark 4.3. If the observation $y(x,T;v)$ belongs to V (implying $D_0=$identity map of $V\to V$), the cost function being defined by

$$J(v)=\int_\Omega\left[(y(x,T;v)-z_d)^2+\sum_{i=1}^n\left(\frac{\partial}{\partial x_i}(y(x,T;v)-z_d)\right)^2\right]dx+(Nv,v), \tag{4.18}$$

then in (4.15) $y(x,T;u)-z_d$ must be replaced by

$$(-\Delta_x+I)\,(y(x,T;u)-z_d(x))$$

and the corresponding problem must be solved in the sense of Theorem 3.1.

Remark 4.4. Let us now suppose that the observation is

$$y'(x,T;v) \quad \text{in} \quad L^2(\Omega), \tag{4.19}$$

implying D = identity from $L^2(\Omega)\to L^2(\Omega)$.

The cost function is given by

$$J(v)=\int_\Omega [y'(x,T;v)-z_d(x)]^2\,dx+(Nv,v)_{\mathscr{U}} \tag{4.20}$$

and the adjoint state $p(u)$ is defined by

$$\left.\begin{aligned} &\frac{\partial^2}{\partial t^2}p(u)+A(t)p(u)=0 \quad \text{in } Q,\\ &p(u)=0 \quad \text{on } \Sigma,\\ &p(x,T;u)=y'(x,T;u)-z_d(x) \quad \text{in } \Omega,\\ &\frac{\partial}{\partial t}p(x,T;u)=0 \quad \text{in } \Omega, \end{aligned}\right\} \tag{4.21}$$

and this problem is *solved in the sense of Theorem 3.1.*[11]

From Theorem 3.2 we conclude that the optimal control is determined by the simultaneous solution of (4.2), (4.21) and of

$$\int_Q (p(u)+Nu)(v-u)\,dx\,dt\geq 0 \quad \forall v\in\mathscr{U}_{\text{ad}}. \tag{4.22}$$

4.3. Petrowsky Type Equation. Distributed Control. Distributed Observation

We now consider a system whose state is given by (cf. Section 1.6)

$$\frac{\partial^2 y(v)}{\partial t^2}+\Delta(a\,\Delta y(v))=f+v, \qquad v\in L^2(Q), \tag{4.23}$$

$$\left.\begin{aligned} &\Delta y(v)=0, \qquad \frac{\partial\Delta}{\partial n}y(v)=0 \quad \text{on } \Sigma,\\ &\Delta y(x,0;v)=y_0(x), \qquad \frac{\partial y}{\partial t}(x,0;v)=y_1(x), \qquad x\in\Omega. \end{aligned}\right\} \tag{4.24}$$

---

[11] Hence

$$\int_Q p(u)\left(\frac{\partial^2\varphi}{\partial t^2}+A(t)\varphi\right)dx\,dt=\int_\Omega (y'(x,T;u)-z_d(x))\varphi'(x,T)\,dx \quad \forall\varphi \text{ such that}$$

$$\frac{\partial^2\varphi}{\partial t^2}+A(t)\varphi\in L^2(Q),\quad \varphi\in H^1(Q),\quad \varphi|_\Sigma=0,\quad \varphi(x,0)=0,\quad \varphi'(x,0)=0.$$

If we assume that

$$J(v)=\int_Q (y(x,t;v)-z_d)^2\,dx\,dt+(Nv,v)_{\mathscr{U}}, \qquad \mathscr{U}=L^2(Q), \tag{4.25}$$

—which means that the observation $y(v)$ belongs to $L^2(Q)$—, then from Theorem 2.1, we see that the optimal control is determined by

$$\left.\begin{aligned} &\frac{\partial^2 p(u)}{\partial t^2}+\Delta(a\,\Delta p(u))=y(u)-z_d \quad \text{in } Q,\\ &\Delta p(u)=0, \quad \frac{\partial}{\partial n}\Delta p(u)=0 \quad \text{on } \Sigma,\\ &p(x,T;u)=0, \quad \frac{\partial p}{\partial t}(x,T;u)=0 \quad \text{on } \Omega, \end{aligned}\right\} \tag{4.26}$$

to which we add (4.23), (4.24) (where $v=u$) and

$$\int_Q (p(u)+Nu)(v-u)\,dx\,dt \geqslant 0 \quad \forall v\in\mathscr{U}_{\text{ad}}. \tag{4.27}$$

Example 4.3. If we again take $\mathscr{U}_{\text{ad}}$ to be given by (4.6), then we are confronted with an unilateral problem of type (4.7). A remark analogous to Remark 4.1 can now be made.

4.4. Petrowsky Type Equation. Distributed Control. Observation of the Final State

Let us consider the same system as in Section 4.3, with the observation being

$$y'(x,T;v) \quad \text{in } L^2(\Omega). \tag{4.28}$$

The cost function being given by (4.20), the adjoint state is defined by

$$\left.\begin{aligned} &\frac{\partial^2}{\partial t^2}p(u)+\Delta(a\,\Delta p(u))=0 \quad \text{in } Q,\\ &\left.\begin{aligned}&\Delta p(u)=0\\ &\frac{\partial}{\partial n}\Delta p(u)=0\end{aligned}\right\} \text{ on } \Sigma,\\ &p(x,T;u)=y'(x,T;u)-z_d(x) \quad \text{in } \Omega,\\ &p'(x,T;u)=0 \quad \text{in } \Omega. \end{aligned}\right\} \tag{4.29}$$

This problem has to be solved in the sense of Theorem 3.1, that is

$$\int_Q p(u)\left(\frac{\partial^2}{\partial t^2} + \Delta(a\,\Delta\varphi)\right)dx\,dt = \int_\Omega (y'(x,T;u)-z_d(x))\,\varphi'(x,T)\,dx$$

$$\forall\varphi \quad \text{such that } \frac{\partial^2\varphi}{\partial t^2} + \Delta(a\,\Delta\varphi)\in L^2(Q), \quad \varphi, \ \frac{\partial\varphi}{\partial t}, \quad \Delta\varphi\in L^2(\Omega),$$

$$\Delta\varphi|_\Sigma=0, \quad \frac{\partial}{\partial n}\Delta\varphi|_\Sigma=0, \quad \varphi(x,0)=0, \quad \varphi(x,0)=0.$$

The optimal control is determined by the simultaneous solution of (4.23), (4.24) (where $v=u$), (4.29) (or 4.30) and (4.22).

4.5. Orientation

We shall now investigate to what extent and how the results relative to decoupling (Chapter 3, Sections 4 and 5) can be adapted to the situation we are confronted with in this chapter. In the sequel, we shall also be concerned with situations where the control and observation are defined in a manner which is different from the cases we have so far studied. Finally we shall study the case where both the control and observation are on the boundary.

5. Decoupling

5.1. Problem Statement. Rewriting as a System of First Order Equations

We shall consider the case where there are no constraints ($\mathscr{U}_{\text{ad}}=\mathscr{U}$) in the framework of Section 2.2 and the sections following and see how we may extend the decoupling results (Chapter 3, Sections 4, 5). For this purpose it is convenient to write the equations defining the state of the system as a *system of first order equations*, which will be done in this section.

Remark 5.1. Formally the problem has been reduced to that we considered in Chapter 3, but the proofs are technically different. It is clear that just by rewriting we cannot pass from the hyperbolic to the parabolic case.

Assumptions. In order not to multiply the technical difficulties, we assume that

$$\left.\begin{aligned} a(t;\varphi,\psi)=a(\varphi,\psi) \quad &\text{does not depend on } t,\\ a(\varphi,\varphi)\geqslant\alpha\|\varphi\|_V^2=\alpha\|\varphi\|^2, \quad &\alpha>0, \quad \forall\varphi\in V.\end{aligned}\right\} \tag{5.1}$$

Notation. We set

$$\mathfrak{h}=V\times H; \tag{5.2}$$

if $\boldsymbol{\varphi},\boldsymbol{\psi}\in\mathfrak{h}$, $\boldsymbol{\varphi}=\{\varphi_0,\varphi_1\}$, $\boldsymbol{\psi}=\{\psi_0,\psi_1\}$, we define the scalar product on $\mathfrak{h}$ by

$$[\boldsymbol{\varphi},\boldsymbol{\psi}]=a(\varphi_0,\psi_0)+(\varphi_1,\psi_1) \tag{5.3}$$

(this is valid since $a(\varphi_0,\psi_0)=a(\psi_0,\varphi_0)$ and (5.1) holds). We then introduce

$$\mathscr{V}=V\times V. \tag{5.4}$$

We identify $\mathfrak{h}$ with its dual. Let $\mathscr{V}'$ be the dual of $\mathscr{V}$. Then

$$\mathscr{V}\subset\mathfrak{h}\subset\mathscr{V}'$$

and

$$\mathscr{V}'=V\times V'. \tag{5.5}$$

The *state* $y(v)$ is given by

$$\left.\begin{aligned} &\frac{d^2}{dt^2}y(v)+Ay(v)=f+Bv,\\ &y(v)\in L^2(0,T;V), \quad \frac{dy}{dt}(v)\in L^2(0,T;H),\end{aligned}\right\} \tag{5.6}$$

and

$$y(0;v)=\xi_0, \quad y'(0;v)=\xi_1.\,[12] \tag{5.7}$$

We assume that (notation analogous to that of Chapter 3, Section 4)

$$\mathscr{U}=L^2(0,T;E), \quad \mathscr{H}=L^2(0,T;F), \tag{5.8}$$

and that

$$Bv=\text{"}t\to B(t)v(t)\text{"}, \quad B(t)\in\mathscr{L}(E;H). \tag{5.9}$$

We now introduce

$$\boldsymbol{y}=\{y_0,y_1\}\in L^2(0,T;\mathfrak{h}). \tag{5.10}$$

---

[12] Here we change notation (y_i becomes ξ_i, $i=0,1$) in order to avoid confusion with $\boldsymbol{y}=\{y_0,y_1\}$ defined in the sequel.

We define, for $\boldsymbol{\varphi}\in\mathscr{V}$,

$$\mathscr{A}\boldsymbol{\varphi}=\{-\varphi_1, A\varphi_0\}, \qquad \mathscr{A}=\begin{pmatrix}0 & -1\\ A & 0\end{pmatrix} \tag{5.11}$$

and note that

$$\mathscr{A}\in\mathscr{L}(\mathscr{V};\mathscr{V}'). \tag{5.12}$$

The operator $\mathscr{A}$ is also in $\mathscr{L}(\mathfrak{h}; H\times V')$.

Thus formulated, problem (5.6), (5.7) is equivalent to[13]

$$\left.\begin{aligned}&\frac{d\boldsymbol{y}(v)}{dt}+\mathscr{A}\boldsymbol{y}(v)=\boldsymbol{f}+\mathscr{B}v, \qquad \boldsymbol{y}(v)\in L^2(0,T;\mathfrak{h}),\\ &\boldsymbol{y}(0;v)=\boldsymbol{\xi}=\{\xi_0,\xi_1\}\in\mathfrak{h},\end{aligned}\right\} \tag{5.13}$$

where

$$\boldsymbol{f}=\{0,f\}, \tag{5.14}$$

$$\mathscr{B}v=\{0,Bv\}. \tag{5.15}$$

We shall now study the (equivalent) formulation (5.13).

Let us remark that

$$\mathscr{A}+\mathscr{A}^*=0. \tag{5.16}$$

Indeed if $\boldsymbol{\varphi}\in\mathscr{V}$, $\boldsymbol{\psi}\in\mathscr{V}$,

$$[\mathscr{A}^*\boldsymbol{\varphi},\boldsymbol{\psi}]=[\boldsymbol{\varphi},\mathscr{A}\boldsymbol{\psi}]=a(\varphi_0,-\psi_1)+(\varphi_1,A\psi_0)=a(\varphi_1,\psi_0)+(-A\varphi_0,\psi_1),$$

which implies that $\mathscr{A}^*\boldsymbol{\varphi}=\{\varphi_1,-A\varphi_0\}$, whence (5.16).

5.2. Rewriting of the Set of Equations Determining the Optimal Control

We consider the case of Section 2.2 (and of Section 3.4). Hence $J(v)$ is given by (2.9) which is now written as (under the hypothesis of Section 5.1):

$$J(v)=\int_0^T\|C(t)y(t;v)-z_d\|_F^2\,dt+\int_0^T(N(t)v(t),v(t))_E\,dt \tag{5.17}$$

assuming that

$$Nv=\text{“}t\to N(t)\,v(t)\text{”}, \qquad N(t)\in\mathscr{L}(E;E).$$

---

[13] $\boldsymbol{y}(v)=\{y_0(v),y_1(v)\}$.

Let us introduce

$$\begin{aligned}&\mathscr{C}(t)\in\mathscr{L}(\mathfrak{h};F),\\&\mathscr{C}(t)\varphi=C(t)\varphi_0;\end{aligned}\tag{5.18}$$

then (5.17) may be written as

$$J(v)=\int_0^T\|\mathscr{C}(t)y(v)-z_d\|_F^2dt+\int_0^T(N(t)v(t),v(t))_E dt.\tag{5.19}$$

The optimal control u is characterized by

$$\int_0^T(\mathscr{C}(t)y(t;u)-z_d,\mathscr{C}(t)(y(t;v)-y(t;u)))_F dt+(Nu,v-u)_{\mathscr{U}}\geqslant 0,\qquad\forall v\in\mathscr{U}_{\mathrm{ad}},\tag{5.20}$$

or also by

$$\int_0^T[\mathscr{C}^*(t)(\mathscr{C}(t)y(u)-z_d),y(v)-y(u)]dt+(Nu,v-u)_{\mathscr{U}}\geqslant 0\ \ \forall v\in\mathscr{U}_{\mathrm{ad}}.\tag{5.21}$$

Calculation of $\mathscr{C}^*(t)$:

$$\begin{aligned}&\mathscr{C}^*(t)\in\mathscr{L}(F';\mathfrak{h})\quad\text{is defined by}\ (f'\in F')\\&[\mathscr{C}^*(t)f',\varphi]=[f',\mathscr{C}(t)\varphi]=(f',C(t)\varphi_0)=(C^*(t)f',\varphi_0),\\&C(t)\in\mathscr{L}(V;F),\qquad C^*(t)\in\mathscr{L}(F';V').\end{aligned}\tag{5.22}$$

Hence

$$[\mathscr{C}^*(t)f',\varphi]=a(A^{-1}C^*(t)f',\varphi_0),$$

and consequently

$$\mathscr{C}^*(t)f'=\{A^{-1}C^*(t)f',0\},\qquad f'\in F'.\tag{5.23}$$

The adjoint state $p(u)$ is then naturally introduced (by analogy with Chapter 3) by

$$\begin{aligned}&-\frac{d}{dt}p(u)+\mathscr{A}^*p(u)=\mathscr{C}^*(t)\Lambda_F(\mathscr{C}(t)y(u)-z_d),\\&p(T;0)=0.\end{aligned}\tag{5.24}$$

Remark 5.2. We have

$$\mathscr{C}^*(t)\Lambda_F(\mathscr{C}(t)y(u)-z_d)=\{A^{-1}C^*(t)\Lambda_F(C(t)y_0(u)-z_d),0\}\in L^2(0,T;\mathfrak{h})\tag{5.25}$$

which implies that (5.24) admits a unique solution

$$p(u)\in L^2(0,T;\mathfrak{h}).$$

We now transform (5.21). From (5.24) we deduce that

$$\int_0^T (\mathscr{C}(t)\,y(u)-z_d, \mathscr{C}(t)(y(v)-y(u)))_F\,dt$$

$$= \int_0^T \left[-\frac{d}{dt}\,p(u)+\mathscr{A}^*\,p(u),\ y(v)-y(u)\right]dt$$

$$= \int_0^T \left[p(u),\ \left(\frac{d}{dt}+\mathscr{A}\right)(y(v)-y(u))\right]dt$$

$$= \int_0^T [p(u), \mathscr{B}(v-u)]\,dt = \int_0^T (\mathscr{B}^*\,p(u), v-u)\,dt$$

$$= \int_0^T (\Lambda_E^{-1}\,\mathscr{B}^*\,p(u), v-u)_E\,dt$$

and consequently (5.21) is equivalent to

$$(\Lambda_E^{-1}\,\mathscr{B}^*\,p(u)+N\,u, v-u)_{\mathscr{U}} \geqslant 0, \qquad \forall\, v \in \mathscr{U}_{\mathrm{ad}}. \tag{5.26}$$

But we may verify that

$$\mathscr{B}^*\,p(u) = B^*\,p_1(u)$$

which implies that (5.26) is finally equivalent to

$$(B^*\,p_1(u)+N u, v-u)_{\mathscr{U}} \geqslant 0, \qquad \forall\, v \in \mathscr{U}_{\mathrm{ad}}. \tag{5.27}$$

Let us write the system of equations (5.24) explicitly:

$$\left.\begin{aligned} &-\frac{d}{dt}\,p_0(u)+p_1(u)=A^{-1}C^*(t)\,\Lambda_F(C(t)\,y_0(u)-z_d),\\ &-\frac{d}{dt}\,p_1(u)-A\,p_0(u)=0,\\ &\qquad p_0(T;u)=0, \quad p_1(T;u)=0. \end{aligned}\right\} \tag{5.28}$$

From the second equation of (5.28) we deduce that

$$p_0(u) = -\frac{d}{dt}\,A^{-1}\,p_1(u)$$

and therefore

$$A^{-1}\frac{d^2}{dt^2}p_1(u)+p_1(u)=A^{-1}C^*(t)\Lambda_F(C(t)y_0(u)-z_d),$$

which gives us

$$\left.\begin{aligned}&\frac{d^2}{dt^2}p_1(u)+A\,p_1(u)=C^*(t)\Lambda_F(C(t)y_0(u)-z_d),\\&p_1(T;u)=0\quad\text{and}\quad\frac{d}{dt}p_1(T;u)=0.\end{aligned}\right\}\tag{5.29}$$

Hence $p_1(u)$ is equal to $p(u)$ defined by (2.16) or (3.16) according as $C(t)\in\mathscr{L}(H;F)$ or $\mathscr{L}(V;F)$.

Remark 5.3. The method we have presented here does not require that we distinguish between the two cases "$C(t)\in\mathscr{L}(H;F)$" or "$C(t)\in\mathscr{L}(V;F)$" up to (5.24), (5.26). But the distinction comes into play when we want to interpret (5.29). In the first case, the solution of (5.29) is the usual solution, whereas in the second case we have to use transposition as in section 3.

Remark 5.4. The system of equations (5.13) is formally analogous to that we have studied in Chapter 3. But the operator $\mathscr{A}$ is not coercive on $\mathscr{V}$. From (5.16) we have

$$[\mathscr{A}\,\varphi,\varphi]=0$$

and hence the results of Chapter 3 are not applicable.

We may however "approximate" $\mathscr{A}$ by means of perturbed operators which are coercive on $\mathscr{V}$ and therefore approximate the control problems considered in this chapter by parabolic control problems. This procedure termed "*parabolic regularization*" will be studied in the following chapter.

5.3. Decoupling

In the case where there are no constraints ($\mathscr{U}_{\text{ad}}=\mathscr{U}$), (5.26) becomes

$$\Lambda_E^{-1}\mathscr{B}^*p(u)+N\,u=0.\tag{5.30}$$

and the system of equations determining the optimal control is:

$$\left.\begin{aligned}&\frac{dy}{dt}+\mathscr{A}\,y+\mathscr{D}_1\,p=f,\\&-\frac{dp}{dt}+\mathscr{A}^*p-\mathscr{D}_2\,y=-\mathscr{C}^*(t)\Lambda_F z_d=g,\\&y(0)=\xi,\qquad p(T)=0,\end{aligned}\right\}\tag{5.31}$$

where

$$\left.\begin{aligned}\mathscr{D}_1 &= \mathscr{B} N^{-1} \Lambda_E^{-1} \mathscr{B}^*,\\ \mathscr{D}_2 &= \mathscr{C}^* \Lambda_F \mathscr{C}.\end{aligned}\right\} \tag{5.32}$$

The operator $\mathscr{P}(s)$ *and the function* $r(s)$. We argue as in Chapter 3, Section 4. Let $s \in]0, T[$ and $\boldsymbol{h} \in \mathfrak{h}$. We consider the system (compare with (4.12), (4.13), Chapter 3)

$$\left.\begin{aligned}&\left.\begin{aligned}\frac{d\boldsymbol{\varphi}}{dt} + \mathscr{A}\boldsymbol{\varphi} + \mathscr{D}_1\boldsymbol{\psi} &= \boldsymbol{f}\\ -\frac{d\boldsymbol{\psi}}{dt} + \mathscr{A}^*\boldsymbol{\psi} - \mathscr{D}_2\boldsymbol{\varphi} &= \boldsymbol{g}\end{aligned}\right\} \text{in }]s, T[,\\ &\boldsymbol{\varphi}(s) = \boldsymbol{h}, \qquad \boldsymbol{\psi}(T) = 0.\end{aligned}\right\} \tag{5.33}$$

Lemma 5.1. *The system of equations* (5.33) *admits a unique solution and*

$$\boldsymbol{\varphi}, \boldsymbol{\psi} \in L^2(s, T; \mathfrak{h}).$$

Proof. Indeed, the system of equations (5.33) is precisely that which determines the optimal control of a system whose state is given by

$$\frac{d\boldsymbol{\varphi}(v)}{dt} + \mathscr{A}\boldsymbol{\varphi}(v) = \boldsymbol{f} + \mathscr{B}v, \qquad \boldsymbol{\varphi}(s) = \boldsymbol{h} \tag{5.34}$$

and the cost function by

$$J_s^h(v) = \int_s^T \|\mathscr{C}(t)\boldsymbol{\varphi}(v) - z_d\|_F^2 dt + \int_s^T (N(t)v(t), v(t))_E dt \tag{5.35}$$

as v ranges over $\mathscr{U}(s, T) = L^2(s, T; E)$.

In a manner analogous to the proof of Lemma 4.2, Chapter 3, we may verify that

$$\begin{aligned}&\text{the mapping } \boldsymbol{h} \to \{\boldsymbol{\varphi}, \boldsymbol{\psi}\} \quad \text{is an affine, continuous map of}\\ &\mathfrak{h} \to L^2(s, T; \mathfrak{h}) \times L^2(s, T; \mathfrak{h}).\end{aligned} \tag{5.36}$$

But from (5.33), $\dfrac{d\boldsymbol{\psi}}{dt} \in L^2(s, T; H \times V')$ implying that $\boldsymbol{\psi}(s)$ has a meaning. If we return to scalar notation for ψ_0, ψ_1, we shall obtain an equation analogous to (5.27) for ψ_1.

Let us then assume that[14]

$$C(t) \in \mathscr{L}(H; F). \tag{5.37}$$

[14] If $C(t) \in \mathscr{L}(V; F)$ the following calculations are formal.

Under these conditions

$$\psi_1 \text{ is continuous from } [s,T] \to V,$$
$$\psi_1' \text{ is continuous from } [s,T] \to H$$

and $\psi_0 = -A^{-1}\psi_1'$ is continuous from $[s,T] \to D(A)$.
Therefore

Lemma 5.2. *If* (5.37) *holds, the mapping*

$$\boldsymbol{h} \to \boldsymbol{\psi}(s) \tag{5.38}$$

is a continuous affine map of $\mathfrak{h} \to D(A) \times V$.[15]

Corollary 5.1. *If* (5.37) *holds, we have*

$$\left.\begin{aligned} &\boldsymbol{\psi}(s) = \mathscr{P}(s)\boldsymbol{h} + \boldsymbol{r}(s), \\ &\mathscr{P}(s) \in \mathscr{L}(\mathfrak{h}; D(A)\times V), \qquad \boldsymbol{r}(s) \in \mathfrak{h}. \end{aligned}\right\} \tag{5.39}$$

As in Lemma 4.3, Chapter 3, we obtain the "*fundamental identity*"

$$\boldsymbol{p}(t) = \mathscr{P}(t)\boldsymbol{y}(t) + \boldsymbol{r}(t), \tag{5.40}$$

where $\mathscr{P}(t)$ and $\boldsymbol{r}(t)$ are given as follows.
(i) we solve

$$\left.\begin{aligned} &\left.\begin{aligned} &\frac{d\boldsymbol{\beta}}{dt} + \mathscr{A}\boldsymbol{\beta} + \mathscr{D}_1\boldsymbol{\gamma} = 0 \\ &\frac{d\boldsymbol{\gamma}}{dt} + \mathscr{A}^*\boldsymbol{\gamma} - \mathscr{D}_2\boldsymbol{\beta} = 0 \end{aligned}\right\} \text{ in }]s,T[, \\ &\boldsymbol{\beta}(s) = \boldsymbol{h}, \qquad \boldsymbol{\gamma}(T) = 0, \end{aligned}\right\} \tag{5.41}$$

and then

$$\mathscr{P}(s)\boldsymbol{h} = \boldsymbol{\gamma}(s); \tag{5.42}$$

(ii) we solve

$$\left.\begin{aligned} &\left.\begin{aligned} &\frac{d\boldsymbol{\xi}}{dt} + \mathscr{A}\boldsymbol{\eta} + \mathscr{D}_1\boldsymbol{\xi} = \boldsymbol{f} \\ -&\frac{d\boldsymbol{\xi}}{dt} + \mathscr{A}^*\boldsymbol{\xi} - \mathscr{D}_2\boldsymbol{\xi} = \boldsymbol{g} \end{aligned}\right\} \text{ in }]s,T[, \\ &\boldsymbol{\eta}(s) = 0, \qquad \boldsymbol{\xi}(T) = 0 \end{aligned}\right\} \tag{5.43}$$

and then

$$\boldsymbol{r}(s) = \boldsymbol{\xi}(s). \quad \square \tag{5.44}$$

We then obtain

Lemma 5.3. *We have*

$$\mathscr{P}(s)^* = \mathscr{P}(s) \quad (\text{adjoint taken in } \mathscr{L}(\mathfrak{h};\mathfrak{h})). \tag{5.45}$$

---

[15] If $C(t) \in \mathscr{L}(V;F)$ we may show that (5.38) is continuous from $\mathfrak{h} \to \mathfrak{h}$.

Proof. Let $\boldsymbol{h}\in\mathfrak{h}, \boldsymbol{\beta}, \gamma$ be the solution corresponding to (5.41). We have

$$\int_s^T \left[-\frac{d\gamma}{dt}+\mathscr{A}^*\gamma-\mathscr{D}_2\boldsymbol{\beta}, \underline{\boldsymbol{\beta}}\right]dt=0$$

$$=[\gamma(s), \underline{\boldsymbol{\beta}}(s)]-\int_s^T [\mathscr{D}_2\underline{\boldsymbol{\beta}}, \boldsymbol{\beta}]\,dt+\int_s^T \left[\gamma, \frac{d}{dt}\underline{\boldsymbol{\beta}}+\mathscr{A}\underline{\boldsymbol{\beta}}\right]dt$$

whence

$$[\mathscr{P}(s)\boldsymbol{h}, \underline{\boldsymbol{h}}]\int_s^T ([\mathscr{D}_1\gamma, \underline{\gamma}]+[\mathscr{D}_2\boldsymbol{\beta}, \underline{\boldsymbol{\beta}}])\,dt; \tag{5.46}$$

the right hand side of (5.46) being symmetric in $\boldsymbol{h}, \underline{\boldsymbol{h}}$ we deduce (5.45). □

Lemma 5.4. *There exists a constant* c_1 *such that*

$$[\mathscr{P}(s)\boldsymbol{h}]\leqslant c_1[\boldsymbol{h}] \quad \forall\, \boldsymbol{h}\in\mathfrak{h}, \qquad s\in[0,T], \tag{5.47}$$

where $[\boldsymbol{h}]=[\boldsymbol{h},\boldsymbol{h}]^{\frac{1}{2}}$.

Proof. Consider the system whose state is given by (5.34) with $f=0$. Let the cost function be

$$\mathring{J}_s^{\boldsymbol{h}}(v)=\int_s^T \|\mathscr{C}(t)\varphi(v)\|_F^2\,dt+\int_s^T (N(t)v(t), v(t))_E\,dt. \tag{5.48}$$

If u is the corresponding optimal control

$$\mathring{J}_s^{\boldsymbol{h}}(u)\leqslant J_s^{\boldsymbol{h}}(0)\leqslant c_1[\boldsymbol{h}]^2$$

and furthermore u is given by (5.41), with

$$\Lambda_E^{-1}\mathscr{B}^*\gamma+Nu=0. \tag{5.49}$$

But then

$$\mathring{J}_s^{\boldsymbol{h}}(u)=\int_s^T [\mathscr{D}_2\boldsymbol{\beta}, \boldsymbol{\beta}]\,dt+\int_s^T (\Lambda_E N u, u)\,dt$$

$$=\int_s^T \left[-\frac{d\gamma}{dt}+\mathscr{A}^*\gamma, \beta\right]dt+\int_s^T (\Lambda_E N u, u)\,dt$$

$$=[\gamma(s), \boldsymbol{\beta}(s)]$$

(by taking into account the first equation of (5.41) and of (5.49)). Thus

$$[\mathscr{P}(s)\boldsymbol{h},\boldsymbol{h}]=\mathring{J}_s^{\boldsymbol{h}}(u) \tag{5.50}$$

and therefore

$$[\mathscr{P}(s)\boldsymbol{h},\boldsymbol{h}]\leqslant C_1[\boldsymbol{h}]^2 \quad \text{whence (5.47).} \quad \square$$

Further, using a proof similar to that of Lemma 4.4, we may show that the function $t\rightarrow[\mathscr{P}(t)\boldsymbol{h},\boldsymbol{h}]$ is continuous on $[0,T]$, $\forall\,\boldsymbol{h},\boldsymbol{h}\in\mathfrak{h}$.

5.4. Riccati Integro-differential Equation

We now use identity (5.40)[16]. Let us first carry out a formal calculation. In (5.31) replacing $\boldsymbol{p}(t)$ by (5.40), we obtain

$$-\mathscr{P}'(t)\boldsymbol{y}(t)-\mathscr{P}(t)\frac{d}{dt}\boldsymbol{y}(t)-\boldsymbol{r}'+\mathscr{A}^*(\mathscr{P}(t)\boldsymbol{y}(t)+\boldsymbol{r}(t))-\mathscr{D}_2\boldsymbol{y}=\boldsymbol{g}, \tag{5.51}$$

and using the first equation of (5.31) in (5.40), we get,

$$\left.\begin{aligned}&-\mathscr{P}'(t)\boldsymbol{y}(t)-\mathscr{P}(t)(-\mathscr{A}\boldsymbol{y}-\mathscr{D}_1(\mathscr{P}\boldsymbol{y}+\boldsymbol{r})+\boldsymbol{f})-\\&-\boldsymbol{r}'+\mathscr{A}^*(\mathscr{P}(t)\boldsymbol{y}(t)+\boldsymbol{r}(t))-\mathscr{D}_2\boldsymbol{y}=\boldsymbol{g}\end{aligned}\right\} \tag{5.52}$$

and (5.52) being an identity in $y(t)$ (t fixed but arbitrary), we deduce that

$$\mathscr{P}'(t)+\mathscr{P}(t)\mathscr{A}+\mathscr{A}^*\mathscr{P}(t)+\mathscr{P}(t)\mathscr{D}_1\mathscr{P}(t)=\mathscr{D}_2(t), \tag{5.53}$$

$$-\boldsymbol{r}'+\mathscr{A}^*\boldsymbol{r}-\mathscr{P}(t)\boldsymbol{r}=\mathscr{P}(t)\boldsymbol{f}+\boldsymbol{g}, \tag{5.54}$$

to which we must adjoin the boundary conditions

$$\mathscr{P}(T)=0, \tag{5.55}$$

$$\boldsymbol{r}(T)=0, \tag{5.56}$$

(corresponding to $\boldsymbol{p}(T)=0$).

Remark 5.5. From (5.16), this equation may also be written as

$$-\mathscr{P}'(t)+\mathscr{P}(t)\mathscr{A}+\mathscr{A}\mathscr{P}(t)+\mathscr{P}(t)\mathscr{D}_1\mathscr{P}(t)=\mathscr{D}_2(t). \tag{5.57}$$

The structure of equations (5.57), (5.55) is formally the same as that of the Riccati equations (4.55), (4.57) of Chapter 3; but in (4.55), Chapter 3, A and A^* are coercive whereas here $\mathscr{A}+\mathscr{A}^*=0$.

Our objective now is

(i) to justify the formal calculations

(ii) to study the structure of (5.53) is greater detail and to give examples.

---

[16] Let us reiterate as in Chapter 3 that we know the existence of $\mathscr{P}(s)$ and $\boldsymbol{r}(s)$.

Step (i). We may proceed in two different ways:

1. We follow a development analogous to Chapter 3, Section 4. We first work in finite dimensions and we then pass to the limit as in Chapter 3, Section 4.5; the technical details are tedious but not fundamentally different from that of Chapter 3.

2. We use parabolic regularization—for this we refer the reader to Chapter 5, Section 2.

Step (ii). We decompose the operator in the form

$$\mathscr{P}(s) = \begin{pmatrix} P_0(s) & P_1(s) \\ P_2(s) & P_3(s) \end{pmatrix} \tag{5.58}$$

where from Lemma 5.1 and Corollary 5.1:

$$\left.\begin{aligned} &P_0(s) \in \mathscr{L}(V; D(A)),\\ &P_1(s) \in \mathscr{L}(H; D(A)),\\ &P_2(s) \in \mathscr{L}(V; V),\\ &P_3(s) \in \mathscr{L}(H; V).^{17} \end{aligned}\right\} \tag{5.59}$$

We have

Lemma 5.5. *The operators* $P_i(s)$ *satisfy the following relations*:

$$P_0(s)^* A = A P_0(s), \tag{5.60}$$

$$P_2(s) = P_1(s)^* A, \tag{5.61}$$

$$P_3(s)^* = P_3(s), \tag{5.62}$$

where the adjoints are taken in the duality of H (resp. V) *with itself* (resp. V').

Proof. Indeed from (5.45) we have

$$a(P_0(s)h_0 + P_1(s)h_1, \underline{h}_0) + (P_2(s)h_0 + P_3(s)h_1, \underline{h}_1) = a(h_0, P_0(s)\underline{h}_0 + P_1(s)\underline{h}_1)$$
$$+ (h_1, P_2(s)\underline{h}_0 + P_3(s)\underline{h}_1) \quad \forall\, \boldsymbol{h} = \{h_0, h_1\}, \quad \underline{\boldsymbol{h}} = \{\underline{h}_0, \underline{h}_1\} \in \mathfrak{h}.$$

We thus deduce the relations of the Lemma.

If we note that

$$\mathscr{D}_1 = \begin{pmatrix} 0 & 0 \\ 0 & BN^{-1}\Lambda_E^{-1}B^* \end{pmatrix} = \begin{pmatrix} 0 & 0 \\ 0 & D_1 \end{pmatrix},$$

and

$$\mathscr{D}_2 = \begin{pmatrix} A^{-1}C^*\Lambda_F C & 0 \\ 0 & 0 \end{pmatrix} = \begin{pmatrix} A^{-1}D_2 & 0 \\ 0 & 0 \end{pmatrix},$$

then equation (5.53) is equivalent to

[17] If $C(t) \in \mathscr{L}(V; F)$, then $P_0(s) \in \mathscr{L}(V; V)$, $P_1(s) \in \mathscr{L}(H; V)$, $P_2(s) \in \mathscr{L}(V; H)$, $P_3(s) \in \mathscr{L}(H; H)$.

$$-P_0'+P_1A+P_2+P_1D_1P_2=A^{-1}D_2, \tag{5.63}$$

$$-P_1'-P_0+P_3+P_1D_1P_3=0, \tag{5.64}$$

$$-P_2'+P_3A-AP_0+P_3D_1P_2=0, \tag{5.65}$$

$$-P_3'-P_2-AP_1+P_3D_1P_3=0, \tag{5.66}$$

together with $P_i(T)=0,\ i=0,1,2,3$.

But equation (5.65) is a consequence of (5.60), (5.61), (5.62), (5.64); in fact, by passing to adjoints, (5.64) gives us

$$-(P_1^*)'-P_0^*+P_3^*+P_3^*D_1P_1^*=0,$$

and multiplying on the right by A:

$$-(P_1^*A)'-P_0^*A+P_3^*A+P_3^*D_1P_1^*A=0$$

when by taking into account (5.60), (5.61), (5.62) we get (5.65). □

Summarizing

Proposition 5.1. *The operators* P_0, P_1, P_3 *satisfy* (5.60), (5.62), (5.63) (where we replace P_2 by P_1^*A), (5.64) *and* (5.65) (where we replace P_2 by P_1^*A), *with* $P_i(T)=0$. *Then* $\mathscr{P}(t)$ *is given by*

$$\begin{pmatrix} P_0(t) & P_1(t) \\ P_1^*(t)A & P_3(t) \end{pmatrix}.$$

Remark 5.6. The $P_i(t)$ satisfy, as they should, a system of *Riccati integro-differential equations.* Equation (5.53) (or (5.57)) is more symmetric than the set of equations of Proposition 5.1. For decoupling, it is therefore advantageous to write the equations as a system of first order equations in t.

Remark 5.7. In examples, each operator $P_i(t)$ is represented by a kernel $P_i(x,\xi,t)$ which is a distribution.

We thus obtain classes of integro-partial differential Riccati equations whose direct study (i. e. without any reference to a control problem) is largely an open problem.

5.5. Another Optimal Control Problem. Decoupling

Let us consider the case of Section 2.3; then

$$J(v)=\int_0^T \|C(t)y'(t;v)-z_d\|_F^2\,dt+\int_0^T (N(t)v(t),v(t))_E\,dt. \tag{5.67}$$

We introduce

$$\left.\begin{aligned} &\mathscr{C}(t)\in\mathscr{L}(\mathfrak{h};F),\\ &\mathscr{C}(t)\varphi=C(t)\varphi_1,\quad \varphi=\{\varphi_0,\varphi_1\}\in\mathfrak{h},\quad C(t)\in\mathscr{L}(H;F).\end{aligned}\right\} \tag{5.68}$$

Then (5.67) may be written as

$$J(v)=\int_0^T \|\mathscr{C}(t)\,y(v)-z_d\|_F^2\,dt+\int_0^T (N(t)v(t),v(t))_E\,dt. \tag{5.69}$$

In this case the operator $\mathscr{C}(t)^*$ is given by

$$\mathscr{C}(t)^* f'=\{0,C(t)^* f'\}. \tag{5.70}$$

The adjoint state $p(u)$ is defined by

$$\left.\begin{aligned} -\frac{d}{dt}p(u)+\mathscr{A}^*p(u)&=\mathscr{C}^*(t)\Lambda_F(\mathscr{C}\,y(u)-z_d),\ {}^{18}\\ p(T;u)&=0.\end{aligned}\right\} \tag{5.71}$$

The optimal control u is characterized by (cf. (5.26))

$$(\Lambda_E^{-1}\mathscr{B}^*p(u)+Nu,v-u)_{\mathscr{U}}\geqslant 0\quad \forall v\in\mathscr{U}_{\mathrm{ad}}, \tag{5.72}$$

and hence in the case where there are no constraints, by

$$\Lambda_E^{-1}\mathscr{B}^*p(u)+Nu=0. \tag{5.73}$$

Let us write the system of equations (5.71) explicitly:

$$\left.\begin{aligned} &-\frac{d}{dt}p_0(u)+p_1(u)=0,\\ &-\frac{d}{dt}p_1(u)-A\,p_0(u)=C(t)^*\Lambda_F(C(t)\,y_1-z_d),\qquad p_i(T;u)=0,\end{aligned}\right\} \tag{5.74}$$

whence

$$\left.\begin{aligned} \frac{d^2}{dt^2}p_0(u)+A\,p_0(u)&=-C(t)^*\Lambda_F(C(t)\,y_1-z_d),\\ p_0(T;u)=0,&\qquad p_0'(T;u)=0.\end{aligned}\right\} \tag{5.75}$$

Comparing with (2.27), (2.28) (where we considered the case with A depending on t) we see that $p_0=-p$.

Noting that (5.73) is equivalent to

$$(\Lambda_E^{-1}B^*p_1+Nu,v-u)_{\mathscr{U}}\geqslant 0\ \ \forall v\in\mathscr{U}_{\mathrm{ad}}, \tag{5.76}$$

we see that this condition is equivalent to (2.33).

Remark 5.8. To obtain (5.72) we have to integrate by parts and utilize Green's formulae which lead to the same technical difficulties as in Section 2.3.

---

[18] We write in the same way as in (5.24), but the $\mathscr{C}(t)$ are different.

The decoupling is done in exactly the same manner as in Sections 5.3 and 5.4. We again have the identity

$$p(t) = \mathscr{P}(t)\, y(t) + r(t) \tag{5.77}$$

which leads to (5.53), (5.54) (with new operators $\mathscr{D}_1$ and $\mathscr{D}_2$). We again obtain a system of *Riccati integro-differential equations.*

Remark 5.9. Clearly (as in Remark 4.3, Chapter 3) once $\mathscr{P}(t)$ and $r(t)$ determined, we deduce (synthesis or feedback) that

$$u(t) = -N^{-1}(t)\Lambda_E^{-1}\mathscr{B}^*[\mathscr{P}(t)\,y(t) + r(t)]. \tag{5.78}$$

We also have remarks analogous to that of Chapter 3, Sections 4.7 and 4.8 (the remarks are purely formal as far as Section 4.8 is concerned).

6. Control via Initial Conditions. Estimation

6.1. Problem Statement

Consider a system whose state is given by

$$y''(v) + A(t)\,y(v) = f \quad \text{in }]0, T[, \tag{6.1}$$

where $f \in L^2(0, T; H)$ and

$$y(0; v) = 0, \qquad y'(0; v) = v, \tag{6.2}$$

where v is given in H. Thus

$$\mathscr{U} = H. \tag{6.3}$$

We then consider the cost function

$$J(v) = \|y(T; v) - z_d^0\|^2 + |y'(T; v) - z_d^1|^2 \; {}^{19} \tag{6.4}$$

where

$$z_d^0 \in V, \qquad z_d^1 \in H.$$

Let

$$\mathscr{U}_{\text{ad}} = \text{closed, convex subset of } H. \tag{6.5}$$

We search for

$$\operatorname*{Inf}_{v \in \mathscr{U}_{\text{ad}}} J(v).$$

Remark 6.1. The cost function $J(v)$ does not contain the factor $(Nv, v)_{\mathscr{U}} = (Nv, v)$. What follows is also valid if we introduce this term with $N \geqslant 0$, but the case (6.4) is more interesting—cf. Theorem 6.1 in the sequel.

---

[19] Let us recapitulate that $\| \; \|$ (resp. $| \; |$) denotes the norm on V (resp. H).

6.2. Coercivity of $J(v)$

Theorem 6.1. *Let us assume that* (1.4), (1.5), (1.6) *hold. The cost function being given by* (6.4), *there exists an optimal control u which is unique.*

Proof. Let us consider the homogeneous quadratic part of $J(v)$, denoted by $\pi(v,v)$. We have,

$$\pi(v,v)=\|y(T;v)-y(T;0)\|^2+|y'(T;v)-y'(T;0)|^2.$$

Let us set $\psi=y(t;v)-y(t;0)$; we have,

$$\psi''+A(t)\psi=0, \qquad \psi(0)=0, \qquad \psi'(0)=v, \tag{6.6}$$

and

$$\pi(v,v)=\|\psi(T)\|^2+|\psi'(T)|^2. \tag{6.7}$$

We shall verify that

$$\pi(v,v)\geqslant c\,|v|^2, \qquad c>0, \qquad \forall v\in H. \tag{6.8}$$

Indeed, consider the backward problem

$$\begin{aligned}&\psi''+A(t)\psi=0;\\ &\psi(T) \quad \text{given in } V,\\ &\psi'(T) \quad \text{given in } H.\end{aligned}$$

It is *well posed* and *in particular* the mapping

$$\{\psi(T),\psi'(T)\}\to\psi'(0)$$

is continuous from $V\times H\to H$ and hence

$$|\psi'(0)|^2\leqslant c_1(\|\psi(T)\|^2+|\psi'(T)|^2),$$

which is equivalent to (6.8). □

Remark 6.2. If we take

$$J(v)=|y(T;v)-z_d^0|^2+|y'(T;v)-z_d^1|^2,$$

the analogue of Theorem 6.1 is *no longer true*.

The *existence* of an optimal control (generally non-unique) is no longer assured unless $\mathscr{U}_{\text{ad}}$ is *bounded* in H.

From the "estimation" point of view, this has the following significance:

(i) without a priori information on $v(\mathscr{U}_{\text{ad}}=\mathscr{U})$ or if $\mathscr{U}_{\text{ad}}$ is not bounded, we have to observe $y(T;v)$ in V and $y'(T;v)$ in H in order to estimate u;

(ii) if we have a priori information on v allowing us to conclude that "$\mathscr{U}_{\text{ad}}$ is bounded", it may be sufficient to observe $y(T;v)$ in H and $y'(T;v)$ in H, or even only $y(T;v)$ in H.

Remark 6.3. The result of Theorem 6.1 is peculiar to *reversible systems: it is necessary to be able to reverse the direction of time in the equation specifying the state* of the system. There is no analogue of this result in the context of problems considered in Chapter 3.

6.3. System of Equations Determining the Optimal Control

Let us again take the cost function (6.4). Then the optimal control u is characterized by

$$\left.\begin{aligned}&(y(T;u)-z_d^0, y(T;v)-y(T;u))_V\\&+(y'(T;u)-z_d^1, y'(T;v)-y'(T;u))\geqslant 0\\&\forall v\in\mathcal{U}_{ad}.\end{aligned}\right\}\tag{6.9}$$

We define the adjoint state $p(u)$ by using the analogue of Theorem 3.1 (in the backward sense): there exists a unique $p(u)\in L^2(0,T;H)$ such that

$$\left.\begin{aligned}&\int_0^T (p(u),\varphi''+A\varphi)dt=(y'(T;u)-z_d^1,\varphi'(T))+(y(T;u)-z_d^0,\varphi(T))_V,\\&\forall\varphi\in L^2(0,T;V),\quad \varphi'\in L^2(0,T;H),\quad \varphi''+A\varphi\in L^2(0,T;H),\\&\varphi(0)=0,\quad \varphi'(0)=0.\end{aligned}\right\}\tag{6.10}$$

We may show that $p(u)$ is continuous from $[0,T]\to H$ and we have Green's formula cf. C. Baiocchi [1], [2])

$$\begin{aligned}\int_0^T (p(u),\psi''+A\psi)dt=&(y'(T;u)-z_d^1,\psi'(T))+(y(T;u)-z_d^0,\psi(T))_V\\&-(p(0;u),\psi'(0)),\end{aligned}\tag{6.11}$$

if ψ has the same properties as φ in (6.16) but $\psi'(0)\neq 0$.

In (6.11) let us take $\psi=y(v)-y(u)$; we deduce that

$$\begin{aligned}&(y(T;u)-z_d^0,y(T;v)-y(T;u))_V+(y'(T;u)-z_d^1,y'(T;v)-y'(T;u))\\&=(p(0;u),v-u),\end{aligned}$$

and consequently (6.9) is equivalent to

$$(p(0;u),v-u)\geqslant 0\ \ \forall v\in\mathcal{U}_{ad}.\tag{6.12}$$

We therefore have

Theorem 6.2. *Let the hypotheses of Theorem* 6.1 *hold. The optimal control u is given by the simultaneous solution of*

$$\left.\begin{aligned}&y''(u)+A(t)y(u)=f,\\&\qquad y(0;u)=0,\quad y'(0;u)=u,\end{aligned}\right\}\tag{6.13}$$

together with equation (6.10) *and equation* (6.12).

Remark 6.4. Let Λ be the canonical isomorphism of V onto V'. Equation (6.10) may be interpreted as

$$\left.\begin{aligned} p''(u)+A(t)p(u)&=0,\\ p(T;u)&=y'(T;u)-z_d^1,\\ p'(T;u)&=-\Lambda(y(T;u)-z_d^0).\end{aligned}\right\}\tag{6.14}$$

Remark 6.5. We may eliminate u and thus obtain the following system of equations:

$$\left.\begin{aligned} &y''+A(t)y=f, \quad p''+A(t)p=0 \quad \text{in }]0,T[,\\ &y(0)=0, \quad p(T)=y'(T)-z_d^1,\\ &p'(T)=-\Lambda(y(T)-z_d^0),\\ &(p(0),v-y'(0))\geqslant 0 \ \ \forall v\in\mathscr{U}_{\mathrm{ad}}.\end{aligned}\right\}\tag{6.15}$$

Problem (6.15) is a nonlinear problem of unilateral type.

Example 6.1. Let A be a second order operator as in 4.1.
Let

$$V=H_0^1(\Omega) \text{ and } \mathscr{U}_{\mathrm{ad}}=\{v\,|\,v\geqslant 0 \text{ a.e. on } \Omega\}.$$

Then (6.15) becomes

$$\left.\begin{aligned} &\frac{\partial^2 y}{\partial t^2}+A\left(x,t,\frac{\partial}{\partial x}\right)y=f, \quad \frac{\partial^2 p}{\partial t^2}+A\left(x,t,\frac{\partial}{\partial x}\right)p=0 \quad \text{in } Q,\\ &\left.\begin{aligned} &y(x,0)=0, \quad p(x,T)=\frac{\partial y}{\partial t}(x,T)-z_d^1(x),\\ &\frac{\partial p}{\partial t}(x,T)=-(-\Delta_x+1)\left(\frac{\partial y}{\partial t}(x,T)-z_d^1\right)^{20}\end{aligned}\right\}\text{in } \Omega\\ &y=0 \quad \text{on } \Sigma, \quad p=0 \quad \text{on } \Sigma,\\ &\frac{\partial y}{\partial t}(x,0)\geqslant 0, \quad p(x,0)\geqslant 0, \quad p(x,0)\frac{\partial y}{\partial t}(x,0)=0 \quad \text{a.e. in } \Omega.\end{aligned}\right\}\tag{6.16}$$

Example 6.2. If we take $V=H^1(\Omega)$ in the preceding example, we have to replace the conditions "$y=0$ on Σ" and "$p=0$ on Σ" by

$$\frac{\partial y}{\partial \nu_A}=0, \quad \frac{\partial p}{\partial \nu_A}=0 \quad \text{on } \Sigma.\tag{6.17}$$

[20] We assume that

$$(\varphi,\psi)_V=(\varphi,\psi)_{H_0^1(\Omega)}=\int_\Omega\left(\varphi\psi+\sum_{i=1}^n\frac{\partial\varphi}{\partial x_i}\frac{\partial\psi}{\partial x_i}\right)dx,$$

giving us $\Lambda=-\Delta_x+I$.

Remark 6.6. Everything that we have done in this chapter is valid for the case where A is not *necessarily symmetric*, but only the second order principal part of A is symmetric. It suffices to replace A by A^* in the equations determining the adjoint state p.

Remark 6.7. Results analogous to the preceding can be obtained if the state $y(v)$ is given by

$$\left.\begin{aligned} y''(v)+A(t)y(v)=f,& \\ y(0;v)=v, \quad y'(0;v)=0.& \end{aligned}\right\} \tag{6.18}$$

If $\mathscr{U}=V$, the solution of (6.18) is to be taken in the sense of Section 1.

If $\mathscr{U}=H$, the solution of (6.18) is to be taken in the sense of Section 3, and hence

$$\left.\begin{aligned} &\int_0^T (y(v),\varphi''+A(t)\varphi)dt=\int_0^T (f,\varphi)dt+(v,\varphi'(T)),\\ &\forall\varphi\in L^2(0,T;V),\quad \varphi'\in L^2(0,T;H),\quad \varphi''+A(t)\varphi\in L^2(0,T;H),\\ &\varphi(0)=0,\quad \varphi'(0)=0. \end{aligned}\right\} \tag{6.19}$$

In this case $y(T;v)\in H$ and $y'(T;v)\in V'$ and we may obtain results which are analogous to the preceding if we take

$$J(v)=|y(T;v)-z_d^0|^2+\|y'(T;v)-z_d^1\|_{V'}^2. \tag{6.20}$$

7. Boundary Control (I)

7.1. Problem Statement

We shall now examine the situation where control is exercised through the boundary.

We first study the case where the state is given by[21]

$$\frac{\partial^2 y(v)}{dt^2}+Ay(v)=f \quad \text{in } Q, \tag{7.1}$$

$$\frac{\partial y(v)}{\partial \nu_A}=v \quad \text{on } \Sigma, \tag{7.2}$$

$$\left.\begin{aligned} y(x,0)=y_0(x),\quad x\in\Omega,\\ \frac{\partial y(x,0)}{dt}=y_1(x),\quad x\in\Omega, \end{aligned}\right\} \tag{7.3}$$

[21] $A=$ second order operator as in Section 4.1.

and in the following section we shall consider the case where we replace (7.2) (Neumann condition) by the Dirichlet condition:

$$y(v) = v \quad \text{on } \Sigma.$$

Clearly, problem (7.1), (7.2), (7.3) must first be correctly formulated. We shall do this under the assumption

$$v \in \mathscr{U} = L^2(\Sigma). \tag{7.4}$$

Remark 7.1. Initially the difficulties associated with the formulation given below in Section 7.2 can be avoided by taking v to be more regular (cf. Remarks (i), (ii) of Chapter 3, Section 9.1). However when writing the optimality conditions (cf. Chapter 3, Section 9.5) we are confronted with other difficulties. Here we shall restrict ourselves to case (7.4).

7.2. Definition of the State of the System

Lemma 7.1. *There exists a unique element $y(v)$ in $L^2(Q)$ such that*

$$\left.\begin{aligned}\int_Q y(v)\left(\frac{\partial^2\varphi}{\partial t^2} + A\varphi\right)dx\,dt = \int_Q f\varphi\,dx\,dt - \int_\Omega y_0(x)\frac{\partial\varphi}{\partial t}(x,0)\,dx \\ + \int_\Omega y_1(x)\varphi(x,0)\,dx + \int_\Sigma v\varphi\,d\Sigma\end{aligned}\right\} \tag{7.5}$$

for all functions φ such that $\varphi \in X$, where

$$\left.\begin{aligned}X = \Big\{\varphi \mid \varphi \in L^2(0,T; H^1(\Omega)), \varphi' \in L^2(Q), \varphi'' + A\varphi \in L^2(Q), \\ \frac{\partial\varphi}{\partial\nu_A} = 0 \text{ on } \Sigma,\ \varphi(x,T) = 0, \varphi'(x,T) = 0\Big\}.\end{aligned}\right\} \tag{7.6}$$

Proof. Indeed, if $v \in L^2(\Sigma)$ the linear form

$$\varphi \to \int_\Sigma v\varphi\,d\Sigma$$

in continuous on X,[22] allowing us to conclude that the right hand side of (7.5) is continuous on X and hence from (3.2) we get the desired result. □

Let us also note that

$$\text{the mapping } v \to y(v) \text{ is a continuous affine map of } L^2(\Sigma) \to L^2(Q). \tag{7.7}$$

[22] This is true even if we take $v \in H^{-\frac{1}{2}}(\Sigma)$.

Remark 7.2. By applying Green's formula we may verify without difficulty that the solution $y(v)$ of (7.5) satisfies equations (7.1), (7.2), (7.3).

We shall now study the system whose state is given by (7.5).

Remark 7.3. Expansion in eigen-functions. Assume that

$$A(t) = A \quad \text{does not depend on } t. \tag{7.8}$$

Let $w_1, \dots, w_m, \dots,$ be the eigen-functions of A for the *Neumann problem:*

$$\left.\begin{aligned} &A w_j = \lambda_j w_j, \\ &\frac{\partial w_j}{\partial v_A} = 0 \quad \text{on } \Gamma, \qquad \int_\Omega w_j w_k dx = o_j, \qquad j = 1, \dots. \end{aligned}\right\} \tag{7.9}$$

Then $y(v) = y$, the solution of (7.5) may be represented uniquely by

$$y(v) = y = \sum_{j=1}^{\infty} g_j(t) w_j, \tag{7.10}$$

$$g_j \in L^2(0,\mathrm{T}), \qquad \sum_{j=1}^{\infty} \int_0^T |g_j(t)|^2 dt < \infty. \tag{7.11}$$

We compute the g_j as follows (compare with Chapter 3, Remarks 8.7 and 9.3).

In (7.5) let us take

$$\varphi = \psi(t) w_j(x), \qquad \psi \in C^2[0,T], \qquad \psi(T) = 0, \qquad \psi'(T) = 0. \tag{7.12}$$

We get

$$\int_0^T (g_j \psi'' + \lambda_j g_j \psi) dt = \int_0^T f_j \psi dt - y_{0j} \psi'(0) + y_{1j} \psi(0) + \int_0^T v_j \psi dt \quad \forall \psi, \tag{7.13}$$

where

$$\left.\begin{aligned} f_j(t) &= \int_\Omega f(x,t) w_j(x) dx, \\ y_{0j} &= \int_\Omega y_0(x) w_j(x) dx, \\ y_{1j} &= \int_\Omega y_1(x) w_j(x) dx, \\ v_j(t) &= \int_\Omega v(x,t) w_j(x) d\Gamma. \end{aligned}\right\} \tag{7.14}$$

Hence

$$g_j'' + \lambda_j g_j = f_i + v_j, \qquad g_j(0) = y_{0j}, \qquad g_j'(0) = y_{1j}. \tag{7.15}$$

7.3. Distributed Observation

Let us first assume that we observe $y(v)$ in Q and let the cost function be

$$J(v)=\int_Q (y(v)-z_d)^2\,dx\,dt+(N v,v)_{L^2(\Sigma)}, \tag{7.16}$$

$$N\in\mathscr{L}(L^2(\Sigma);L^2(\Sigma)),\qquad N\geqslant \nu\,(\text{identity}),\qquad \nu>0.$$

Then there exists an optimal control u in $\mathscr{U}_{\text{ad}}$ (closed convex subset of $\mathscr{U}=L^2(\Sigma)$) which is characterized by

$$\int_Q (y(u)-z_d)(y(v)-y(u))\,dx\,dt+(N u,v-u)_{L^2(\Sigma)}\geqslant 0\quad \forall v\in\mathscr{U}_{\text{ad}}. \tag{7.17}$$

We introduce the adjoint state by means of

$$\left.\begin{aligned}
&\frac{\partial^2}{\partial t^2}p(u)+A\,p(u)=y(u)-z_d \quad \text{in } Q,\\
&\frac{\partial}{\partial \nu_A}p(u)=0 \quad \text{on } \Sigma,\\
&p(x,T;u)=0,\qquad \frac{\partial p}{\partial t}(x,T;u)=0,\qquad x\in\Omega.
\end{aligned}\right\} \tag{7.18}$$

This problem which admits a unique solution.

(7.17) may now be transformed in the following way: from (7.5) (which is applicable with $v=u$) we deduce that

$$\int_Q (y(v)-y(u))\left(\frac{\partial^2\varphi}{\partial t^2}+A\varphi\right)dx\,dt=\int_\Sigma (v-u)\varphi\,d\Sigma\quad \forall\varphi\in X. \tag{7.19}$$

But we may take $\varphi=p(u)$, and then (7.19) gives us

$$\int_Q (y(v)-y(u))(y(u)-z_d)\,dx\,dt=\int_\Sigma (v-u)p(u)\,d\Sigma. \tag{7.20}$$

We then have

Theorem 7.1. *Let us assume that the state is given by* (7.5) *and the cost function by* (7.16). *Then the optimal control u is determined by the simultaneous solution of* (7.5) *(where $v=u$),* (7.18) *and*

$$\int_\Sigma (p(u)+N u)(v-u)\,d\Sigma\geqslant 0\quad \forall v\in\mathscr{U}_{\text{ad}}. \tag{7.21}$$

Example 7.1. Case where there are no constraints: $\mathscr{U}_{\text{ad}}=\mathscr{U}$.
Then (7.21) reduces to

$$p(u)+N u=0. \tag{7.22}$$

Eliminating u, we obtain the system of equations

$$\left.\begin{aligned}
&\frac{\partial^2 y}{\partial t^2} + A y = f, \quad \frac{\partial^2 p}{\partial t^2} + A p = y - z_d \quad \text{in } Q,\\
&\frac{\partial y}{\partial \nu_A} + N^{-1} p = 0, \quad \frac{\partial p}{\partial \nu_A} = 0 \quad \text{on } \Sigma,\\
&\left.\begin{aligned}
&y(x,0) = y_0(x) \quad \frac{\partial y}{\partial t}(x,0) = y_1(x),\\
&p(x,T) = 0, \quad \frac{\partial p}{\partial t}(x,T) = 0
\end{aligned}\right\} \text{on } \Omega.
\end{aligned}\right\} \tag{7.23}$$

Example 7.2. $\mathscr{U}_{\text{ad}} = \{v \mid v \geqslant 0 \text{ on } \Sigma\}$.
The optimal control u is obtained by solving the unilateral problem

$$\left.\begin{aligned}
&\frac{\partial^2 y}{\partial t^2} + A y = f, \quad \frac{\partial^2 p}{\partial t^2} + A p = y - z_d \quad \text{on } Q,\\
&\left.\begin{aligned}
&\frac{\partial y}{\partial \nu_A} = 0, \quad p + N\frac{\partial y}{\partial \nu_A} \geqslant 0,\\
&\frac{\partial y}{\partial \nu_A}\left(p + N\frac{\partial y}{\partial \nu_A}\right) = 0, \quad \frac{\partial p}{\partial \nu_A} = 0
\end{aligned}\right\} \text{on } \Sigma,\\
&\left.\begin{aligned}
&y(x,0) = y_0(x), \quad \frac{\partial y}{\partial t}(x,0) = y_1(x),\\
&p(x,T) = 0, \quad \frac{\partial p}{\partial t}(x,T) = 0
\end{aligned}\right\} \text{on } \Omega,
\end{aligned}\right\} \tag{7.24}$$

and then

$$u = \frac{\partial y}{\partial \nu_A}. \tag{7.25}$$

7.4. Boundary Observation

Let us now assume that we observe $y(v)$ on Σ. Let us admit that the following lemma is true.

Lemma 7.2. *We have: $y(v) \in L^2(\Sigma)$ and the mapping $v \to y(v)|_\Sigma$ is a continuous, affine map of $L^2(\Sigma)$ into itself.*

Remark 7.4. The main idea of a proof of Lemma 7.2 is the following: we remarked that if $v \in H^{-\frac{1}{2}}(\Sigma)$ then $y(v) \in L^2(Q) = L^2(0, T; L^2(\Omega))$ and we may directly verify that if $v \in H^{\frac{1}{2}}(\Sigma)$ then $y(v) \in L^2(0, T; H^2(\Omega))$. Then by using interpolation[23] theory (cf. Lions-Magenes [1], Chapter 1) we deduce that if $v \in L^2(\Sigma)$ then $y(v) \in L^2(0, T; H^1(\Omega))$ and hence in particular $y(v) \in L^2(\Sigma)$ whence the Lemma.

We choose as cost function

$$\left.\begin{aligned} &J(v) = \int_\Sigma (y(v) - z_d)^2 \, d\Sigma + (N v, v)_{L^2(\Sigma)}, \\ &N \in \mathscr{L}(L^2(\Sigma); L^2(\Sigma)), N \geqslant \nu \text{ (identity)}, \qquad \nu > 0. \end{aligned}\right\} \tag{7.26}$$

Let $\mathscr{U}_{\text{ad}}$ be a closed, convex subset of $\mathscr{U}$. Then the optimal control u is characterized by

$$\int_\Sigma (y(u) - z_d)(y(v) - y(u)) \, d\Sigma + (N u, v - u)_{L_2(\Sigma)} \geqslant 0 \quad \forall v \in \mathscr{U}_{\text{ad}}. \tag{7.27}$$

We introduce the adjoint state $p(u)$ by[24]

$$\left.\begin{aligned} &\frac{\partial^2}{\partial t^2} p(u) + A p(u) = 0 \quad \text{in } Q, \\ &\frac{\partial}{\partial \nu_A} p(u) = y(u) - z_d \quad \text{on } \Sigma, \\ &p(x, T; u) = 0, \quad \frac{\partial p}{\partial t}(x, T; u) = 0 \quad \text{in } \Omega. \end{aligned}\right\} \tag{7.28}$$

By using Green's formula we transform (7.27) as follows: formally, we multiply the first equation in (7.28) by $y(v) - y(u)$ and integrate by parts to obtain

$$\begin{aligned} 0 &= -\int_\Sigma \frac{\partial p}{\partial \nu_A}(u)(y(v) - y(u)) \, d\Sigma + \int_\Sigma p(u) \left(\frac{\partial y}{\partial \nu_A}(v) - \frac{\partial y}{\partial \nu_A}(u) \right) d\Sigma \\ &= -\int_\Sigma (y(u) - z_d)(y(v) - y(u)) \, d\Sigma + \int_\Sigma p(u)(v - u) \, d\Sigma. \end{aligned}$$

We justify the formal calculation by approximating y by regular functions (for example) and by passing to the limit. Condition (7.27)

[23] We reduce the affine case to the linear situation.

[24] We utilize the backward analogue of Lemma 7.1 to define $p(u)$.

then becomes

$$\int_{\Sigma} (p(u)+Nu)(v-u)\,d\Sigma \geqslant 0 \quad \forall v \in \mathscr{U}_{ad}. \tag{7.29}$$

We thus finally have

Theorem 7.2. *We assume that the state is given by* (7.5) *and the cost function by* (7.26). *Then the unique optimal control u is given by the simultaneous solution of* (7.5) *(with $v=u$)* (7.28) *and* (7.29).

Example 7.3. Case where there are no constraints: $\mathscr{U}_{ad}=\mathscr{U}$. Then (7.29) reduces to

$$p(u)+Nu=0 \quad \text{on } \Sigma, \tag{7.30}$$

and eliminating u we obtain the following system of equations

$$\left.\begin{array}{l}
\dfrac{\partial^2 y}{\partial t^2}+Ay=f, \quad \dfrac{\partial^2 p}{\partial t^2}+Ap=0 \quad \text{in } Q,\\[2ex]
\dfrac{\partial y}{\partial v_A}+N^{-1}p=0, \quad \dfrac{\partial p}{\partial v_A}=y-z_d \quad \text{on } \Sigma,\\[2ex]
\left.\begin{array}{l}
y(x,0)=y_0(x), \quad \dfrac{\partial y}{\partial t}(x,0)=y_1(x),\\[2ex]
p(x,T)=0, \quad \dfrac{\partial p}{\partial t}(x,T)=0
\end{array}\right\} \text{in } \Omega.
\end{array}\right\} \tag{7.31}$$

Example 7.4. Assume that

$$\mathscr{U}_{ad}=\{v \mid v>0 \text{ a.e. on } \Sigma\}.$$

Then u is obtained by solving the *unilateral problem*

$$\left.\begin{array}{l}
\dfrac{\partial^2 y}{\partial t^2}+Ay=f, \quad \dfrac{\partial^2 p}{\partial t^2}+Ap=0 \quad \text{in } Q,\\[2ex]
p+N\dfrac{\partial y}{\partial v_A}=\geqslant 0, \quad \dfrac{\partial y}{\partial v_A}\geqslant 0,\\[2ex]
\left(p+N\dfrac{\partial y}{\partial v_A}\right)\dfrac{\partial y}{\partial v_A}=0, \quad \dfrac{\partial p}{\partial v_A}=y-z_d \quad \text{on } \Sigma,\\[2ex]
\left.\begin{array}{l}
y(x,0)=y_0(x), \quad \dfrac{\partial y}{\partial t}(x,0)=y_1(x),\\[2ex]
p(x,T)=0, \quad \dfrac{\partial p}{\partial t}(x,T)=0
\end{array}\right\} \text{in } \Omega,
\end{array}\right\} \tag{7.32}$$

and then

$$u=\frac{\partial y}{\partial \nu_A}.$$

Remark 7.5. The preceding may be extended to the case where the system is governed by non-hyperbolic operators, as for example those of Section 4.3.

8. Boundary Control (II)

8.1. Problem Statement

We use the same notation as in the previous section. We assume that the state $y(v)$ is given by

$$\frac{\partial^2 y(v)}{\partial t^2}+A\,y(v)=f \text{ in } Q, \tag{8.1}$$

$$y(v)=v \quad \text{on } \Sigma, \tag{8.2}$$

$$y(x,0;v)=y_0(x), \quad \frac{\partial y}{\partial t}(x,0;v)=y_1(x). \tag{8.3}$$

The control is therefore boundary control and the mixed problem (8.1), (8.2), (8.3) is of Dirichlet type.

If we assume that $v\in L^2(\Sigma)$, then we must define the solution of the problem by transposition (as in the previous section) but we do not consider whether $y(v)\in L^2(Q)$ or not.

Clearly we may consider control problems where $y(v)\notin L^2(Q)$; for example if $y(v)\in H^{-1}(Q)$ (which happens to be the case—cf. Lions-Magenes [1]) we may take as cost function

$$J(v)=\|y(v)-z_d\|^2_{H^{-1}(Q)}+(N\,v,v)_{L^2(\Sigma)}. \tag{8.4}$$

Here we shall assume that the control v is more "regular".

8.2. Control v Regular

We now make a very strong assumption:

$$\mathscr{U}=H_0^2(\Sigma) \tag{8.5}$$

(i.e. $v\in\mathscr{U}$ if and only if all the second derivatives of v on Σ are in $L^2(\Sigma)$ and

$$v(x,0)=v(x,T)=v'(x,0)=v'(x,T)=0,\ x\in\Gamma).$$

Then problem (8.1), (8.2), (8.3) admits a unique solution $y(v) \in H^1(Q)$. Indeed we may first construct ψ such that

$$\psi \in H^2(Q), \quad \psi|_\Sigma = v, \quad \psi(x,0)=0, \quad \psi'(x,0)=0,$$

and $z = y(v) - \psi$ must satisfy

$$\begin{aligned} z'' + Az &= f - (\psi'' + A\psi) = \tilde{f} \in L^2(Q), \\ z &= 0 \quad \text{on } \Sigma, \\ z(x,0) &= y_0(x), \quad z'(x,0) = y_1(x), \quad x \in \Omega. \end{aligned}$$

Let us take as observation the final state $y(x,T;v)$ and let

$$J(v) = |y(x,T;v) - z_d|^2_{L^2(\Omega)} + \nu \|v\|^2_{H^2_0(\Sigma)}, \quad \nu > 0, \tag{8.6}$$

where we have taken

$$\|v\|^2_{H^2_0(\Sigma)} = \int_\Sigma (\Delta_\Gamma v)^2 \, d\Gamma \, dt + \int_\Sigma (v'')^2 \, d\Gamma \, dt, \tag{8.7}$$

Δ_Γ being the Laplace-Beltrami operator on Γ.

The optimal control u which exists and is unique is given by

$$(y(x,T;u) - z_d, \, y(x,T;v) - y(x,T;u)) + \nu(u, v-u)_{H^2_0(\Sigma)} \geq 0 \quad \forall v \in \mathscr{U}_{ad}, \tag{8.8}$$

($\mathscr{U}_{ad}$ being a closed, convex subset of $H^2_0(\Sigma)$).

The adjoint state $p(u)$ is given by

$$\left.\begin{aligned} p''(u) + Ap(u) &= 0 \quad \text{in } Q, \\ p(x,T;u) &= 0, \\ p'(x,T;u) &= y(x,T;u) - z_d(x), \\ p(u) &= 0 \quad \text{on } \Sigma. \end{aligned}\right\} \tag{8.9}$$

It may be shown (cf. Lions-Magenes [1]) that

$$\frac{\partial p}{\partial \nu_A}(u) \in H^{-2}(\Sigma) \quad \text{(this result is not the best possible)} \tag{8.10}$$

and we may deduce from (8.9) by using Green's formula that

$$(y(x,T;u) - z_d, \, y(x,T;v) - y(x,T;u)) = \int_\Sigma \frac{\partial p}{\partial \nu_A^*}(v-u)\, d\Sigma,$$

and then (8.8) becomes

$$\int_\Sigma \frac{\partial p}{\partial \nu_A}(u)\,(v-u)\, d\Sigma + \nu(u, v-u)_{H^2_0(\Sigma)} \geq 0 \quad \forall v \in \mathscr{U}_{ad}, \tag{8.11}$$

and finally

$$\int_{\Sigma}\left(\frac{\partial p}{\partial \nu_A}(u)+\nu\left(\Delta_\Gamma^2+\frac{\partial^4}{\partial t^4}\right)u\right)(v-u)\,d\Sigma \geqslant 0 \quad \forall v \in \mathscr{U}_{\mathrm{ad}}. \tag{8.12}$$

Therefore

Theorem 8.1. *We assume that* (8.5) *holds and that the cost function is given by* (8.6). *Then the optimal control u is determined by the simultaneous solution of the system of equations* (8.1), (8.2), (8.3) *(with $v=u$) and* (8.9), (8.12).

8.3. Examples

Example 8.1. $\mathscr{U}_{\mathrm{ad}}=\mathscr{U}$.

We then solve the system

$$\left.\begin{aligned}
&y''+Ay=f, \quad p''+Ap=0 \quad \text{in } Q,\\
&y(x,0)=y_0(x), \quad y'(x,0)=y_1(x),\\
&p(x,T)=0, \quad p'(x,T)=y(x,T)-z_d(x) \quad \text{in } \Omega,\\
&p=0 \quad \text{on } \Sigma,\\
&\frac{\partial p}{\partial \nu_A}+\nu\left(\Delta_\Gamma^2+\frac{\partial^4}{\partial t^4}\right)y=0 \quad \text{on } \Sigma,\\
&y(x,0)=y(x,T)=y'(x,0)=y'(x,T)=0 \quad \text{if } x\in\Gamma.
\end{aligned}\right\} \tag{8.13}$$

Then

$$u=y|_{\Sigma}. \tag{8.14}$$

Example 8.2. $\mathscr{U}_{\mathrm{ad}}=\{v \mid v\in H_0^2(\Sigma),\, v\geqslant 0 \text{ on } \Sigma\}$.

We then solve the unilateral problem

$$\left.\begin{aligned}
&y''+Ay=f, \quad p''+Ap=0 \quad \text{in } Q,\\
&y(x,0)=y_0(x), \quad y'(x,0)=y_1(x),\\
&p(x,T)=0, \quad p'(x,T)=y(x,T)-z_d(x) \quad \text{in } \Omega,\\
&p=0 \quad \text{on } \Sigma,\\
&y\geqslant 0 \quad \text{on } \Sigma,\\
&\frac{\partial p}{\partial \nu_A}+\nu\left(\Delta_\Gamma^2+\frac{\partial^4}{\partial t^4}\right)y\geqslant 0 \quad \text{on } \Sigma,\\
&y\left[\frac{\partial p}{\partial \nu_A}+\left(\Delta_\Gamma^2+\frac{\partial^4}{\partial t^4}\right)y\right]=0 \quad \text{on } \Sigma,\\
&y(x,0)=y(x,T)=y'(x,0)=y'(x,T)=0 \quad \text{if } x\in\Gamma
\end{aligned}\right\} \tag{8.15}$$

The optimal control u is then given by (8.14).

9. Parabolic-Hyperbolic Systems

9.1. Recapitulation of Some General Results

Let $\mathscr{V}$ and $\mathfrak{h}$ be two Hilbert spaces[25] over $\mathbb{R}$, $\mathscr{V} \subset \mathfrak{h}$, $\mathscr{V}$ dense in $\mathfrak{h}$; identifying $\mathfrak{h}$ with its dual we have

$$\mathscr{V} \subset \mathfrak{h} \subset \mathscr{V}'.$$

Let $\mathscr{A}$ be given, with

$$\mathscr{A} \in \mathscr{L}(\mathscr{V}; \mathscr{V}'), \tag{9.1}$$

$$\langle \mathscr{A}\varphi, \varphi \rangle \geqslant \alpha \|\varphi\|_{\mathscr{V}}^2, \quad \alpha > 0, \quad \forall \varphi \in \mathscr{V}, \text{ [26]} \tag{9.2}$$

and let Λ be an unbounded operator in $\mathfrak{h}$ such that

$$\left.\begin{array}{l} -\Lambda \text{ is the infinitesimal generator of a semi-group } G(s) \text{ in } \mathfrak{h} \\ \text{and in } \mathscr{V} \text{ and in } \mathscr{V}', \text{ such that } G(s) \text{ is a contraction in } \mathfrak{h}. \text{ [27]} \end{array}\right\} \tag{9.3}$$

We denote by $D(\Lambda; \mathfrak{h})$ (resp. $D(\Lambda; \mathscr{V})$, resp. $D(\Lambda, \mathscr{V}')$) the domain of Λ in $\mathfrak{h}$ (resp. $\mathscr{V}$, resp. $\mathscr{V}'$).

In this setting, it has been proved in Lions-Magenes [1], Chapter 3, that

Theorem 9.1. *Let f be given in $\mathscr{V}'$. There exists a unique y such that*

$$y \in \mathscr{V} \cap D(\Lambda; \mathscr{V}'), \tag{9.4}$$

$$\Lambda y + \mathscr{A} y = f. \tag{9.5}$$

The mapping $f \to y$ is continuous from $\mathscr{V}' \to \mathscr{V} \cap D(\Lambda; \mathscr{V}')$.

Application. Consider the real variables $\{x_0, t\}$, with

$$\{x_0, t\} \in \mathcal{O} =]0, a_0[\times]0, T[, \quad a_0 < \infty, \quad T < \infty. \tag{9.6}$$

Let $V \subset H \subset V'$ as in Section 1 and let us consider

$$\mathscr{V} = L^2(\mathcal{O}; V), \quad \mathfrak{h} = L^2(\mathcal{O}; H). \text{ [28]} \tag{9.7}$$

Then $\mathscr{V}' = L^2(\mathcal{O}; V')$.

---

[25] These spaces play a role which is different from the spaces $\mathscr{V}_i$ introduced in Section 5.

[26] The angular brackets denote the scalar product between $\mathscr{V}'$ and $\mathscr{V}$.

[27] I.e. $\|G(s)\|_{\mathscr{L}(\mathfrak{h};\mathfrak{h})} \leqslant 1$.

[28] Hence for example $\mathscr{V} = \{\psi \mid \psi(x_0, t) \in V, x_0, t \to \psi(x_0, t)$, measurable from

$$\mathcal{O} \to V, \int_{\mathcal{O}} \|\psi(x_0, t)\|_{\mathscr{V}}^2 \, dx_0 \, dt < \infty\}.$$

Let $a(x_0,t;\varphi,\psi)$ be a family of bi-linear forms on V with

$$\left.\begin{array}{l} x_0,t\to a(x_0,t;\varphi,\psi) \text{ bounded measurable on } \mathcal{O}, \\ a(x_0,t;\varphi,\psi)\geq\alpha\|\varphi\|_V^2 \text{ a. e. in } \mathcal{O}, \quad \forall\varphi\in V, \quad \alpha>0. \text{ [29]} \end{array}\right\} \tag{9.8}$$

Then if $\psi\in\mathscr{V}$

$$\mathscr{A}\psi = \text{“}x_0,t\to A(x_0,t)\psi(x_0,t)\text{”}, \tag{9.9}$$

where

$$A(x_0,t)\in\mathscr{L}(V;V') \tag{9.10}$$

is defined by

$$(A(x_0,t)\varphi_1,\varphi_2)=a(x_0,t;\varphi_1,\varphi_2) \;\; \forall\varphi_1,\varphi_2\in V.$$

Assumption (9.2) holds.

The operator Λ is defined by

$$\Lambda\varphi=\frac{\partial\varphi}{\partial t}+\frac{\partial\varphi}{\partial x_0}, \tag{9.11}$$

with[30]

$$\left.\begin{array}{l} D(\Lambda;\mathfrak{h})=\left\{\varphi\,|\,\varphi\in\mathfrak{h},\dfrac{\partial\varphi}{\partial t}+\dfrac{\partial\varphi}{\partial x_0}\in\mathfrak{h},\right. \\ \left.\varphi(x_0,0)=0,\varphi(0,t)=0\right\}. \text{ [31]} \end{array}\right\} \tag{9.12}$$

The operator Λ satisfies (9.3); indeed it suffices to write $G(s)$ explicitly:

$$G(s)\varphi(x_0,t)=\begin{cases}\varphi(x_0-s,t-s) & \text{if } x_0>s, \quad t>s,\\ 0 \quad \text{otherwise.}\end{cases} \tag{9.13}$$

We are therefore in a position to apply Theorem 9.1. Consequently,

Theorem 9.2. *We assume that* (9.8) *holds*[32]. *There exists a unique function* $y=y(x_0,t)$ *having the following properties:*

$$\left.\begin{array}{l} y\in L^2(\mathcal{O};V), \\ \dfrac{\partial y}{\partial t}+\dfrac{\partial y}{\partial x_0}\in L^2(\mathcal{O};V'), \end{array}\right\} \tag{9.14}$$

---

[29] It would suffice if $a(x_0,t;\varphi,\psi)+\lambda\|\varphi\|_H^2\geq\alpha\|\varphi\|_V^2$ for appropriate λ.

[30] $D(\Lambda;\mathscr{V})$ or $D(\Lambda;\mathscr{V}')$ are similarly defined, replacing $\mathfrak{h}$ by $\mathscr{V}$ or $\mathscr{V}'$.

[31] These conditions have meaning.

[32] Note that $a(x_0,t;\varphi,\psi)$ is not necessarily symmetric in φ,ψ.

$$\left.\begin{aligned}&\frac{\partial y}{\partial t}+\frac{\partial y}{\partial x_0}+A(x_0,t)y=f,\\ &f \text{ given in } L^2(\mathcal{O};V'),\end{aligned}\right\}\tag{9.15}$$

$$y(x_0,0)=0, \qquad x_0\in]0,a_0[, \tag{9.16}$$

$$y(0,t)=0, \qquad t\in]0,T[. \tag{9.17}$$

Example 9.1. Let us take

$$V=H_0^1(\Omega), \qquad H=L^2(\Omega),$$

$$A(x_0,t)\varphi=-\sum_{i,j=1}^{n}\frac{\partial}{\partial x_i}\left(a_{ij}(x,x_0,t)\frac{\partial\varphi}{\partial x_j}\right),$$

$$a_{ij}(x,x_0,t)\in L^\infty(\Omega\times\mathcal{O}),$$

$$\sum_{i,j=1}^{n} a_{ij}(x,x_0,t)\xi_i\xi_j \geqslant \alpha(\xi_1^2+\cdots+\xi_n^2) \ \ \forall \xi\in\mathbb{R}^n.$$

Then the problem solved by Theorem 9.2 is the following:

$$\frac{\partial y}{\partial t}+\frac{\partial y}{\partial x_0}+A\left(x,x_0,t,\frac{\partial}{\partial x}\right)y=f \quad \text{in } \Omega\times\mathcal{O},$$

$$y(x,x_0,t)=0 \quad \text{if } x\in\Gamma\ (=\text{boundary of } \Omega),$$

$$y(x,0,t)=0, \qquad x\in\Omega, \qquad t\in]0,T[,$$

$$y(x,x_0,0)=0, \qquad x\in\Omega, \qquad x_0\in]0,a_0[,$$

and

$$y, \frac{\partial y}{\partial x_i}\ (i=1,2,\dots,n)\in L^2(\Omega\times\mathcal{O}).$$

We shall consider other examples later.

Remark 9.1. The operator $\frac{\partial}{\partial t}+\frac{\partial}{\partial x_0}+A\left(x_0,t\frac{\partial}{\partial x}\right)$ is the "combination" of the *hyperbolic operator* $\frac{\partial}{\partial t}+\frac{\partial}{\partial x_0}$ and the *parabolic operator* $\frac{\partial}{\partial t}+A\left(x_0,t,\frac{\partial}{\partial x}\right)$; this justifies the title of this section.

Remark 9.2. It is clear that we may consider far more general examples of the preceding type by taking Λ to be a first order hyperbolic operator—or a first order hyperbolic system in $\mathcal{O}=G\times]0,T[$, G open in $\mathbb{R}^m$, $x_0\in\mathbb{R}^m$.

Remark 9.3. In Theorem 9.2, we may equally well take $A\left(x_0,t,\frac{\partial}{\partial x}\right)$ to be an operator or a system of order $2m$, $m>1$.

9.2. Complement

We shall now extend Theorem 9.2 to the case where the initial condition (9.17) is non-zero. We shall need the following *trace result:*

Lemma 9.1. *Let* $z \in L^2(\mathcal{O}; V)$, $\frac{\partial z}{\partial x_0} + \frac{\partial z}{\partial t} \in L^2(\mathcal{O}; V')$. *We may then define* $z(x_0, 0) \in L^2(0, a_0; H)$; *conversely if* $\varphi(x_0)$ *is given in* $L^2(0, a_0; H)$, *we can find* z *(non-unique) with the above property and continuously depending on* φ *such that* $z(x_0, 0) = \varphi(x_0)$.

Proof. 1. By extension and by reflection we shall reduce the problem to the case where $\mathcal{O} = \mathbb{R}^2$.

We use the "diagonalization" (cf. J. Dixmier [1]) procedure: there exists a measurable sum[33]

$$\mathfrak{h} = \overset{\oplus}{\int} \mathfrak{h}(\lambda)\, d\mu(\lambda), \qquad 0 < \lambda_0 \leqslant \lambda < \infty,$$

such that the canonical isomorphism Λ of $V \to V'$ becomes the multiplication operator λ^2. More precisely, there exists a unitary operator J of H such that

$$J \Lambda \varphi = \lambda^2 J \varphi \quad \forall \varphi \in V.$$

We further carry out the operation of Fourier Transform $\tilde{J}$ in the variables x_0 and t (ξ_0 and τ being the dual variables).

If

$$\hat{z} = \tilde{\mathcal{F}} J z,$$

we have

$$\left.\begin{aligned} &\lambda \hat{z} \in L^2(\mathbb{R}^2; \mathfrak{h}), \\ &\frac{1}{\lambda}(\xi_0 + \tau)\hat{z} \in L^2(\mathbb{R}^2; \mathfrak{h}). \end{aligned}\right\} \tag{9.18}$$

Let us set,

$$B = B(\xi_0, \tau, \lambda) = \left(\lambda^2 + \frac{1}{\lambda^2}(\xi_0 + \tau)^2\right)^{\frac{1}{2}}. \tag{9.19}$$

We shall prove that $z(x_0, 0) \in L^2(\mathbb{R}_t; H)$. Since J is unitary and also $\tilde{J}$ is unitary, this is equivalent to proving

$$\int_{-\infty}^{+\infty} \hat{z}(\xi_0, \tau, \lambda)\, d\tau = \psi(\xi_0, \lambda) \in L^2(\mathbb{R}_{\xi_0}; \mathfrak{h}). \tag{9.20}$$

---

[33] We assume that V is separable.

Now

$$\psi(\xi_0,\lambda)=\int_{-\infty}^{+\infty} B\hat{z}\frac{1}{B}\,d\tau,$$

and hence

$$\|\psi(\xi_0,\lambda)\|^2_{\mathfrak{h}(\lambda)}\leqslant \int_{-\infty}^{+\infty} B^2\|\hat{z}(\xi_0,\tau,\lambda)\|^2_{\mathfrak{h}(\lambda)}\,d\tau \int_{-\infty}^{+\infty}\frac{1}{B^2}\,d\tau$$

and consequently

$$\|\psi(\xi_0,\lambda)\|^2_{\mathfrak{h}(\lambda)}\leqslant c\int_{-\infty}^{+\infty} B^2\|\hat{z}(\xi_0,\tau,\lambda)\|^2_{\mathfrak{h}(\lambda)}\,d\tau,$$

from which by virtue of (9.18) we obtain

$$\int_{\lambda}^{\infty}\int_{-\infty}^{+\infty}\|\psi(\xi_0,\lambda)\|^2_{\mathfrak{h}(\lambda)}\,d\xi_0\,d\mu(\lambda)<\infty$$

whence (9.20).

2. Conversely, let φ be given in $L^2(0,T;H)$. We shall construct z.

We again reduce the problem to the line $\mathbb{R}_t$ and to the plane. We then use the technique introduced in 1., to construct z satisfying (9.18) and

$$\int_{-\infty}^{\infty}\hat{z}(\xi_0,\tau,\lambda)\,d\tau=\psi(\xi_0,\lambda) \tag{9.21}$$

given $\psi\in L^2(\mathbb{R}_t;\mathfrak{h})$.

We take

$$\hat{z}(\xi_0,\tau,\lambda)=F(\xi_0,\tau,\lambda)\,\psi(\tau,\lambda),$$

with

$$F(\xi_0,\tau,\lambda)=c\,\frac{1}{\lambda^2+(1/\lambda^2)(\xi_0+\tau)^2},$$

$$c^{-1}=\int_{-\infty}^{+\infty}\frac{d\rho}{1+\rho^2}.$$

It may be verified that (9.18), (9.21) hold. □

We can now prove

Theorem 9.3. *Let us assume that* (9.8) *holds. We are given*

$$f\in L^2(\mathcal{O};V'), \tag{9.22}$$

$$y_0(x_0)\in L^2(0,a_0;H). \tag{9.23}$$

Then there exists a function $y=y(x_0,t)$ which is unique having the following properties: y satisfies (9.14), (9.15), (9.17) *and*

$$y(x_0,0)=y_0(x_0). \tag{9.24}$$

Proof. From Lemma 9.1, there exists a $z \in L^2(\mathcal{O}; V)$ such that

$$\frac{\partial z}{\partial x_0} + \frac{\partial z}{\partial t} \in L^2(\mathcal{O}; V') \quad \text{and} \quad z(x_0,0) = y_0(x_0).$$

Then $y - z = \tilde{y}$ must satisfy

$$\frac{\partial \tilde{y}}{\partial t} + \frac{\partial \tilde{y}}{\partial x_0} + A\tilde{y} = f - \left(\frac{\partial z}{\partial t} + \frac{\partial z}{\partial x_0}\right) - Az = \tilde{f},$$

and conditions analogous to (9.16), (9.17). But since

$$\frac{\partial z}{\partial x_0} + \frac{\partial z}{\partial t} \quad \text{and} \quad Az \in L^2(\mathcal{O}; V'), \quad \text{we have } \tilde{f} \in L^2(\mathcal{O}; V')$$

and the result is a consequence of Theorem 9.2. □

9.3. Control Problems

We formulate the problem in the context of Theorem 9.3.
Therefore, let $\mathcal{U}$ be the space of controls and

$$B \in \mathcal{L}(\mathcal{U}; L^2(\mathcal{O}; V')). \tag{9.25}$$

For $v \in \mathcal{U}$, the state

$$y(v) = y(x_0, t; v)$$

is given by (f and y_0 being given and satisfying (9.22), (9.23))

$$\frac{\partial y(v)}{\partial x_0} + \frac{\partial y(v)}{\partial t} + A\left(x_0, t, \frac{\partial}{\partial x}\right) y(v) = f + Bv \quad \text{in } \Omega \times \mathcal{O}, \tag{9.26}$$

$$y(x_0, 0; v) = y_0(x_0), \qquad x_0 \in]0, a_0[, \tag{9.27}$$

$$y(0, t; v) = 0, \qquad t \in]0, T[. \tag{9.28}$$

Let the observation be distributed[34] in $\Omega \times \mathcal{O}$, and let the cost function be given by

$$J(v) = \int_{\mathcal{O}} |y(x_0, t; v) - z_d|_H^2 \, dx_0 \, dt + (Nv, v)_{\mathcal{U}}, \tag{9.29}$$

where

$$\left.\begin{aligned} &z_d \text{ is given in } L^2(\Omega \times \mathcal{O}; H), \\ &N \in \mathcal{L}(\mathcal{U}; \mathcal{U}), \qquad N \geqslant \gamma \text{ (identity)}, \qquad \gamma > 0. \end{aligned}\right\}$$

Let $\mathcal{U}_{ad}$ = closed, convex subset of $\mathcal{U}$. We look for $\operatorname{Inf} J(v)$, $v \in \mathcal{U}_{ad}$.

[34] Clearly other cases are possible: we may observe $y(x_0, T; v)$ in $L^2(0, a_0; H)$, or $y(a_0, t; v) \in L^2(0, T; H)$ etc. We shall not give a detailed account of the various possibilities. They may be treated using the same principles.

The optimal control exists and it is unique. It is characterized by

$$\int_{\Omega} (y(u)-z_d, y(v)-y(u))\,dx_0\,dt + (Nu, v-u)_{\mathcal{U}} \geq 0 \quad \forall v \in \mathcal{U}_{\text{ad}}. \tag{9.30}$$

The adjoint state $p(u)$ is defined by

$$\left.\begin{aligned} -\frac{\partial}{\partial x_0}p(u) - \frac{\partial}{\partial t}p(u) + A^*\left(x_0, t, \frac{\partial}{\partial x}\right)p(u) &= y(u) - z_d \quad \text{in } \Omega \times \mathcal{O}, \\ p(x_0, T; u) = 0, &\quad x_0 \in]0, a_0[, \\ p(a_0, t; u) = 0, &\quad t \in]0, T[, \end{aligned}\right\} \tag{9.31}$$

where

$$\left.\begin{aligned} &p(u) \in L^2(\mathcal{O}; V), \\ &\left(\frac{\partial}{\partial x_0} + \frac{\partial}{\partial t}\right)p(u) \in L^2(\mathcal{O}; V'). \end{aligned}\right\} \tag{9.32}$$

Let us admit for the moment

Lemma 9.2. *We have*

$$\left.\begin{aligned} &\iint_{\mathcal{O}} \left(\left(\frac{\partial}{\partial x_0} + \frac{\partial}{\partial t}\right)p(u), y(v) - y(u)\right)dx_0\,dt \\ &\qquad = -\iint_{\mathcal{O}} \left(p(u), \left(\frac{\partial}{\partial x_0} + \frac{\partial}{\partial t}\right)(y(v) - y(u))\right)dx_0\,dt. \end{aligned}\right\} \tag{9.33}$$

Then from (9.31) and (9.33), we deduce that

$$\begin{aligned} &\int_{\mathcal{O}} (y(u) - z_d, y(v) - y(u))\,dx_0\,dt \\ &\quad = \int_{\mathcal{O}} \left(p(u), \left(\frac{\partial}{\partial x_0} + \frac{\partial}{\partial t} + A(x_0, t)\right)(y(v) - y(u))\right)dx_0\,dt \\ &\quad = \int_{\mathcal{O}} (p(u), B(v-u))\,dx_0\,dt \\ &\quad = (B^* p(u), v-u) \;{}^{35} = (\Lambda_{\mathcal{U}}^{-1} B^* p(u), v-u)_{\mathcal{U}} \end{aligned}$$

and consequently we have (provided Lemma 9.2 is verified to be true):

Theorem 9.4. *Let us assume that the state* $y(v)$ *is given by* (9.26), (9.27), (9.28) *and the cost function is given by* (9.29). *The optimal control* u *exists and it is unique. It is determined by the simultaneous solution of*

---

[35] The bracket here denotes the scalar product between $\mathcal{U}'$ and $\mathcal{U}$.

(9.26), (9.27), (9.28) *(where* $v=u$*)*, (9.31) *and*

$$(\Lambda_{\mathcal{U}}^{-1} B^* p(u)+Nu, v-u)_{\mathcal{U}} \geqslant 0 \quad \forall v \in \mathcal{U}_{\mathrm{ad}}. \tag{9.34}$$

Proof of Lemma 9.2. Indeed

$$y(v)-y(u) \in \mathcal{V} \cap D(\Lambda; \mathcal{V}'), \qquad \Lambda = \frac{\partial}{\partial x_0} + \frac{\partial}{\partial t},$$

since

$$y(x_0, 0; v) - y(x_0, 0; u) = 0;$$

moreover $p(u) \in D(\Lambda^*; \mathcal{V}') \cap \mathcal{V}$. The equality (9.33) is therefore equivalent to

$$(\Lambda \varphi, \psi) = (\varphi, \Lambda^* \psi) \ \forall \varphi, \psi,$$

with $\varphi \in \mathcal{V} \cap D(\Lambda; \mathcal{V}')$ and $\psi \in \mathcal{V} \cap D(\Lambda^*; \mathcal{V}')$. Thus, this equality is true if $\varphi \in D(\Lambda; \mathcal{V})$ and since $D(\Lambda; \mathcal{V})$ is dense in $\mathcal{V} \cap D(\Lambda; \mathcal{V}')$, by passing to the limit the desired result is proved. □

9.4. Example (I)

In the framework of Section 9.3, let us take

$$V = H_0^1(\Omega),$$
$$\mathcal{U} = L^2(\mathcal{O}; V') = L^2(\mathcal{O}; H^{-1}(\Omega)),$$
$$B = \text{identity}.$$

Then the optimal control u is determined by the simultaneous solution of the system of equations and inequality

$$\left.\begin{array}{l}
\left.\begin{array}{l}
\dfrac{\partial}{\partial x_0} y(u) + \dfrac{\partial}{\partial t} y(u) + A\left(x_0, t, \dfrac{\partial}{\partial x}\right) y(u) = f + u \\
-\left(\dfrac{\partial}{\partial x_0} + \dfrac{\partial}{\partial t}\right) p(u) + A^*\left(x_0, t, \dfrac{\partial}{\partial x}\right) p(u) = y(u) - z_d
\end{array}\right\} \text{in } \Omega \times \mathcal{O}, \\
y(x, x_0, t; u) = 0, \quad p(x, x_0, t; u) = 0 \quad \text{if } x \in \Gamma, \quad \{x_0, t\} \in \mathcal{O}, \\
y(x, x_0, 0; u) = y_0(x, x_0), \quad x_0 \in]0, a_0[, \quad x \in \Omega, \\
y(x, 0, t; u) = 0, \quad t \in]0, T[, \quad x \in \Omega, \\
p(x, x_0, T; u) = 0, \quad x_0 \in]0, a_0[, \quad x \in \Omega, \\
p(x, a_0 t; u) = 0, \quad t \in]0, T[, \quad x \in \Omega,
\end{array}\right\} \tag{9.35}$$

$$\int_{\Omega \times \mathcal{O}} (p(u) + (-\Delta_x + 1) N u)(v-u) \, dx \, dx_0 \, dt \geqslant 0 \quad \forall v \in \mathcal{U}_{\mathrm{ad}}. \tag{9.36}$$

If we take

$$\mathscr{U} = L^2(\Omega \times \mathcal{O}) \tag{9.37}$$

then (9.35) remains unchanged and (9.36) is replaced by

$$\int_{\Omega \times \mathcal{O}} (p(u) + Nu)(v - u)\,dx\,dx_0\,dt \geqslant 0 \quad \forall v \in \mathscr{U}_{\mathrm{ad}}. \tag{9.38}$$

9.5. Example (II)

Let us again consider the situation of Section 9.3:

$$\begin{aligned} &V = H^1(\Omega), \\ &B \in \mathscr{L}(\mathscr{U}; L^2(\mathcal{O} \times \Gamma)). \end{aligned} \tag{9.39}$$

By taking scalar product with $\psi \in V$, let us write (9.26) in the form

$$\begin{aligned} \left(\left(\frac{\partial}{\partial x_0} + \frac{\partial}{\partial t}\right) y(v), \psi\right) + a(x_0, t; y(v), \psi) &= \int_\Omega f \psi(x)\,dx \\ &+ \int_\Gamma Bv(x, x_0, t)\psi(x)\,d\Gamma \quad \forall \psi \in H^1(\Omega). \end{aligned} \tag{9.40}$$

To simplify somewhat, let us assume

$$\mathscr{U} = L^2(\mathcal{O} \times \Gamma), \qquad B = \text{identity}. \tag{9.41}$$

Then (9.40) may be interpreted as

$$\left(\frac{\partial}{\partial x_0} + \frac{\partial}{\partial t}\right) y(v) + A\left(x_0, t; \frac{\partial}{\partial x}\right) y(v) = f \quad \text{in } \Omega \times \mathcal{O}, \tag{9.42}$$

$$\frac{\partial y}{\partial v_A}(x, x_0, t; v) = v \quad \text{if } x \in \Gamma, \quad \{x_0, t\} \in \mathcal{O}, \tag{9.43}$$

to which (9.27), (9.28) are to be added.

The state is therefore given by *the solution of a parabolic-hyperbolic problem with boundary control.*

Let us assume that

$$\mathscr{U}_{\mathrm{ad}} = \{v \mid v \geqslant 0 \text{ a.e. on } \Gamma \times \mathcal{O}\}. \tag{9.44}$$

Then the optimal control u is determined by the solution of the unilateral problem:

$$\left.\begin{aligned} &\left(\frac{\partial}{\partial x_0} + \frac{\partial}{\partial t}\right) y + A\left(x_0, t, \frac{\partial}{\partial x}\right) y = f \\ -&\left(\frac{\partial}{\partial x_0} + \frac{\partial}{\partial t}\right) p + A^*\left(x_0, t, \frac{\partial}{\partial x}\right) p = y - z_d \end{aligned}\right\} \text{ in } \Omega \times \mathcal{O}, \tag{9.45}$$

$$\left.\begin{aligned} &\frac{\partial y}{\partial \nu_A}(x,x_0,t)\geq 0, \quad \frac{\partial p}{\partial \nu_{A^*}}(x,x_0,t)=0\\ &p(x,x_0,t)+N\frac{\partial y}{\partial \nu_A}(x,x_0,t)\geq 0\\ &\frac{\partial y}{\partial \nu_A}\left[p+N\frac{\partial y}{\partial \nu_A}\right]=0 \end{aligned}\right\} \text{ on } \Gamma\times\mathcal{O}, \tag{9.46}$$

and

$$\left.\begin{aligned} &\left.\begin{aligned} y(x,x_0,0)&=y_0(x_0)\\ p(x,x_0,T)&=0 \end{aligned}\right\} x_0\in\,]0,a_0[, \quad x\in\Omega,\\ &\left.\begin{aligned} y(x,0,t)&=0\\ p(x,a_0,t)&=0 \end{aligned}\right\} t\in\,]0,T[, \quad x\in\Omega. \end{aligned}\right\} \tag{9.47}$$

9.6. Decoupling

In Theorem 9.4, let $\mathcal{U}_{\text{ad}}=\mathcal{U}$, that is, there are no constraints. Then (9.34) reduces to

$$\Lambda_{\mathcal{U}}^{-1}B^*p(u)+Nu=0. \tag{9.48}$$

Therefore, the optimal control is determined by the solution of the systems of equations

$$\left.\begin{aligned} &\frac{\partial y}{\partial x_0}+\frac{\partial y}{\partial t}+Ay+BN^{-1}\Lambda_{\mathcal{U}}^{-1}B^*p=f,\\ &-\frac{\partial p}{\partial x_0}-\frac{\partial p}{\partial t}+A^*p-y=-z_d, \end{aligned}\right\} \tag{9.49}$$

and (9.47).

We shall briefly study the decoupling of the above system of equations by adapting the methods of Chapter 3, Sections 4 and 5 and of Section 5 of this chapter.

We assume that

$$\left.\begin{aligned} &\mathcal{U}=L^2(\mathcal{O};E),\\ &B=B(x_0,t), \quad B(x_0,t)\in\mathcal{L}(E;V'). \end{aligned}\right\} \tag{9.50}$$

We consider the following problem (compare with Lemma 4.1, Chapter 3):

$$\left.\begin{aligned} &\frac{\partial \varphi}{\partial x_0}+\frac{\partial \varphi}{\partial t}+A\varphi+BN^{-1}\Lambda_{\mathcal{U}}^{-1}B^*\psi=f, \quad x_0\in\,]0,a_0[, \quad t\in\,]s,T[,\\ &-\frac{\partial \psi}{\partial x_0}-\frac{\partial \psi}{\partial t}+A^*\psi-\varphi=-z_d, \quad x_0\in\,]0,a_0[, \quad t\in\,]s,T[, \end{aligned}\right\} \tag{9.51}$$

with

$$\left.\begin{aligned} \varphi(x_0,s)&=h(x_0), \quad h\in L^2(0,a_0;H),\\ \psi(x_0,T)&=0,\\ \varphi(0,t)&=0,\\ \psi(a_0,t)&=0. \end{aligned}\right\} \tag{9.52}$$

This problem admits a unique solution. Indeed it corresponds to an optimal control without constraints of a system whose state $\varphi(v)$ is given by

$$\left(\frac{\partial}{\partial x_0}+\frac{\partial}{\partial t}\right)\varphi(v)+A\varphi(v)=f+B(x_0,t)v \quad \text{in } x_0\in]0,a_0[, \quad t\in]s,T[, \tag{9.53}$$

with

$$\varphi(x_0,s)=h(x_0), \quad \varphi(0,t)=0,$$

and cost function

$$\int_{]0,a_0[\times]s,T[} |y(v)-z_d|^2\,dx_0\,dt+\int_{]0,a_0[\times]s,T[}(Nv,v)_E\,dx_0\,dt.$$ [36]

Then from Lemma 9.1, $\psi(x_0,s)$ is uniquely defined and belongs to $L^2(0,a_0;H)$. Hence the mapping

$$h\to\psi(x_0,s) \text{ is a continuous affine map of } L^2(0,a_0;H) \text{ into itself.} \tag{9.54}$$

Let us denote by $\psi(s)$ the function "$x_0\to\psi(x_0,s)$"$\in L^2(0,a_0;H)$. Then

$$\psi(s)=P(s)h+r(s) \tag{9.55}$$

where

$$P(s)\in\mathscr{L}(L^2(0,a_0;H);L^2(0,a_0;H)), \tag{9.56}$$

$$r(s)\in L^2(0,a_0;H). \tag{9.57}$$

From this we deduce the "fundamental identity":

$$p(t)=P(t)y(t)+r(t), \tag{9.58}$$

where

$$p(t)=\text{"}x_0\to p(x_0,t)\text{"}$$

(and analogous notation for $y(t)$, $r(t)$).

The operator $P(t)$ is given by (compare with Chapter 3, Lemma 4.3)

$$\left.\begin{aligned} &\left(\frac{\partial}{\partial x_0}+\frac{\partial}{\partial t}\right)\beta+A\beta+BN^{-1}\Lambda_{\mathscr{U}}^{-1}B^*\gamma=0,\\ &-\left(\frac{\partial}{\partial x_0}+\frac{\partial}{\partial t}\right)\gamma+A^*\gamma-\beta=0, \quad \text{for } x_0\in]0,a_0[, \quad t\in]s,T[, \end{aligned}\right\} \tag{9.59}$$

[36] We assume that $N=N(x_0,t)\in\mathscr{L}(E;E)$.

$$\left.\begin{aligned}\beta(x_0,s)&=h, & \gamma(x_0,T)&=0,\\ \beta(0,t)&=0, & \beta(a_0,t)&=0,\end{aligned}\right\} \tag{9.60}$$

and

$$\gamma(x_0,s)=P(s)h. \tag{9.61}$$

$r(s)$ is given by

$$\left.\begin{aligned}&\left(\frac{\partial}{\partial x_0}+\frac{\partial}{\partial t}\right)\eta+A\eta+BN^{-1}\Lambda_{\mathscr{U}}^{-1}B^*\xi=f,\\ &-\left(\frac{\partial}{\partial x_0}+\frac{\partial}{\partial t}\right)\xi+A^*\xi-\eta=-z_d,\\ &\eta(x_0,s)=0, \quad \xi(x_0,T)=0,\\ &\eta(0,t)=0, \quad \xi(a_0,t)=0,\end{aligned}\right\} \tag{9.62}$$

and

$$r(s)=\xi(x_0,s). \tag{9.63}$$

We may prove (in a manner analogous to the proof of Lemma 4.4, Chapter 3) that

$$P(s)^*=P(s). \tag{9.64}$$

Let us now carry out a *formal* calculation to find a representation for $P(s)$ (we do not undertake to justify the formal calculations; the justification is technically extremely complicated). The operator $P(s)$ is defined by a *distribution kernel* (cf. L. Schwartz [3], [4]):

$$\left.\begin{aligned}&P(s)h=\int_0^{a_0}P(x_0,\xi_0,s)h(\xi_0)\,d\xi_0,\\ &P(x_0,\xi_0,s) \text{ being a distribution on }]0,a_0[\times]0,a_0[\\ &\text{with values in } \mathscr{L}(H;H).\end{aligned}\right\} \tag{9.65}$$

Then

$$p(x_0,t)=\int_0^{a_0}P(x_0,\xi_0,t)y(\xi_0,t)\,d\xi_0+r(x_0,t),$$

and hence

$$-\int_0^{a_0}\frac{\partial P}{\partial x_0}(x_0,\xi_0,t)y(\xi_0,t)\,d\xi_0-\frac{\partial r}{\partial x_0}-\int_0^{a_0}\frac{\partial P}{\partial t}(x_0,\xi_0,t)y(\xi_0,t)\,d\xi_0$$

$$-\frac{\partial r}{\partial t}-\int_0^{a_0}P(x_0,\xi_0,t)\frac{\partial y}{\partial t}(\xi_0,t)\,d\xi_0+\int_0^{a_0}A^*Py\,d\xi_0+A^*r-y=-z_d,$$

whence, by taking into account the first equation in (9.49):

$$-\int_0^{a_0} \frac{\partial P}{\partial x_0} y\, d\xi_0 - \frac{\partial r}{\partial x_0} - \int_0^{a_0} \frac{\partial P}{\partial t} y\, d\xi_0$$

$$+ \int_0^{a_0} P\left(\frac{\partial y}{\partial \xi_0} + Ay + BN^{-1}\Lambda_{\mathscr{U}}^{-1}B^* p - f\right) d\xi_0 - \frac{\partial r}{\partial t}$$

$$+ \int_0^{a_0} A^* P y\, d\xi_0 + A^* r - y = -z_d,$$

whence we deduce that

$$\left.\begin{aligned} &-\frac{\partial P}{\partial t} - \left(\frac{\partial P}{\partial x_0} + \frac{\partial P}{\partial \xi_0}\right) + PA + A^*P \\ &+ \int_0^{a_0} P(x_0,\eta_0,t) BN^{-1}\Lambda_{\mathscr{U}}^{-1}B^* P(\eta_0,\xi_0,t)\, d\eta_0 = \text{identity} \otimes \delta(\xi_0 - x_0), \end{aligned}\right\} \quad (9.66)$$

and

$$\left.\begin{aligned} &-\frac{\partial r}{\partial t} - \frac{\partial r}{\partial x_0} + A^* r + \int_0^{a_0} P(x_0,\xi_0,t) BN^{-1}\Lambda_{\mathscr{U}}^{-1}B^* r(\xi_0,t)\, d\xi_0 \\ &= \int_0^{a_0} P(x_0,\xi_0,t) f(\xi_0,t)\, d\xi_0 - z_d, \end{aligned}\right\} \quad (9.67)$$

with initial and boundary conditions

$$\left.\begin{aligned} &P(a_0,\xi_0,t) = 0, \\ &P(x_0,a_0,t) = 0, \quad \text{and } P(x_0,\xi_0,T) = 0, \end{aligned}\right\} \quad (9.68)$$

$$r(a_0,t) = 0 \quad \text{and } r(x_0,T) = 0. \quad (9.69)$$

Example. In the context of Section 9.4, let $P(x,\xi,x_0,\xi_0,t)$ be the kernel of P.

To simplify matters, assume that $\mathscr{U} = L^2(\Omega \times \mathcal{O})$ and $N = \nu$ (identity). Then the kernel $P(x,\xi,x_0,\xi_0,t)$ is characterized by

$$\left.\begin{aligned} &-\frac{\partial P}{\partial t} - \frac{\partial P}{\partial x_0} - \frac{\partial P}{\partial \xi_0} - (A_x^* + A_\xi^*)P \\ &+ \nu^{-1} \int_{\Omega \times]0,a_0[} P(x,\eta,x_0,\eta_0,t) P(\eta,\xi,\eta_0,\xi_0,t)\, d\eta\, d\eta_0 \\ &= \delta(x-\xi)\delta(x_0-\xi_0), \end{aligned}\right\} \quad (9.70)$$

$$\left.\begin{aligned}&P(x,\xi,x_0,\xi_0,t)=P(\xi,x,\xi_0,x_0,t),\\&P(x,\xi,x_0,\xi_0,t)=0 \quad\text{if } x\in\Gamma \text{ or } \xi\in\Gamma,\\&\qquad\qquad\{x_0,\xi_0,t\}\in\,]0,a_0[\times]0,a_0[\times]0,T[,\\&P(x,\xi,a_0,t)=0,\qquad P(x,\xi,x_0,a_0,t)=0,\\&P(x,\xi,x_0,\xi_0,T)=0.\end{aligned}\right\}\tag{9.71}$$

We thus obtain an integro-differential equation of Riccati type, which is partially parabolic and partially hyperbolic.

Existence of a *weak* solution of (9.70), (9.71) is known *a priori.*

As we have already indicated, when we were considering analogous parabolic or hyperbolic cases, a direct study of problems of the type (9.70), (9.71) appears to be very interesting.

10. Existence Theorems

10.1. Orientation

In this section we shall present some existence theorems for non-linear optimal control problems. We shall be concerned with results in the context of hyperbolic problems. These results are analogous to those obtained in Section 15, Chapter 3 for parabolic problems.

10.2. Example. Introduction of a "Viscosity" Term

Let us consider as space of controls

$$\mathcal{U}=L^\infty(Q),\qquad Q=\Omega\times]0,T[,\tag{10.1}$$

and to fix ideas, a system, whose state $y(v)$ is given by

$$\frac{\partial^2 y}{\partial t^2}(v)+A\,y(v)+v\frac{\partial y}{\partial t}=f \quad\text{in } Q,\tag{10.2}$$

(A = second order operator, as in Section 4.1),

$$y(v)=0 \quad\text{on } \Sigma,\tag{10.3}$$

$$\left.\begin{aligned}&y(x,0;v)=y_0(x),\qquad x\in\Omega,\\&\frac{\partial y}{\partial t}(x,0;v)=y_1(x)\ ^{37}.\end{aligned}\right\}\tag{10.4}$$

[37] Problem (10.2), (10.3), (10.4) admits a unique solution $y(v)\in H^1(Q)$.

Let the cost function be given, for example, by

$$J(v)=\int_Q |y(v)-z_d|^2\,dx\,dt+\nu\,\|v\|_{L^\infty(Q)},\qquad \nu>0. \tag{10.5}$$

We wish to find $\operatorname{Inf} J(v)$, $v\in\mathscr{U}_{ad}$ = convex subset of $L^\infty(Q)$ which is closed in the weak topology of the dual of $L^1(Q)$.

Even in this simple case we are faced with the following difficulty (analogous to that encountered in Chapter 3, Section 5.3). Let v_n be a minimizing sequence. Then from the form of the cost function in (10.5)[38], v_n ranges in a bounded set of $L^\infty(Q)$. Therefore we may extract a subsequence, again denoted by v_n, such that

$$v_n\to v \quad \text{in } L^\infty(Q) \quad \text{in the “weak star” sense,}\qquad v\in\mathscr{U}_{ad}. \tag{10.6}$$

On the other hand, multiplying (10.2) (for $v=v_n$) by $\frac{\partial}{\partial t}(y(v_n))$ and setting $y(v_n)=y_n$, we obtain

$$\frac{1}{2}\frac{d}{dt}\left[|y_n'|^2+a(t;y_n,y_n)\right]+\int_\Omega v_n y_n'(t)^2\,dx-\frac{1}{2}a'(t;y_n,y_n)=\int_\Omega f y_n'\,dx \tag{10.7}$$

(where $|y_n'|^2=\int_\Omega (y_n')^2\,dx$), from which we deduce that

$$\left.\begin{array}{l} y_n \ \text{(respectively } y_n'\text{) ranges in a bounded subset of}\\ L^\infty(0,T;H_0^1(\Omega)) \ \text{(respectively } L^\infty(0,T;L^2(\Omega))\text{).}\end{array}\right\} \tag{10.8}$$

From these facts we may deduce that we can extract a subsequence, again denoted by y_n, such that

$$\left.\begin{array}{ll} y_n\to y & \text{weakly (weak star) in } L^\infty(0,T;H_0^1(\Omega)),\\ y_n'\to y' & \text{weakly in } L^\infty(0,T;L^2(\Omega)).\end{array}\right\} \tag{10.9}$$

But this is not sufficient to be able to conclude

$$v_n y_n'\to v\,y' \quad \text{(for example in the sense of } \mathscr{D}'(\Omega)\text{)}. \tag{10.10}$$

Remark 10.1. However if we can obtain a stronger convergence of the sequence v_n then conclusion (10.10) is true. This will be the case if we take our controls to be more regular. If, for example, we assume that

$$\mathscr{U}=L^\infty(Q)\cap H^1(Q), \tag{10.11}$$

and in (10.5) we replace $J(v)$ by

$$\left.\begin{array}{l} J_\varepsilon(v)=\int_Q |y(v)-z_d|^2\,dx\,dt+\nu\|v\|_{L^\infty(Q)}+\varepsilon\|v\|_{H^1(Q)},\\ \varepsilon>0,\end{array}\right\} \tag{10.12}$$

[38] If $\mathscr{U}_{ad}$ is bounded the term $\nu\|v\|_{L^\infty(Q)}$ is superfluous.

then we can show that $v_n \to v$ strongly[39] in $L^2(Q)$ and then (10.10) is true. Hence for the case (10.11), (10.12) an *optimal control exists.*

We shall obtain a conclusion analogous to that of Remark 10.1 by *modifying the state equation*[40]. For $\varepsilon > 0$ fixed, we assume that the state $y(v)$ is given by

$$\frac{\partial^2}{\partial t^2} y(v) + A\, y(v) - \varepsilon \Delta y'(v) + v\, y'(v) = f \quad \text{in } Q, \tag{10.13}$$

conditions (10.3), (10.4) remaining unchanged.

It may be verified that this problem admits a unique solution such that

$$y(v) \in H^1(\Omega) \quad \text{and} \quad y'(v) \in L^2(0, T; H_0^1(\Omega)).$$

We have

Theorem 10.1. *Let us assume that the state is given by* (10.13) ($\varepsilon > 0$), (10.3) *and* (10.4) *and that the cost function is given by* (10.5). *Then there exists at least one optimal control.*

Proof. Let v_n be a minimizing sequence and let $y(v_n) = y_n$. Then v_n is bounded in $L^\infty(Q)$ and we deduce from (10.13), (10.3) and (10.4) that

$$\frac{1}{2}\frac{d}{dt}\left[|y_n'(t)|^2 + a(t; y_n(t), y_n(t))\right] - \frac{1}{2} a'(t; y_n(t), y_n(t))$$
$$+ \varepsilon \int_\Omega |\operatorname{grad}_x y_n'(x,t)|^2 dx + \int_\Omega v_n y_n'^2 dx = \int_\Omega f y_n' dx.$$

Then

$$\left.\begin{array}{l} y_n \text{ and } y_n' \text{ range in a bounded set of } L^2(0, T; H_0^1(\Omega)) \\ \text{and } y_n' \text{ ranges in a bounded set of } L^\infty(0, T; L^2(\Omega)). \end{array}\right\} \tag{10.14}$$

From (10.13)

$$y_n'' = f - A y_n + \varepsilon \Delta y_n' - v_n y_n' \tag{10.15}$$

ranges in a bounded subset of $L^2(0, T; H^{-1}(\Omega))$ and utilizing the compactness Lemma of Lions [3], Chapter 4, Proposition 4.2, we may extract a subsequence, again denoted by y_n, such that (10.6) and (10.9) hold and further

$$y_n' \to y' \quad \text{strongly in } L^2(Q). \tag{10.16}$$

Therefore (10.10) is satisfied and $y = y(v)$ with v being an optimal control. □

---

[39] Since the injection of $H^1(Q) \to L^2(Q)$ is compact.

[40] This is an example of "*parabolic regularization*" studied in Chapter 5.

Remark 10.2. Let the state $y(v)$ be given by (10.13), (10.3) and (10.4) and let

$$J_\varepsilon(v) = \int_Q |y_\varepsilon(v) - z_d|^2\, dx\, dt + \nu \|v\|_{L^\infty(Q)}.$$

We do not know if

$$\left(\inf_{v \in \mathscr{U}_{ad}} J_\varepsilon(v)\right) \to \left(\inf_{v \in \mathscr{U}_{ad}} J(v)\right) \quad \text{when } \varepsilon \to 0.$$

Remark 10.3. Results similar to that of Theorem 10.1 when A is an operator of order $2m$ in x and also when the equations are nonlinear (an example will be given) can be obtained.

10.3. Time Optimal Control

Let

$$\mathscr{U}_{ad} = \text{closed, bounded, convex subset of } H^1(Q),\ Q = \Omega \times]0, T[. \tag{10.17}$$

Let us set

$$\beta(\varphi) = |\varphi|^{\rho-1}\varphi, \qquad \rho > 1. \tag{10.18}$$

Let the operator A be given as in 10.2 and consider a system whose state is given by

$$y''(v) + Ay(v) + \beta(y'(v)) = f + v \quad \text{in } Q, \tag{10.19}$$

$f \in L^2(Q)$, $v \in \mathscr{U}_{ad}$, with boundary conditions (10.3), (10.4).

It follows from Lions-Strauss [1][41] that there exists a function $y(v)$ which is unique such that (10.19), (10.3) and (10.4) are satisfied and

$$\left.\begin{aligned} &y(v) \in L^\infty(0, T; H_0^1(\Omega)),\\ &y'(v) \in L^\infty(0, T; L^2(\Omega)),\\ &y(v) \in L^{\rho+1}(Q). \end{aligned}\right\} \tag{10.20}$$

Let

$$K = \text{weakly closed subset of } V \times H. \tag{10.21}$$

Let us assume that

$$\left.\begin{aligned} &\text{there exists a } v \in \mathscr{U}_{ad} \text{ such that for some appropriate } \tau \text{ in}\\ &]0, T[, \text{ we have } \{y(\tau; v), y'(\tau, v)\} \in K. \end{aligned}\right\} \tag{10.22}$$

We set

$$\tau_0 = \inf \tau, \qquad \tau \text{ satisfying (10.22)}. \tag{10.23}$$

τ_0 is termed the *optimal time*. We shall prove

[41] In the work cited more general existence and uniqueness results may be found.

Theorem 10.2. *We assume that* (10.17), (10.21), (10.22) *hold. Then there exists a* $v \in \mathcal{U}_{\text{ad}}$ *such that*

$$\{y(\tau_0; v), y'(\tau_0; v)\} \in K, \tag{10.24}$$

τ_0 *being the optimal time defined by* (10.23).

Proof. 1. Let τ_n be such that

$$\{y(\tau_n; v_n), y'(\tau_n; v_n)\} \in K, \quad \tau_n \to \tau_0, \quad v_n \in \mathcal{U}_{\text{ad}}.$$

Let us set

$$y_n = y(v_n).$$

From (10.19), (10.3), (10.4) (with $v = v_n$) we deduce that[42]

$$\frac{1}{2}\frac{d}{dt}[|y_n'(t)|^2 + a(y_n(t), y_n(t))] + \int_\Omega |y_n'|^{\rho+1} dx = \int_\Omega (f + v_n) y_n' dx \tag{10.25}$$

whence we deduce

$$\left.\begin{array}{l} y_n \text{ (resp. } y_n') \text{ ranges in a bounded subset of} \\ L^\infty(0,T; H_0^1(\Omega)) \text{ (resp. of } L^\infty(0,T; L^2(\Omega)) \cap L^{\rho+1}(Q)), \end{array}\right\} \tag{10.26}$$

and that

$$\|y_n(\tau_n)\|_V + |y_n'(\tau_n)| \leqslant \text{constante}. \tag{10.27}$$

Hence we may extract a subsequence, again denoted by v_n, y_n, such that

$$v_n \to v \quad \text{weakly in } H^1(Q) \text{ (and strongly in } L^2(Q)), \ v \in \mathcal{U}_{\text{ad}}, \tag{10.28}$$

$$\left.\begin{array}{ll} y_n \to y & \text{“weak star” in } L^\infty(0,T; H_0^1(\Omega)), \\ y_n' \to y' & \text{“weak star” in } L^\infty(0,T; L^2(\Omega)) \cap L^{\rho+1}(Q), \end{array}\right\} \tag{10.29}$$

$$\beta(y_n') \to g \quad \text{weakly in } L^{(\rho+1)/\rho}(Q). \tag{10.30}$$

Further

$$y_n'' = f + v_n - A y_n - \beta(y_n'),$$

and hence

$$\left.\begin{array}{l} y_n'' \to y'' = f + v - A y - g \quad \text{in } L^\infty(0,T; H^{-1}(\Omega)) \\ \qquad + L^{(\rho+1)/\rho}(Q) \quad \text{“weak star”}. \end{array}\right\} \tag{10.31}$$

From (10.27) we may assume that

$$\left.\begin{array}{l} \{y_n(\tau_n), y_n'(\tau_n)\} \to \{\chi_0, \chi_1\} \quad \text{weakly in } V \times H \\ \text{and } \{\chi_0, \chi_1\} \in K. \end{array}\right\} \tag{10.32}$$

---

[42] Suppose that A does not depend on t just in order to simplify; for the case when A depends on t, see Lions-Strauss, loc. cit., and Strauss [1].

Let us verify that

$$\left.\begin{aligned} y_n(\tau_n) &\to y(\tau_0) \quad \text{weakly in } H, \\ y_n'(\tau_n) &\to y'(\tau_0) \quad \text{weakly in } (H^{-1}(\Omega)+L^{(\rho+1)/\rho}(\Omega)). \end{aligned}\right\} \tag{10.33}$$

Indeed

$$y_n(\tau_n)-y(\tau_0)=y_n(\tau_n)-y_n(\tau_0)+y_n(\tau_0)-y(\tau_0);$$

and from (10.29) we have (in particular) $y_n(\tau_0)\to y(\tau_0)$ weakly in H, and further

$$|y_n(\tau_n)-y_n(\tau_0)| = \left|\int_{\tau_0}^{\tau_n} y_n'(t)\,dt\right| \leqslant c|\tau_n-\tau_0|\to 0.$$

In a similar manner (utilizing (10.31)) we verify the second assertion of (10.33).

Comparing with (10.32), we then have

$$\{y(\tau_0), y'(\tau_0)\} = \{\chi_0,\chi_1\}\in K. \tag{10.34}$$

The theorem will be proved if we can verify that $y=y(v)$, which will result from

$$g=\beta(y'). \tag{10.35}$$

2. To achieve this, we shall use the monotonicity of the operator $y'\to\beta(y')$. Let φ be an arbitrary function having properties analogous to $y=y(v)$ in (10.20) and (10.3). From the monotonicity of $y'\to\beta(y')$, we have

$$\int_\Omega (\beta(y_n')-\beta(\varphi'))(y_n'-\varphi')\,dx\,dt\geqslant 0 \quad \forall\varphi. \tag{10.36}$$

But

$$\begin{aligned}\int_Q \beta(y_n')(y_n'-\varphi')\,dx\,dt &= -\int_Q (y_n''+Ay_n-f-v_n)(y_n'-\varphi')\,dx\,dt \\ &= -\tfrac{1}{2}\left(|y_n'(T)|^2+a(y_n(T),y_n(T))\right)+\tfrac{1}{2}\left(|y_1|^2+a(y_0,y_0)\right) \\ &\quad +\int_Q (f+v_n)(y_n'-\varphi')\,dx\,dt+\int_Q (y_n''+Ay_n)\varphi'\,dx\,dt\end{aligned}$$

whence

$$\left.\begin{aligned} &\liminf \int_Q \beta(y_n')(y_n'-\varphi')\,dx\,dt \\ &\leqslant -\tfrac{1}{2}\left(|y'(T)|^2+a(y(T),y(T))\right)+\tfrac{1}{2}\left(|y_1|^2+a(y_0,y_0)\right) \\ &\quad +\int_Q (f+v)(y'-\varphi')\,dx\,dt\ ^{43}+\int_Q (y''+Ay)\varphi'\,dx\,dt \end{aligned}\right\} \tag{10.37}$$

---

[43] It is here and only here that we use the strong convergence of v_n toward v. We do not know whether the theorem is true without assuming that $\mathscr{U}_{ad}$ is relatively compact in $L^2(Q)$.

and since $y''+Ay+g=f+v$, the right hand side of (10.37) is equal to

$$\int_Q g(y'-\varphi')\,dx\,dt$$

and hence from (10.36) we deduce that

$$\int_Q (g-\beta(\varphi'))(y'-\varphi')\,dx\,dt \geqslant 0. \tag{10.38}$$

We now utilize a method due to G. Minty [2]. In (10.38) we take

$$\varphi = y - \lambda\psi, \quad \lambda > 0,$$

ψ having properties analogous to that of φ.

We obtain

$$\lambda \int_Q (g-\beta(y'-\lambda\psi))\psi\,dx\,dt \geqslant 0,$$

and hence

$$\int_Q (g-\beta(y'-\lambda\psi))\psi\,dx\,dt \geqslant 0,$$

and if we let $\lambda \to 0$ we get

$$\int_Q (g-\beta(y'))\psi\,dx\,dt \geqslant 0 \quad \forall \psi,$$

whence the result (10.35). □

Remark 10.4. In the particular case when β is given by (10.18), the preceding general arguments may be replaced by a uniform convexity argument due to E. de Giorgi (cf. Lions-Strauss [1], p. 94).

Notes

The proof of Section 1.3 is due to O. A. Ladyzenskaya [1] and is identical to that given in Lions-Magenes [1], Chapter 3.

The proof of Section 1.4 utilizes the Faedo-Galerkin method in a manner similar to that used in Chapter 3, Section 1.

Theorems 2.1 to 3.4 (with the exception of 3.1) are extensions to infinite dimensions with *unbounded operators* (the extension to infinite dimensions with bounded operators being trivial) of two point boundary value problems which arise naturally in the finite dimensional case. Therefore this situation is again analogous to that encountered in Chapter 3 where we considered *parabolic operators*. But in this case there are additional technical difficulties which are rather serious. The "transposition" technique turns out to be an essential tool.

The examples (cf. Section 4) we have considered lead to *unilateral problems* (of a type different from that encountered in Chapter 3). A more detailed study (regularity of solution, a study of the switching surface) needs to be carried out and appears to us to be an interesting subject for research.

Decoupling (Section 5) leads to the consideration of integro-differential equations of Riccati type (formally analogous to those obtained in Chapter 3). A complete justification of the derivation of these equations appears to present considerable difficulties. Explicit results have been obtained only in a particular case. The decoupling problems for Examples 7.2 and 7.3 have not been studied here. A systematic study of this equation and in particular a direct study of the Riccati equation, that is, without reference to a control problem, appears to us to be of interest.

The duality theory of Kalman can be adapted to the situation in Section 6; for the finite dimensional case, cf. Kalman [3], Kalman-Bucy [1].

The results of Section 7 may be extended to the case of systems governed by operators of arbitrary order $2m$ by using the methods of this section.

The results of Section 8 are technically complicated and probably not the best possible. One faces the difficulty of regularity in hyperbolic equations. This has been noted and studied in detail in Lions-Magenes [1], Chapter 5. With regard to the above, it should be noted that the "L^q estimates", $q \neq 2$ are imprecise (W. Littman [1]). Therefore the results of Chapter 3, Section 14, *cannot be extended* to the hyperbolic case. We refer the reader to Lions-Magenes [1], Chapter 3 for a proof of Theorem 9.1. A detailed study of unilateral problems (in particular of their regularity) encountered in this section appears to be interesting. Similarly a study of the Riccati equations, briefly introduced in Section 9.6, should also prove to be interesting. The operator $\frac{\partial}{\partial t} + \frac{\partial}{\partial x_0}$ could perhaps be replaced by a *first order hyperbolic system*

$$\frac{\partial}{\partial t} + \sum_{i=1}^{\mu} B_i(x_0,t)\frac{\partial}{\partial x_{0i}},$$

which leads to new Riccati equations. For example, (9.70) then becomes

$$-\frac{\partial P}{\partial t} - \sum_{i=1}^{\mu} \frac{\partial}{\partial x_{0i}}(B_i(x_0,t)P) - \sum_{i=1}^{\mu} \frac{\partial}{\partial \xi_{0i}}(B_i(\xi_0,t)P)$$

$$-(A_x^* + A_\xi^*)P + v^{-1} \int_{\Omega \times]0,T[} P(x,\eta,x_0,\eta_0,t)P(\eta,\xi,\eta_0,\xi_0,t)\,d\eta\,d\eta_0$$

$$= \delta(x-\xi)\delta(x_0-\xi_0),$$

where P now is a *matrix* $\|P_0(x,\xi,x_0,\xi_0,t)\|$, A is a system of elliptic operators and G is the domain spanned by x_0.

The boundary conditions are rather complicated. The theory of boundary value problems for first order hyperbolic systems needs to be used. *Equations of transport* and equations which are "partially elliptic, partially parabolic" may be studied within the same framework—cf. O. A. Oleinik [1] for a study of such equations.

Partial differential equations of the type introduced in Section 9.2 (without a control term) have been studied in A. M. Il'in [1], Lions [9].

The results of Section 10 pose more problems than they solve. In this vein for example, the arguments in Theorem 10.2 use both monotonicity and compactness properties and it is not clear whether it is essential to use both properties. Other existence theorems may be found in Schmaedeke [1].

We have not given explicit details of a certain number of possible adaptations to the hyperbolic case of problems studied in Chapter 3. Examples of these are:

— problems where both control and state constraints are present (cf. Chapter 3, Section 13);

— applications of duality, Chapter 3, Section 12;[44]

— the case of point-wise control (as Chapter 3, Section 8), and for which we refer to Lions-Magenes [1];

— the case of hyperbolic operators or operators well-posed in the sense of Petrovsky, with delay terms.

We refer the reader to Fattorini [5], Russell [3], K. Tsujioka [1] for consideration of controllability problems in the hyperbolic case.

For systems gouverned by hyperbolic equations and the Goursat problem, a variant of the Maximum Principle may be found in A. I. Egorov [1].

Certain control problems may be reduced to moment problems by using explicit representations for the state of the system (in terms of eigenfunctions). Cf. Butkovskii-Poltavskii [1] and Butkovskii [1].

Finally control of systems governed by Schroedinger's equations can be studied using the same methods. In the case of no constraints, when considering decoupling, we are led to new nonlinear partial differential equations of Riccati type.

[44] Along these lines, see A. Bensoussan [3].

CHAPTER V

Regularization, Approximation and Penalization

1. Regularization

1.1. Parabolic Regularization

We consider the same situation as in Chapter 4, Section 1. Hence, let $V \subset H \subset V'$; we assume that the form $a(t; \varphi, \psi)$ does not depend on t—this solely to simplify the exposition. Hence:

$$a(\varphi, \psi) = a(\psi, \varphi) = \text{continuous, symmetric bi-linear form on } V, \tag{1.1}$$

and to further simplify the exposition, we assume that,

$$a(\varphi, \varphi) \geqslant \alpha \|\varphi\|^2, \quad \alpha > 0, \quad \forall \varphi \in V. \tag{1.2}$$

We first consider the evolution equation—without any control

$$y'' + A y = f, \tag{1.3}$$

f given in $L^2(0, T; H)$, with

$$y(0) = y_0, \quad y'(0) = y_1, \quad y_0 \in V, \quad y_1 \in H. \tag{1.4}$$

We have seen (Chapter 4, Section 1) that there exists a unique function y which is a solution of (1.3), (1.4) satisfying:

$$y \in L^2(0, T; V), \quad y' \in L^2(0, T; H). \tag{1.5}$$

We shall approximate Problem (1.3), (1.4), (1.5) by a problem which is of parabolic type. Let $\varepsilon > 0$ be given. We shall prove the following results:

Theorem 1.1. *There exists a function y_ε which is unique and satisfies*

$$y_\varepsilon'' + A y_\varepsilon + \varepsilon A y_\varepsilon' = f, \tag{1.6}$$

$$y_\varepsilon(0) = y_0, \quad y_\varepsilon'(0) = y_1, \tag{1.7}$$

$$y_\varepsilon \in L^2(0, T; V), \quad y_\varepsilon' \in L^2(0, T; V). \tag{1.8}$$

Theorem 1.2. *As* $\varepsilon \to 0$, *we have*

$$y_\varepsilon \to y \quad \text{in } L^2(0,T;V), \tag{1.9}$$

$$y'_\varepsilon \to y' \quad \text{in } L^2(0,T;H), \tag{1.10}$$

and y *is the solution of* (1.3), (1.4), (1.5).

Remark 1.1. Problem (1.6), (1.7), (1.8) is the "regularized parabolic" problem corresponding to problem (1.3), (1.4), (1.5). This definition will be justified in the sequel (by writing the equation in the form of a first order system of equations in t).

Proof of Theorem 1.1. Let us proceed as in Chapter 4, Section 5.1. We thus introduce

$$\mathscr{V} = V \times V, \qquad \mathfrak{h} = V \times H, \qquad \mathscr{V}' = V \times V',$$

the space $\mathfrak{h}$ being endowed with the scalar product

$$[\boldsymbol{\varphi}, \boldsymbol{\psi}] = a(\varphi_0, \psi_0) + (\varphi_1, \psi_1), \qquad \boldsymbol{\varphi} = \{\varphi_0, \varphi_1\} \in \mathfrak{h}, \qquad \boldsymbol{\psi} = \{\psi_0, \psi_1\} \in \mathfrak{h}.$$

Let us set

$$y_\varepsilon = y_{\varepsilon 0}, \qquad y'_\varepsilon = y_{\varepsilon 1}, \qquad \mathbf{y}_\varepsilon = \{y_{\varepsilon 0}, y_{\varepsilon 1}\}, \tag{1.11}$$

$$\mathscr{A} = \begin{pmatrix} 0 & -I \\ A & 0 \end{pmatrix}, \qquad \mathscr{A}_\varepsilon = \begin{pmatrix} 0 & -I \\ A & \varepsilon A \end{pmatrix}. \tag{1.12}$$

Problem (1.6), (1.7), (1.8) is equivalent to

$$\frac{d}{dt} \mathbf{y}_\varepsilon + \mathscr{A}_\varepsilon \mathbf{y}_\varepsilon = \mathbf{f}, \qquad \mathbf{f} = \{0, f\}, \tag{1.13}$$

$$\mathbf{y}_\varepsilon(0) = \{y_0, y_1\}, \tag{1.14}$$

$$\mathbf{y}_\varepsilon \in L^2(0,T;\mathscr{V}) \quad (\text{and then } \mathbf{y}'_\varepsilon L^2(0,T;\mathscr{V}')). \tag{1.15}$$

Theorem 1.1 will be proved if we can verify that we are in a position to apply Theorem 1.2, Chapter 3 (this justifies the terminology—see Remark 1.1). Now, we have seen (Chapter 4, (5.16)) that

$$[\mathscr{A} \boldsymbol{\varphi}, \boldsymbol{\varphi}] = 0$$

giving us

$$[\mathscr{A}_\varepsilon \boldsymbol{\varphi}, \boldsymbol{\varphi}] = \varepsilon a(\varphi_1, \varphi_1)$$

and hence for $\lambda > 0$:

$$[\mathscr{A}_\varepsilon \boldsymbol{\varphi}, \boldsymbol{\varphi}] + \lambda [\boldsymbol{\varphi}]^2 = \lambda a(\varphi_0, \varphi_0) + (\lambda |\varphi_1|^2_{\mathscr{V}} + \varepsilon a(\varphi_1, \varphi_1))$$
$$\geqslant \alpha \inf(\lambda, \varepsilon) \|\boldsymbol{\varphi}\|^2_{\mathscr{V}},$$

which proves (1.2), Chapter 3. □

Proof of Theorem 1.2. Let us take the scalar product of the two terms in (1.6) with y'_ε (instead of (1.3) with y_ε). We get

$$\frac{1}{2}\frac{d}{dt}[|y'(t)|^2+a(y_\varepsilon(t),y_\varepsilon(t))]+\varepsilon a(y'_\varepsilon(t),y'_\varepsilon(t))=(f(t),y'_\varepsilon(t)),$$

whence

$$|y'_\varepsilon(t)|^2+a(y_\varepsilon(t),y_\varepsilon(t))+2\varepsilon\int_0^t a(y'_\varepsilon(\sigma),y'_\varepsilon(\sigma))\,d\sigma$$
$$=|y_1|^2+a(y_0,y_0)+2\int_0^t(f(\sigma),y'_\varepsilon(\sigma))\,d\sigma.$$

In particular, we deduce that as $\varepsilon\to 0$

$$\boldsymbol{y}_\varepsilon \quad \text{ranges in a bounded set in } L^2(0,T;\mathfrak{h}),\text{ [1]} \tag{1.16}$$

$$\sqrt{\varepsilon}\,y'_\varepsilon \quad \text{ranges in a bounded set in } L^2(0,T;V). \tag{1.17}$$

Hence we may extract a subsequence, again denoted by y_ε, such that

$$\begin{aligned} y_\varepsilon&\to z \quad \text{weakly in } L^2(0,T;V),\\ y'_\varepsilon&\to z' \quad \text{weakly in } L^2(0,T;H). \end{aligned} \tag{1.18}$$

But, since according to (1.17), $\varepsilon A y'_\varepsilon\to 0$ in $L^2(0,T;V')$, we deduce from (1.6), (1.8) that

$$y''_\varepsilon\to z''=f-Az \quad \text{weakly in } L^2(0,T;V'). \tag{1.19}$$

Then, in particular, $y_\varepsilon(0)\to z(0)$ weakly in H, $y'(0)\to z'(0)$ weakly in V'; hence $z(0)=y_0$, $z'(0)=y_1$. Hence from the uniqueness of the solution of (1.3), (1.4), (1.5) $z=y$.

Thus we have already proved that $\boldsymbol{y}_\varepsilon\to\boldsymbol{y}$ weakly in $L^2(0,T;\mathfrak{h})$ and it remains to prove strong convergence. For this purpose we introduce the quantity

$$\begin{aligned} X_\varepsilon=&|y'_\varepsilon(t)-y'(t)|^2+a(y_\varepsilon(t)-y(t),y_\varepsilon(t)-y(t))\\ &+2\varepsilon\int_0^t a(y'_\varepsilon(\sigma),y'_\varepsilon(\sigma))\,d\sigma. \end{aligned} \tag{1.20}$$

Using (1.6) and (1.3) we see that

$$\begin{aligned} X_\varepsilon=&2\int_0^t(f,y'_\varepsilon)\,d\sigma+2\int_0^t(f,y')\,d\sigma\\ &-2[(y'_\varepsilon(t),y'(t))+a(y_\varepsilon(t),y(t))]+2(|y_1|^2+a(y_0,y_0)). \end{aligned} \tag{1.21}$$

---

[1] In fact in a bounded set in $L^\infty(0,T;\mathfrak{h})$.

But again according to (1.3) and (1.6):

$$(y''(t), y'_\varepsilon(t)) + a(y(t), y'_\varepsilon(t)) = (f(t), y'_\varepsilon(t)),$$
$$(y''_\varepsilon(t), y'(t)) + a(y_\varepsilon(t), y'(t)) + \varepsilon\, a(y'_\varepsilon(t), y'(t)) = (f(t), y'(t)),$$

and thus

$$(y'_\varepsilon(t), y'(t)) + a(y_\varepsilon(t), y(t))$$
$$= \int_0^t (f(\sigma), y'_\varepsilon(\sigma) + y'(\sigma))\, d\sigma - \varepsilon \int_0^t a(y'_\varepsilon(\sigma), y'(\sigma))\, d\sigma + |y_1|^2 + a(y_0, y_0);$$

we now obtain, by substituting in (1.21), that

$$X_\varepsilon \to 0 \tag{1.22}$$

and this is true for every t.

In particular, we obtain the strong convergence of $\boldsymbol{y}_\varepsilon$ toward $\boldsymbol{y}$ in $L^2(0,T;\mathfrak{h})$. □

Remark 1.2. In fact (1.22) also proves

$$y_\varepsilon(t) \to y(t) \quad \text{strongly in } V \ (\forall t \in [0,T]), \tag{1.23}$$

$$y'_\varepsilon(t) \to y'(t) \quad \text{strongly in } H. \tag{1.24}$$

1.2. Application to Optimal Control

Let us now consider a system whose state $y(v)$ is given as in Chapter 4, Section 2 by:

$$y''(v) + A y(v) = f + Bv, \tag{1.25}$$

$$y(0; v) = y_0, y'(0; v) = y_1, \tag{1.26}$$

$$y(v) \in L^2(0,T; V), \qquad y'(v) \in L^2(0,T; H), \tag{1.27}$$

where

$$B \in \mathscr{L}(\mathscr{U}; L^2(0,T; H)). \tag{1.28}$$

We assume (cf. Chapter 4, Section 2.2) that the cost function is given by

$$J(v) = \|Cy(v) - z_d\|^2_{\mathscr{H}} + (Nv, v)_{\mathscr{U}}, \tag{1.29}$$

where

$$C \in \mathscr{L}(L^2(0,T; H); \mathscr{H}), \text{ [2]} \tag{1.30}$$

$$N \in \mathscr{L}(\mathscr{U}; \mathscr{U}), \qquad N \geq \nu \text{ (identity)}, \qquad \nu > 0. \tag{1.31}$$

[2] We may also treat the case where $C \in \mathscr{L}(L^2(0,T; V); \mathscr{H})$ (cf. Chapter 4, Section 3.4) after showing that the parabolic regularisation is again convergent in the "transposed case", as in Chapter 4, Section 3.1. We shall not develop the details.

For given $\varepsilon>0$, we consider the state $y_\varepsilon(v)$ of the *regularized parabolic* system (cf. Section 1.1):

$$y_\varepsilon''(v)+Ay_\varepsilon(v)+\varepsilon Ay_\varepsilon'(v)=f+Bv, \tag{1.32}$$

$$y_\varepsilon(0;v)=y_0, \quad y_\varepsilon'(0;v)=y_1. \tag{1.33}$$

With this we associate:

$$J_\varepsilon(v)=\|Cy_\varepsilon(v)-z_d\|_{\mathscr{H}}^2+(Nv,v)_{\mathscr{U}}. \tag{1.34}$$

If $\mathscr{U}_{\text{ad}}$ = closed, convex subset of $\mathscr{U}$, we set

$$\left.\begin{aligned} j&=\inf J(v), \quad v\in\mathscr{U}_{\text{ad}},\\ j_\varepsilon&=\inf J_\varepsilon(v), \quad v\in\mathscr{U}_{\text{ad}}.\end{aligned}\right\} \tag{1.35}$$

Let u (resp. u_ε) be the optimal control for $J(v)$ (resp. $J_\varepsilon(v)$), that is,

$$u, u_\varepsilon\in\mathscr{U}_{\text{ad}}, \quad J(u)=j, \quad J_\varepsilon(u_\varepsilon)=j_\varepsilon. \tag{1.36}$$

We then have

Theorem 1.3. *We assume that* (1.1), (1.2) *hold. Let* u, u_ε *be defined as in* (1.36) *and* (1.35). *Then as* $\varepsilon\to0$, *we have:*

$$u_\varepsilon\to u \quad \text{in } \mathscr{U}, \tag{1.37}$$

$$\begin{aligned} y_\varepsilon(u_\varepsilon)&\to y(u) \quad \text{in } L^2(0,T;V),\\ y_\varepsilon'(u_\varepsilon)&\to y'(u) \quad \text{in } L^2(0,T;H),\end{aligned} \tag{1.38}$$

$$j_\varepsilon\to j. \tag{1.39}$$

Proof. 1. From Theorem 1.2, we have

$$J_\varepsilon(v)\to J(v) \ \forall v \quad \text{fixed in } \mathscr{U}. \tag{1.40}$$

Hence in particular

$$j_\varepsilon\leqslant J_\varepsilon(u)\to J(u)=j,$$

and hence

$$\limsup j_\varepsilon\leqslant j. \tag{1.41}$$

2. But

$$j_\varepsilon=J_\varepsilon(u_\varepsilon)\geqslant\nu\|u\|_{\mathscr{U}}^2, \tag{1.42}$$

which combined with (1.41) shows that $\|u_\varepsilon\|_{\mathscr{U}}$ is bounded.

We may then extract a subsequence, again denoted by u_ε, such that

$$\left.\begin{aligned} &u_\varepsilon\to w \quad \text{weakly in } \mathscr{U},\\ &w\in\mathscr{U}_{\text{ad}}.\end{aligned}\right\} \tag{1.43}$$

Then, from Theorem 1.2, $y_\varepsilon(u_\varepsilon)\to y(w)$ weakly in $L^2(0,T;V)$ and $y'_\varepsilon(u_\varepsilon)\to y'(w)$ weakly in $L^2(0,T;H)$ and consequently

$$\liminf J_\varepsilon(u_\varepsilon)\geqslant J(w), \tag{1.44}$$

which when compared with (1.41) shows that $w=u$. Hence we have (1.39).

We thus have (1.37), (1.38) the convergence being weak convergence.

3. But as

$$J_\varepsilon(u_\varepsilon)=(Cy_\varepsilon(u_\varepsilon),Cy_\varepsilon(u_\varepsilon))_{\mathscr{H}}+(Nu_\varepsilon,u_\varepsilon)_{\mathscr{U}}-2(Cy_\varepsilon(u_\varepsilon),z_d)$$
$$-\|z_d\|^2\to J(u) \quad\text{and}\quad J_\varepsilon(0)\to J(0),$$

we deduce that (setting

$$(C(y_\varepsilon(u_\varepsilon)-y_\varepsilon(0)),C(y_\varepsilon(u)-y_\varepsilon(0)))_{\mathscr{H}}+(Nu_\varepsilon,u_\varepsilon)_{\mathscr{U}}=|||u_\varepsilon|||^2_{\mathscr{U}}),$$
$$|||u_\varepsilon|||\to|||u|||_{\mathscr{U}}$$

and since $|||\ |||$ is a norm equivalent to $\|\ \|_{\mathscr{U}}$, (1.37) is proved, from which (1.38) follows according to Theorem 1.2. □

Remark 1.3. By virtue of Remark 1.2, we have results analogous to Theorem 1.2 for the following cost functions:

$$J(v)=\|\mathscr{C}y'(v)-z_d\|^2_{\mathscr{H}}+(Nv,v)_{\mathscr{U}}, \tag{1.45}$$

with

$$\mathscr{C}\in\mathscr{L}(L^2(0,T;H);\mathscr{H}) \quad \text{(cf. Chapter 4, Section 2.3);} \tag{1.46}$$

also (cf. Chapter 4, Section 2.4)

$$J(v)=\|D_0y(T;v)-z_d\|^2_{\mathscr{H}}+(Nv,v)_{\mathscr{U}}, \tag{1.47}$$

with

$$D_0\in\mathscr{L}(H;\mathscr{H})\ ^3 \tag{1.48}$$

or again (cf. Chapter 4, Section 2.5)

$$J(v)=\|Dy'(T;v)-z_d\|^2_{\mathscr{H}}+Nv,v)_{\mathscr{U}}, \tag{1.49}$$

with

$$D\in\mathscr{L}(H;\mathscr{H}). \tag{1.50}$$

Remark 1.4. Clearly, we may write down the system of equations and inequalities which determine u_ε by applying the results of Chapter 3. We shall state these explicity in the framework of Theorem 1.3.

Let us write (1.32) in the form of a system of first order equations in t:

$$\frac{d}{dt}y_\varepsilon(v)+\mathscr{A}_\varepsilon y_\varepsilon(v)=f+\mathscr{B}v, \tag{1.51}$$

---

[3] If $D_0\in\mathscr{L}(V;\mathscr{H})$, we may obtain similar results after "transposition of the parabolic regularization".

where (as in Chapter 4, Section 5.1)

$$\mathscr{B} v = \{0, Bv\}. \tag{1.52}$$

Let us introduce (again as in Chapter 4, Section 5.1)

$$\mathscr{C} \in \mathscr{L}(L^2(0,T;\mathfrak{h}); \mathscr{H})$$

by

$$\mathscr{C} y = C y_0, \quad y = \{y_0, y_1\}.$$

We then have

$$J_\varepsilon(v) = \|\mathscr{C} y_\varepsilon(v) - z_d\|_{\mathscr{H}}^2 + (Nv, v)_{\mathscr{U}}. \tag{1.53}$$

We introduce the *adjoint state* $p_\varepsilon(u_\varepsilon) = p_\varepsilon$ by

$$\left.\begin{aligned} -\frac{d}{dt} p_\varepsilon + \mathscr{A}_\varepsilon^* p_\varepsilon &= \mathscr{C}^* \Lambda_{\mathscr{H}}(C y_\varepsilon(u_\varepsilon) - z_d), \\ p_\varepsilon(T) &= 0. \end{aligned}\right\} \tag{1.54}$$

Then the *optimal control* u_ε is determined by (1.51) (where $v = u$) and $y_\varepsilon(0; u_\varepsilon) = \{y_0, y_1\}$ and (1.54) and the condition

$$(\Lambda^{-1} \mathscr{B}^* p_\varepsilon + N u_\varepsilon, v - u_\varepsilon)_{\mathscr{U}} \geq 0 \quad \forall v \in \mathscr{U}_{\text{ad}}. \tag{1.55}$$

Let us return to the "scalar" case. Set:

$$p_\varepsilon = \{p_{\varepsilon 0}, p_{\varepsilon 1}\} \quad \text{and} \quad p_\varepsilon = p_{\varepsilon 1}. \tag{1.56}$$

Then (comparing with Chapter 4, Section 5.1)

$$\left.\begin{aligned} -p_{\varepsilon 0}' + p_{\varepsilon 1} &= A^{-1} \mathscr{C}^* \Lambda_{\mathscr{H}}(C y_\varepsilon(u_\varepsilon) - z_d), \\ -p_{\varepsilon 1}' - A p_{\varepsilon 0} + \varepsilon A p_{\varepsilon 1} &= 0, \\ p_{\varepsilon 0}(T) = 0, \quad p_{\varepsilon 1}(T) &= 0, \end{aligned}\right\} \tag{1.57}$$

whence

$$\left.\begin{aligned} p_\varepsilon'' + A p_\varepsilon - \varepsilon A p_\varepsilon' &= \mathscr{C}^* \Lambda(C y_\varepsilon(u_\varepsilon) - z_d), \\ p_\varepsilon(T) = 0, \quad p_\varepsilon'(T) &= 0. \end{aligned}\right\} \tag{1.58}$$

Condition (1.55) reduces to

$$(\Lambda^{-1} B^* p_\varepsilon + N u_\varepsilon, v - u_\varepsilon)_{\mathscr{U}} \geq 0 \quad \forall v \in \mathscr{U}_{\text{ad}}. \tag{1.59}$$

Summarising,

> the optimal control u_ε of the regularized parabolic problem is determined by the solution of (1.32), (1.33) (where $v = u_\varepsilon$), (1.58) and (1.59). (1.60)

Let us note that

Theorem 1.4. *Hypotheses as in Theorem* 1.3. *Let* p_ε *be the adjoint state defined by* (1.58). *Then*

$$\begin{aligned} p_\varepsilon &\to p \quad \text{in } L^2(0,T;V), \\ p'_\varepsilon &\to p' \quad \text{in } L^2(0,T;H), \end{aligned} \tag{1.61}$$

where p is the adjoint state of the original problem; that is,

$$\begin{aligned} p'' + Ap &= C^* \Lambda(Cy_{\mathscr{H}}(u) - z_d), \\ p(T) &= 0, \quad p'(T) = 0. \end{aligned} \tag{1.62}$$

(cf. (2.19), Chapter 4).

Proof. Indeed from Theorem 1.3, $y_\varepsilon(u_\varepsilon) \to y(u)$ in $L^2(0,T;H)$ and hence we obtain the result by applying Theorem 1.2 to the situation (1.58), (1.62) (and reversing time—which is allowed here). □

Remark 1.5. In the case of no constraints $\mathscr{U}_{\text{ad}} = \mathscr{U}$, the system of equations determining the optimal control for the original problem as well as for the regularized parabolic problem are:

$$\left.\begin{aligned} &y'' + Ay + BN^{-1}\Lambda_{\mathscr{U}}^{-1}B^*p = f, \\ &\quad p'' + Ap - C^*\Lambda_{\mathscr{H}}Cy = -C^*\Lambda_{\mathscr{H}}z_d \quad \text{(original problem)}, \\ &\quad y(0) = y_0, \quad y'(0) = y_1, \\ &\quad p(T) = 0, \quad p'(T) = 0, \end{aligned}\right\} \tag{1.63}$$

and

$$\left.\begin{aligned} &y''_\varepsilon + Ay_\varepsilon + \varepsilon Ay'_\varepsilon + BN^{-1}\Lambda_{\mathscr{U}}^{-1}B^*p_\varepsilon = f, \\ &\quad p''_\varepsilon + Ap_\varepsilon - \varepsilon Ap'_\varepsilon - C^*\Lambda_{\mathscr{H}}Cy_\varepsilon = -C^*\Lambda_{\mathscr{H}}z_d \\ &\qquad\qquad \text{(regularized parabolic problem)}, \\ &\qquad y_\varepsilon(0) = y_0, \quad y'_\varepsilon(0) = y_1, \\ &\qquad p_\varepsilon(T) = 0, \quad p'_\varepsilon(T) = 0. \end{aligned}\right\} \tag{1.64}$$

The optimal control is then determined by

$$\begin{aligned} u &= N^{-1}\Lambda_{\mathscr{U}}^{-1}B^*p, \\ u_\varepsilon &= N^{-1}\Lambda_{\mathscr{U}}^{-1}B^*p_\varepsilon. \end{aligned} \tag{1.65}$$

1.3. Application to Decoupling

Assume that (cf. Chapter 4, Section 5)

$$\mathscr{U} = L^2(0,T;E), \qquad \mathscr{H} = L^2(0,T;F)$$

and that the operators B, C, N are "local"

$$Bv = \text{“}t \to B(t)v\text{”}, \quad \text{etc.}$$

As in Chapter 4, Section 5, we introduce $\mathscr{D}_1$ and $\mathscr{D}_2$ by (5.32). Then (1.64) may be written as

$$\left.\begin{aligned} y_\varepsilon' + \mathscr{A}_\varepsilon y_\varepsilon + \mathscr{D}_1 p_\varepsilon &= f, \\ -p_\varepsilon' + \mathscr{A}_\varepsilon^* p_\varepsilon - \mathscr{D}_2 y_\varepsilon &= -\mathscr{C}^* \Lambda_{\mathscr{H}} z_d, \end{aligned}\right\} \tag{1.66}$$

$$y_\varepsilon(0) = \{y_0, y_1\}, \quad p_\varepsilon(T) = 0. \tag{1.67}$$

We are now in a position to apply the theory of Chapter 3, Section 4. We have the identity

$$p_\varepsilon(t) = \mathscr{P}_\varepsilon(t) y_\varepsilon(t) + r_\varepsilon(t), \tag{1.68}$$

$$\left.\begin{aligned} &\mathscr{P}_\varepsilon(t) \in \mathscr{L}(\mathfrak{h}; \mathfrak{h}), \\ &\mathscr{P}_\varepsilon(t)^* = \mathscr{P}_\varepsilon(t). \end{aligned}\right\} \tag{1.69}$$

The operator $\mathscr{P}_\varepsilon$ is given as follows: we solve

$$\left.\begin{aligned} &\frac{d}{dt}\beta_\varepsilon + \mathscr{A}_\varepsilon \beta_\varepsilon + \mathscr{D}_1 \gamma_\varepsilon = 0 \quad \text{in }]s, T[, \\ &-\frac{d}{dt}\gamma_\varepsilon + \mathscr{A}_\varepsilon^* \gamma_\varepsilon - \mathscr{D}_2 \beta_\varepsilon = 0 \quad \text{in }]s, T[, \\ &\beta_\varepsilon(s) = h \in \mathfrak{h}, \quad \gamma_\varepsilon(T) = 0; \end{aligned}\right\} \tag{1.70}$$

then

$$y_\varepsilon(s) = \mathscr{P}_\varepsilon(s) h. \tag{1.71}$$

The operator $\mathscr{P}_\varepsilon$ satisfies the *Riccati integro-differential equation:*

$$\left.\begin{aligned} -\mathscr{P}_\varepsilon' + \mathscr{P}_\varepsilon \mathscr{A}_\varepsilon + \mathscr{A}_\varepsilon^* \mathscr{P}_\varepsilon + \mathscr{P}_\varepsilon \mathscr{D}_1 \mathscr{P}_\varepsilon = \mathscr{D}_2 \quad \text{in }]0, T[, \\ \mathscr{P}_\varepsilon(T) = 0, \end{aligned}\right\} \tag{1.72}$$

and again, writing $\mathscr{A}_\varepsilon$ explicitly:

$$\left.\begin{aligned} -\mathscr{P}_\varepsilon' + (\mathscr{P}_\varepsilon \mathscr{A} + \mathscr{A} \mathscr{P}_\varepsilon) + \varepsilon \begin{pmatrix} 0 & 0 \\ 0 & A \end{pmatrix} \mathscr{P}_\varepsilon + \mathscr{P}_\varepsilon \mathscr{D}_1 \mathscr{P}_\varepsilon = \mathscr{D}_2 \quad \text{in }]0, T[, \\ \mathscr{P}_\varepsilon(T) = 0. \end{aligned}\right\} \tag{1.73}$$

We also have the identity (cf. Chapter 4, Section 5.3)

$$p(t) = \mathscr{P}(t) y(t) + r(t), \tag{1.74}$$

and we deduce from Theorem 1.4 [4] that as $\varepsilon \to 0$

$$\mathscr{P}_\varepsilon(t) h \to \mathscr{P}(t) h \quad \text{in } \mathfrak{h}, \quad \forall h \in \mathfrak{h}. \tag{1.75}$$

---

[4] We apply the result analogous to the system of equations (1.70).

This allows us to prove that $\mathscr{P}$ satisfies

$$\begin{aligned} -\mathscr{P}'+(\mathscr{P}\mathscr{A}-\mathscr{A}\mathscr{P})+\mathscr{P}\mathscr{D}_1\mathscr{P}=\mathscr{D}_2 \quad \text{in }]0,T[, \\ \mathscr{P}(T)=0. \end{aligned} \tag{1.76}$$

1.4. Various Remarks

Remark 1.6. We may also approximate *parabolic* systems by *elliptic* systems. If the state $y(v)$ is given as in Chapter 3, Section 2 by

$$\left.\begin{aligned} \frac{dy(v)}{dt} + A(t)y(v)=f+Bv, \\ y(0;v)=0, \end{aligned}\right\} \tag{1.77}$$

we may approximate the solution of this system by $y_\varepsilon(v)$ the solution of $(\varepsilon>0)$

$$\left.\begin{aligned} -\varepsilon\frac{d^2}{dt^2}y_\varepsilon(v) + \frac{dy_\varepsilon}{dt}(v)+A(t)y_\varepsilon(v)=f+Bv, \\ y_\varepsilon(0;v)=0, \quad \frac{d}{dt}y_\varepsilon(T;v)=0 \end{aligned}\right\} \tag{1.78}$$

and we may obtain results analogous to that of Theorem 1.3. The passage from (1.77) to (1.78) is the so-called procedure of *elliptic regularization* (Lions [10], Oleinik [1]).

Remark 1.7. In the same spirit (but without changing the nature of the system) we have *stability theorems* of the type shown below. Let us consider, for example, a system whose state is given by (1.25), (1.26), (1.27).

Let A_m be a sequence of operators such that

$$\left.\begin{aligned} (A_m\varphi,\psi)=a_m(\varphi,\psi)=a_m(\psi,\varphi)\to a(\varphi,\psi) \quad \forall\varphi,\psi\in V, \\ a_m(\varphi,\varphi)\geqslant\alpha\|\varphi\|^2, \quad \alpha>0 \text{ independent of } m. \end{aligned}\right\} \tag{1.79}$$

Let $y_m(v)$ be the state corresponding to A_m. Then

$$\left.\begin{aligned} y_m''(v)+A_m y_m(v)=f+Bv, \\ y_m(0;v)=y_{0m}, \quad y_m'(0;v)=y_{1m}, \end{aligned}\right\} \tag{1.80}$$

where

$$y_{0m}(\text{resp. } y_{1m})\to y_0(\text{resp. } y_1) \quad \text{in } V(\text{resp. } H). \tag{1.81}$$

The cost function being given by (1.29), we introduce

$$J_m(v)=\|Cy_m(v)-z_d\|_{\mathscr{H}}^2+(Nv,v)_{\mathscr{U}}. \tag{1.82}$$

We then have results analogous to that of Theorem 1.3 with J_ε replaced by J_m and y_ε by y_m.

This allows us to obtain *approximation theorems:*

1. Using the method of Faedo-Galerkin (to approximate the state);
2. Using finite difference techniques again to approximate the state (in this case there are additional technical difficulties[5]—cf. Bossavit [1], Nedelec [1].

Remark 1.8. We have analogous (convergence) results for certain systems governed by non-linear equations.

Remark 1.9. The methods described above should be distinguished from the method of *regularizing the control*, an example of which is presented in Section 1.5.

Remark 1.10. One has to be careful: controllability is *unstable* under parabolic regularization and approximation.

1.5. Regularization of the Control

Let us consider the situation of Chapter 3, Section 8.3. The state is given by

$$\left.\begin{aligned} &\frac{\partial y(v)}{\partial t} + A\,y(v) = f \quad \text{in } Q, \quad f \in L^2(Q),\\ &\frac{\partial y(v)}{\partial \nu_A} = v \quad \text{on } \Sigma,\\ &y(0;v) = y_0 \quad \text{on } \Omega, \quad y_0 \in L^2(\Omega), \end{aligned}\right\} \tag{1.83}$$

where

$$v \in \mathscr{U} = L^2(\Sigma) = L^2(0,T;L^2(\Gamma)). \tag{1.84}$$

The cost function is given by

$$J(v) = \int_\Sigma (y(v) - z_d)^2\, d\Sigma + (N\,v, v)_{L^2(\Sigma)}. \tag{1.85}$$

We assume that

$$\mathscr{U}_{\text{ad}} = \{v \mid v \in \mathscr{U},\ v(t) \in K \text{ a.e., } K = \text{closed, convex subset of } L^2(\Gamma)\}. \tag{1.86}$$

[5] These difficulties are serious as far as approximating the kernel $\mathscr{P}(x,\xi,t)$ of the operator $\mathscr{P}(t)$ by finite-difference methods are concerned.

We now introduce a control space which is more regular:

$$\mathcal{U}^1 = H^1(0,T;L^2(\Gamma)) = \left\{ v \mid v \in L^2(\Sigma),\ v' = \frac{\partial v}{\partial t} \in L^2(\Sigma) \right\}, \tag{1.87}$$

which is a Hilbert space when endowed with the norm

$$\|v\|_{\mathcal{U}^1} = \left(\int_\Sigma (|v|^2 + |v'|^2) \, d\Sigma \right)^{\frac{1}{2}}.$$

We then introduce

$$\mathcal{U}^1_{\text{ad}} = \mathcal{U}^1 \cap \mathcal{U}_{\text{ad}} \quad \text{(closed, convex set in } \mathcal{U}^1) \tag{1.88}$$

and the *regularized cost function:*

$$J_\varepsilon(v) = \int_\Sigma (y(v) - z_d)^2 \, d\Sigma + (N v, v)_{L^2(\Sigma)} + \varepsilon \|v'\|^2_{L^2(\Sigma)}, \qquad \varepsilon > 0. \tag{1.89}$$

The problem of *regularized control* is to find

$$\operatorname*{Inf}_{v \in \mathcal{U}_{\text{ad}}} J_\varepsilon(v).$$

Clearly, there exists a unique $u_\varepsilon \in \mathcal{U}^1_{\text{ad}}$ such that

$$J_\varepsilon(u_\varepsilon) = \operatorname*{Inf}_{v \in \mathcal{U}_{\text{ad}}} J_\varepsilon(v). \tag{1.90}$$

We have,

Theorem 1.5. *The state of the system being given by* (1.83), *the cost function by* (1.85) *and the regularised cost function by* (1.89), *and* $\mathcal{U}_{\text{ad}}$ *and* $\mathcal{U}^1_{\text{ad}}$ *being given by* (1.86) *and* (1.88) *respectively, we have:*

$$u_\varepsilon \to u \quad \text{in } L^2(\Sigma), \tag{1.91}$$

$$\operatorname*{Inf}_{v \in \mathcal{U}_{\text{ad}}} J_\varepsilon(v) \to \operatorname*{Inf}_{v \in \mathcal{U}_{\text{ad}}} J(v). \tag{1.92}$$

Proof. 1. We have

$$J_\varepsilon(u_\varepsilon) \geqslant \operatorname*{Inf}_{v \in \mathcal{U}_{\text{ad}}} J(v) \geqslant \operatorname*{Inf}_{v \in \mathcal{U}_{\text{ad}}} J(v) = J(u), \quad \text{since } \mathcal{U}_{\text{ad}} \supset \mathcal{U}^1_{\text{ad}}. \tag{1.93}$$

Let us then show that

$$\left. \begin{array}{l} \forall \eta > 0, \quad \text{there exists } \varepsilon(\eta) \text{ such that for } \varepsilon \leqslant \varepsilon(\eta) \quad \text{we have} \\ J_\varepsilon(u_\varepsilon) \leqslant J(u) + \eta. \end{array} \right\} \tag{1.94}$$

For this, we first choose $w \in \mathcal{U}^1_{\text{ad}}$ such that

$$J(w) \leqslant J(u) + \tfrac{1}{2}\eta. \tag{1.95}$$

This is possible since $\mathcal{U}^1_{\text{ad}}$ is dense in $\mathcal{U}_{\text{ad}}$. In fact, if we extend u to $\tilde{u}$ defined on R_t with

$$\tilde{u} \in L^2(\mathbb{R}_t; L^2(\Gamma)), \qquad \tilde{u}(t) \in K \quad \text{a.e.}$$

(this is allowed), and then regularize $\tilde{u}$ in t:

$$\left.\begin{aligned}&\tilde{u}*\rho_n=\tilde{w}_n,\\&\rho_n\in\mathscr{D}(\mathbb{R}_t),\qquad \rho_n\geqslant 0,\qquad \int_{-\infty}^{+\infty}\rho_n\,dt=1,\end{aligned}\right\}\tag{1.96}$$

then (convexity theorem) $\tilde{w}_n(t)\in K$ a.e. If we now set

$$w_n=\text{restriction of } \tilde{w}_n \text{ to }]0,T[,$$

we have:

$$w_n\in\mathscr{U}_{\mathrm{ad}}^1 \text{ and } w_n\to u \text{ in } L^2(0,T;L^2(\Gamma)) \text{ as } n\to\infty.$$

Hence we have (1.95).

Then

$$J_\varepsilon(u_\varepsilon)\leqslant J_\varepsilon(w)\leqslant(\text{from } 1.95)\ J(u)+\tfrac{1}{2}\eta-\varepsilon\|w'\|^2_{L^2(\Sigma)}$$

from which we obtain (1.94) by taking

$$\varepsilon(\eta)\|w'\|^2_{L^2(\Sigma)}\leqslant\frac{\eta}{2}.$$

From (1.94) it follows that

$$\limsup J_\varepsilon(u_\varepsilon)\leqslant J(u),\tag{1.97}$$

from which by comparing with (1.93) we obtain the result (1.92).

2. But then we have

$$C\geqslant J_\varepsilon(u_\varepsilon)\geqslant\nu\|u_\varepsilon\|^2_{L^2(\Sigma)}+\varepsilon\|u'_\varepsilon\|^2_{L^2(\Sigma)},$$

which implies that we can extract a sequence, again denoted by u_ε, such that

$$u_\varepsilon\to u^0 \text{ weakly in } \mathscr{U} \text{ and hence } u^0\in\mathscr{U}_{\mathrm{ad}}.$$

But

$$\liminf J_\varepsilon(u)\geqslant\liminf J(u_\varepsilon)\geqslant J(u^0)$$

which when combined with (1.97) shows that $u^0=u$ and hence we have (1.91) where the convergence is weak convergence.

But setting (cf. Part 3 of the proof of Theorem 1.3)

$$|||v|||^2=\int_\Sigma(y(v)-y(0))^2d\Sigma+(Nv,v)_{L^2(\Sigma)}$$

we see that

$$|||u_\varepsilon|||^2+\varepsilon|||u'_\varepsilon|||^2_{L^2(\Sigma)}\to|||u|||^2$$

and hence

$$|||u_\varepsilon-u|||^2+\varepsilon\|u'_\varepsilon\|^2_{L^2(\Sigma)}\to 0$$

whence (1.91). □

Remark 1.11. Let us write the system of equations and inequalities which determine u_ε: the adjoint state $p(u_\varepsilon)$ is given by

$$\left.\begin{aligned} -\frac{\partial}{\partial t}p(u_\varepsilon)+A^*p(u_\varepsilon)=0 \quad \text{in } Q,\\ \frac{\partial p}{\partial \nu_{A^*}}(u_\varepsilon)=y(u_\varepsilon)-z_d \quad \text{on } \Sigma,\\ p(x,T;u_\varepsilon)=0 \quad \text{in } \Omega, \end{aligned}\right\} \tag{1.98}$$

and u_ε is determined by the simultaneous solution of (1.83) (with $v=u_\varepsilon$), (1.98) and

$$\begin{aligned} &\int_\Sigma [(p(u_\varepsilon)+Nu_\varepsilon)(v-u_\varepsilon)+\varepsilon(u_\varepsilon')(v'-u_\varepsilon')]\,dt\,d\Gamma \geqslant 0,\\ &\forall v\in \mathscr{U}_{ad}^1. \end{aligned} \tag{1.99}$$

From Theorem 1.5 we deduce that $p(u_\varepsilon)\to p(u)$ in $L^2(0,T;H^1(\Omega))$ as $\varepsilon\to 0$.

Example. Take

$$K=\{f\,|\,f\geqslant 0 \text{ a.e. on } \Gamma,\ f\in L^2(\Gamma)\}.$$

Then u_ε is determined by the solution of the unilateral problem:

$$\frac{\partial y_\varepsilon}{\partial t}+Ay_\varepsilon=f, \quad -\frac{\partial p_\varepsilon}{\partial t}+A^*p_\varepsilon=0 \quad \text{in } Q,$$

$$\frac{\partial y_\varepsilon}{\partial \nu_A}\geqslant 0 \quad \text{on } \Sigma, \quad \frac{\partial p_\varepsilon}{\partial \nu_A}=y_\varepsilon-z_d \quad \text{on } \Sigma,$$

$$p_\varepsilon+N\frac{\partial y_\varepsilon}{\partial \nu_A}-\varepsilon\frac{\partial^2}{\partial t^2}\left(\frac{\partial y_\varepsilon}{\partial \nu_A}\right)\geqslant 0 \quad \text{on } \Sigma,$$

$$\frac{\partial y_\varepsilon}{\partial \nu_A}\left(p_\varepsilon+N\frac{\partial y_\varepsilon}{\partial \nu_A}-\varepsilon\frac{\partial^2}{\partial t^2}\frac{\partial}{\partial \nu_A}y_\varepsilon\right)=0 \quad \text{on } \Sigma,$$

$$\frac{\partial}{\partial t}\frac{\partial y_\varepsilon}{\partial \nu_A}=0 \quad \text{for } t=0 \text{ and } t=T, \quad x\in\Gamma,$$

$$y_\varepsilon(x,0)=y_0(x), \quad p(x,T)=0, \quad x\in\Omega.$$

Then

$$u_\varepsilon=\frac{\partial y_\varepsilon}{\partial \nu_A}.$$

2. Approximation in Terms of Systems of Cauchy-Kowaleska Type

2.1. Evolution Equation on a Variety

Let $\Omega \subset \mathbb{R}^n$ be a bounded set with a regular boundary Γ. Let $A(x, \partial/\partial x)$ be defined by

$$A\psi = -\sum_{i,j=1}^{n} \frac{\partial}{\partial x_i}\left(a_{ij}(x)\frac{\partial \psi}{\partial x_j}\right) + a_0 \psi, \tag{2.1}$$

where

$$\left.\begin{aligned} & a_{ij} \in C^2(\bar{\Omega}), \quad a_0 \in L^\infty(\bar{\Omega}), \\ & a(\varphi, \psi) = \sum_{i,j=1}^{n} \int_\Omega a_{ij} \frac{\partial \varphi}{\partial x_j}\frac{\partial \psi}{\partial x_i} dx + \int_\Omega a_0 \varphi \psi \, dx, \\ & a(\varphi, \varphi) \geqslant \alpha \|\varphi\|^2_{H^1(\Omega)}, \quad \alpha > 0, \quad \forall \varphi \in H^1(\Omega). \end{aligned}\right\} \tag{2.2}$$

Remark 2.1. We may also consider problems similar to those treated below when the coefficients of the operator are bounded measurable functions of t, or when the operators are elliptic of arbitrary order. Similarly in (2.2) we may consider the case when $a_0 < 0$ with simple modifications of the techniques we use below.

The state $y(v)$ of the system is given by

$$A y(v) = 0 \quad \text{in } \Omega \quad (\text{and } \forall t \in]0, T[), \tag{2.3}$$

$$\frac{\partial}{\partial t} y(v) + \frac{\partial}{\partial \nu_A} y(v) = v \quad \text{on } \Sigma = \Gamma \times]0, T[, \tag{2.4}$$

$$y(x, 0; v) = y_0(x), \quad x \in \Gamma. \tag{2.5}$$

Clearly we must solve problem (2.3), (2.4), (2.5) in a precise sense. We shall show that this problem lies within the framework of Chapter 3, Section 1 and 2 and we are in fact dealing with a parabolic evolution equation defined on Γ.

Operator $\tilde{A}$. Let h be a given element in $H^{\frac{1}{2}}(\Gamma)$. There exists a $\psi = \psi(h)$ in $H^1(\Omega)$ which is unique such that

$$\left.\begin{aligned} A\psi &= 0 \quad \text{in } \Omega, \\ \psi &= h \quad \text{on } \Gamma. \end{aligned}\right\} \tag{2.6}$$

We may then define (Lions-Magenes [1], Chapter 2) $\partial \psi / \partial \nu_A$ in $H^{-\frac{1}{2}}(\Gamma)$. We define

$$\tilde{A} h = \left\{ \frac{\partial \psi}{\partial \nu_A} \text{ on } \Gamma \right\} \tag{2.7}$$

whence

$$\tilde{A} \in \mathscr{L}(H^{\frac{1}{2}}(\Gamma); H^{-\frac{1}{2}}(\Gamma)). \tag{2.8}$$

Lemma 2.1. *We have,* $\forall h \in H^{\frac{1}{2}}(\Gamma)$:

$$(\tilde{A} h, h)_\Gamma = \int_\Gamma (\tilde{A} h) h \, d\Gamma \geqslant \alpha_1 \|h\|^2_{H^{\frac{1}{2}}(\Gamma)}, \quad \alpha_1 > 0. \tag{2.9}$$

Proof. Indeed,

$$(A\psi, \psi) = \int_\Omega (A\psi)\psi \, dx = -\int_\Gamma \left(\frac{\partial \psi}{\partial \nu_A}\right) \psi \, d\Gamma + a(\psi, \psi) = 0,$$

from which

$$(\tilde{A} h, h)_\Gamma = a(\psi, \psi) \geqslant \alpha \|\psi\|^2_{H^{\frac{1}{2}}(\Omega)} \geqslant \alpha_1 \|h\|^2_{H^{\frac{1}{2}}(\Gamma)}. \quad \square$$

Problem (2.3), (2.4), (2.5) can now be formulated as follows: we first set

$$y(.,t)|_\Gamma = \tilde{y}(t); \quad y(.,t;v) = \tilde{y}(t;v); \tag{2.10}$$

then we look for[6] $\tilde{y}$ as the solution of

$$\frac{d\tilde{y}}{dt} + \tilde{A}\tilde{y} = v \quad \text{in }]0,T[, \tag{2.11}$$

$$\tilde{y}(0) = y_0 \quad \text{on } \Gamma. \tag{2.12}$$

Equation (2.11) with the initial condition (2.12) lies within the framework of Chapter 3, Section 1, provided we identify

$$V = H^{\frac{1}{2}}(\Gamma), \quad H = L^2(\Gamma), \quad V' = H^{-\frac{1}{2}}(\Gamma),$$
$$A \text{ replaced by } \tilde{A}.$$

From (2.9), we see that condition (1.2), Chapter 3 holds (with $\lambda = 0$). Hence

$$\left.\begin{array}{l} \text{if } v \in L^2(0,T; L^2(\Gamma)) = L^2(\Sigma)\ ^7 \text{ and if } y_0 \in L^2(\Gamma), \\ \text{problem (2.11), (2.12) admits a unique solution satisfying} \\ \tilde{y} \in L^2(0,T; H^{\frac{1}{2}}(\Gamma)), \quad \dfrac{d\tilde{y}}{dt} \in L^2(0,T; H^{-\frac{1}{2}}(\Gamma)). \end{array}\right\} \tag{2.13}$$

We return to y by solving:

$$\begin{aligned} A y(x,t) &= 0 \quad \text{in } \Omega, t \text{ fixed a.e.} \\ y(x,t) &= \tilde{y}(x,t) \quad \text{on } \Gamma. \end{aligned} \tag{2.14}$$

---

[6] To simplify the notation, we suppress for the moment "the dependence on v" in $\tilde{y}$.

[7] More generally, we may take v in $L^2(0,T; H^{-\frac{1}{2}}(\Gamma))$.

We have thus proved:

Theorem 2.1. *We assume that* (2.2) *holds and that* v *is given in* $L^2(\Sigma)$. *Then there exists a unique function* $y = y(v)$, *solution of* (2.3), (2.4), (2.5) *satisfying*

$$y \in L^2(0, T; H^1(\Omega)). \tag{2.15}$$

Remark 2.2. Equation (2.11) is a parabolic equation defined on the *variety* Γ, the operator $\tilde{A}$ being a (pseudo-differential) first order elliptic operator on Γ.

The Control Problem. We consider the cost function[8]

$$\left.\begin{aligned} &J(v) = \int_\Sigma (y(v) - z_d)^2 \, d\Sigma + (Nv, v)_{L^2(\Sigma)}, \quad v \in \mathscr{U} = L^2(\Sigma), \\ &N \in \mathscr{L}(L^2(\Sigma); L^2(\Sigma)), \quad N \geqslant v \quad \text{(identity)}. \end{aligned}\right\} \tag{2.16}$$

Let u be the optimal control. It is characterised by

$$\int_\Sigma (y(u) - z_d)\,(y(v) - y(u))\, d\Sigma + (Nu, v-u)_{L^2(\Sigma)} \geqslant 0 \quad \forall v \in \mathscr{U}_{\text{ad}}, \tag{2.17}$$

assuming we seek

$$\operatorname{Inf} J(v), \quad v \in \mathscr{U}_{\text{ad}} = \text{closed, convex set in } \mathscr{U}.$$

The results of Chapter 3, Section 2.3 are now applicable. We introduce the adjoint state $\tilde{p}(u)$ by

$$\left.\begin{aligned} -\frac{d}{dt}\tilde{p}(u) + (\tilde{A})^* \tilde{p}(u) = y(u) - z_d \quad &\text{on } \Sigma, \\ \tilde{p}(T; u) = 0 \quad &\text{on } \Gamma. \end{aligned}\right\} \tag{2.18}$$

Then (2.17) is equivalent to

$$\int_\Sigma (\tilde{p}(u) + Nu)\,(v - u)\, d\Sigma \geqslant 0 \quad \forall v \in \mathscr{U}_{\text{ad}}. \tag{2.19}$$

To specify this, we have to explicitely calculate $(\tilde{A})^*$.

Lemma 2.2. *The operator* $(\tilde{A})^*$ *is defined as follows: let* k *be given in* $H^{\frac{1}{2}}(\Gamma)$; *we solve*

$$\left.\begin{aligned} A^* \chi = 0 \quad &\text{in } \Omega, \\ \chi|_\Gamma = k; & \end{aligned}\right\} \tag{2.20}$$

then

$$(\tilde{A})^* k = \frac{\partial \chi}{\partial \nu_{A^*}}. \tag{2.21}$$

[8] We could have equally well considered a cost function corresponding to observation of the final state.

Proof. This is a consequence of Green's formula:

$$0 = \int_{\Omega} (A\psi)\chi\, dx = -\int_{\Gamma} (\tilde{A}h)k\, d\Gamma + \int_{\Gamma} h \frac{\partial \chi}{\partial \nu_{A^*}} d\Gamma + \int_{\Omega} \psi(A^*\chi)\, dx$$

and hence

$$(\tilde{A}h, k)_\Gamma = \left(h, \frac{\partial \chi}{\partial \nu_{A^*}}\right)_\Gamma, \quad \text{whence (2.21).} \quad \square$$

Then problem (2.18) can be formulated in the following equivalent form:

$$\left.\begin{aligned} A^* p(u) = 0 \quad \text{in } \Omega \ (\forall t \in]0, T[), \\ -\frac{\partial}{\partial t} p(u) + \frac{\partial}{\partial \nu_{A^*}} p(u) = y(u) - z_d \quad \text{on } \Sigma, \\ p(x, T; u) = 0, \qquad x \in \Gamma, \end{aligned}\right\} \tag{2.22}$$

(and $\tilde{p}(u) = p(u)$ on Σ).
We have thus proved:

Theorem 2.2. *Let the hypotheses of Theorem 2.1 hold and let the cost function be given by* (2.16). *Then the optimal control u is determined by the solution of* (2.3), (2.4), (2.5) *(with $v = u$), of* (2.22) *and of*[9]

$$\int_{\Sigma} (p(u) + Nu)(v - u)\, d\Sigma \geqslant 0 \quad \forall v \in \mathcal{U}_{ad}. \tag{2.23}$$

Example 2.1. In the case of no constraints $\mathcal{U}_{ad} = \mathcal{U}$, we have reduced the problem to solving

$$\left.\begin{aligned} Ay = 0, \quad A^* p = 0 \quad \text{in } \Omega \quad \text{and } \forall t, \\ \frac{\partial y}{\partial t} + \frac{\partial y}{\partial \nu_{A^*}} + N^{-1} p = 0 \quad \text{on } \Sigma, \\ \frac{\partial p}{\partial \nu_{A^*}} = y - z_d \quad \text{on } \Sigma, \\ y(x, 0) = y_0(x), \qquad x \in \Gamma, \\ p(x, T) = 0, \qquad x \in \Gamma. \end{aligned}\right\} \tag{2.24}$$

[9] From (2.19).

In the notation of $\tilde{y}$, $\tilde{p}$ (see below), the (equivalent) system may be written as

$$\left.\begin{aligned}\frac{\partial\tilde{y}}{\partial t}+\tilde{A}\tilde{y}+N^{-1}\tilde{p}=0,\qquad -\frac{\partial\tilde{p}}{\partial t}+(\tilde{A})^{*}\tilde{p}=\tilde{y}-z_d,\\ \tilde{y}(0)=y_0,\qquad \tilde{p}(T)=0.\end{aligned}\right\}\tag{2.25}$$

Then

$$u=-N^{-1}(p)_{\Sigma}=-N^{-1}\tilde{p}.\tag{2.26}$$

In the form of (2.25) we may use the decoupling theory of Chapter 3, Sections 4 and 5. We then have

$$\tilde{p}(t)=\tilde{P}(t)\,\tilde{y}(t)+\tilde{r}(t),\tag{2.27}$$

with

$$\left.\begin{aligned}-\frac{d}{dt}\tilde{P}+\tilde{P}\tilde{A}+(\tilde{A})^{*}\tilde{P}+\tilde{P}N^{-1}\tilde{P}=\text{identity},\qquad 0<t<T,\\ \tilde{P}(T)=0,\end{aligned}\right\}\tag{2.28}$$

and

$$\left.\begin{aligned}-\frac{d\tilde{r}}{dt}+(\tilde{A})^{*}\tilde{r}+\tilde{P}N^{-1}\tilde{r}=-z_d,\\ \tilde{r}(T)=0.\end{aligned}\right\}\tag{2.29}$$

Equation (2.28) is a non-linear integro-differential parabolic equation of Riccati type on the variety $\Gamma\times\Gamma$.

The *kernel theorem* of L. Schwartz is valid on the *variety* Γ. Therefore we may represent $\tilde{P}$ with the aid of a *distribution kernel:*

$$\tilde{P}(x,\xi,t)\quad\text{on}\ \Gamma_x\times\Gamma_\xi.$$

Let us note that since $\tilde{A}$ is not a differential operator, $\tilde{A}$ has a kernel $\tilde{A}(x,\xi)$ which is not concentrated on the diagonal $\Gamma_x\times\Gamma_\xi$.

Clearly the relation (2.27) may be "lifted" from Γ to Ω. We may deduce the existence of an operator $P(t)\in\mathscr{L}(L^2(\Omega);L^2(\Omega))$ (in particular) such that

$$p(t)=P(t)\,y(t)+r(t).\tag{2.30}$$

Example 2.2. Let us now consider the case

$$\mathscr{U}_{\text{ad}}=\{v\mid v\geqslant 0\ \text{on}\ \Sigma,\ v\in L^2(\Sigma)\}.\tag{2.31}$$

The optimal control is determined by the solution of the unilateral problem

$$\left.\begin{aligned} &Ay=0, \quad A^*p=0 \quad \text{in } \Omega \text{ and } \forall t \\ &\frac{\partial y}{\partial t}+\frac{\partial y}{\partial \nu_A} \geqslant 0 \quad \text{on } \Sigma, \\ &p+N\left(\frac{\partial y}{\partial t}+\frac{\partial y}{\partial \nu_A}\right) \geqslant 0 \quad \text{on } \Sigma, \\ &\left(\frac{\partial y}{\partial t}+\frac{\partial y}{\partial \nu_A}\right)\left[p+N\left(\frac{\partial y}{\partial t}+\frac{\partial y}{\partial \nu_A}\right)\right]=0 \quad \text{on } \Sigma, \\ &\frac{\partial p}{\partial \nu_{A^*}}=y-z_d \quad \text{on } \Sigma, \\ &y(x,0)=y_0(x) \quad \text{on } \Gamma, \\ &p(x,T)=0 \quad \text{on } \Gamma, \end{aligned}\right\} \tag{2.32}$$

and then

$$u=\frac{\partial y}{\partial t}+\frac{\partial y}{\partial \nu_A} \quad \text{on } \Sigma.$$

We may consider similar problems containing second derivatives of t. For example,

$$\left.\begin{aligned} &Ay(v)=0 \quad \text{in } \Omega, \quad \forall t, \\ &\frac{\partial^2}{\partial t^2} y(v)+\frac{\partial}{\partial \nu_A} y(v)=v \quad \text{on } \Sigma, \\ &y(x,0;v)=y_0(x) \quad \text{on } \Gamma \ (y_0 \text{ given in } H^{\frac{1}{2}}(\Gamma)), \\ &\frac{\partial y}{\partial t}(x,0;v)=y_1(x) \quad \text{on } \Gamma \ (y_1 \text{ given in } L^2(\Gamma)), \end{aligned}\right\} \tag{2.33}$$

under the assumption that

$$A=A^*. \tag{2.34}$$

In this way we transform this problem to lie within the framework of Chapter 4—but on a variety Γ instead of Ω. Thus, we may also apply the parabolic regularization procedure of Section 1.

2.2. Approximation by a System of Cauchy-Kowaleska Type

Problem (2.3), (2.4), (2.5) when considered on the open set Ω (and not on a variety Γ) is not of Cauchy-Kowaleska type. We shall see that we may approximate in a very simple manner this system by a new (parabolic) system of Cauchy-Kowaleska type.

For $\varepsilon>0$, define the state $y_\varepsilon(v)$ by

$$\varepsilon\frac{\partial}{\partial t}y_\varepsilon(v)+A\,y_\varepsilon(v)=0 \quad \text{in} \quad \Omega\times]0,T[, \tag{2.35}$$

$$\frac{\partial}{\partial t}y_\varepsilon(v)+\frac{\partial}{\partial \nu_A}y_\varepsilon(v)=v \quad \text{on } \Sigma, \tag{2.36}$$

$$y_\varepsilon(x,0;v)=\hat{y}_0(x) \quad \text{on } \Omega, \tag{2.37}$$

where $\hat{y}_0$ is defined by

$$\begin{aligned} A\hat{y}_0&=0 \quad \text{in } \Omega,\\ \hat{y}_0&=y_0 \quad \text{on } \Gamma.\ ^{10} \end{aligned} \tag{2.38}$$

Problem (2.35), (2.36), (2.37) which now is of Cauchy-Kowaleska type in Ω admits a unique solution. In fact, it is equivalent to $y_\varepsilon(t;v)=y_\varepsilon(t)$, where $y_\varepsilon(t)$ is given by

$$\left.\begin{aligned} &\varepsilon\frac{d}{dt}(y_\varepsilon(t),\psi)+\frac{d}{dt}(y_\varepsilon(t),\psi)_\Gamma+a(y_\varepsilon(t),\psi)=(v(t),\psi)_\Gamma,\\ &\forall\psi\in H^1(\Omega), \end{aligned}\right\} \tag{2.39}$$

$$y_\varepsilon(0)=\hat{y}_0, \tag{2.40}$$

and we deduce (as in Chapter 3, Section 1) that there exists a unique solution such that

$$\left.\begin{aligned} &y_\varepsilon(v)\in L^2(0,T;H^1(\Omega));\\ &t\to y_\varepsilon(t,v) \quad \text{is continuous from } [0,T]\to L^2(\Omega). \end{aligned}\right\} \tag{2.41}$$

We shall prove,

Theorem 2.3. *Let the hypotheses of Theorem* 2.1 *hold. Let* $y_\varepsilon(v)$ *be the solution of* (2.35), (2.36), (2.37), (2.41). *Then as* $\varepsilon\to 0$, *we have*

$$y_\varepsilon(v)\to y(v) \quad \text{in } L^2(0,T;H^1(\Omega)). \tag{2.42}$$

Proof. We write y_ε (resp. y) instead of $y_\varepsilon(v)$ (resp. $y(v)$).

By taking $\psi=y_\varepsilon$ in (2.39) and integrating by parts we deduce that

---

[10] A problem which admits (cf. Lions-Magenes [1], Chapter 2) a unique solution in $H^{\frac{1}{2}}(\Omega)$—hence in particular $\hat{y}_0\in L^2(\Omega)$.

$$\frac{\varepsilon}{2}\|y_\varepsilon(T)\|^2_{L^2(\Omega)}+\frac{1}{2}\|y_\varepsilon(T)\|^2_{L^2(\Gamma)}+\int_0^T a(y_\varepsilon,y_\varepsilon)\,dt$$
$$=\int_0^T (v,y_\varepsilon)_\Gamma\,dt+\frac{\varepsilon}{2}\|y_0\|^2_{L^2(\Omega)}+\frac{1}{2}\|y_0\|^2_{L^2(\Gamma)},$$

and hence

$$y_\varepsilon \text{ ranges in a bounded set of } L^2(0,T;H^1(\Omega)), \tag{2.43}$$

and

$$\|y_\varepsilon(T)\|_{L^2(\Gamma)}\leqslant C,\qquad \sqrt{\varepsilon}\,\|y_\varepsilon(T)\|_{L^2(\Omega)}\leqslant C. \tag{2.44}$$

Hence we may extract a subsequence, again denoted by y_ε, such that

$$y_\varepsilon\to z \quad \text{weakly in } L^2(0,T;H^1(\Omega)), \tag{2.45}$$

$$y_\varepsilon(T)\to\chi \quad \text{weakly in } L^2(\Omega). \tag{2.46}$$

Let

$$\varphi\in\mathscr{C}^1([0,T];H^1(\Omega)).$$

From (2.39) we deduce that

$$\left.\begin{aligned}&-\varepsilon\int_0^T(y_\varepsilon(t),\varphi')\,dt-\int_0^T(y_\varepsilon(t),\varphi')_\Gamma\,dt+\int_0^T a(y_\varepsilon,\varphi)\,dt\\&\quad=\int_0^T(v(t),\varphi)_\Gamma\,dt+\varepsilon(\hat{y}_0,\varphi(0))+(y_0,\varphi(0))_\Gamma\\&\qquad-\varepsilon(y_\varepsilon(T),\varphi(T))-(y_\varepsilon(T),\varphi(T))_\Gamma.\end{aligned}\right\} \tag{2.47}$$

By virtue of (2.45) (which implies that $y_\varepsilon|_\Sigma\to z|_\Sigma$ in $L^2(\Sigma)$), (2.46), (2.44) we may let $\varepsilon\to 0$ in (2.47). We get

$$-\int_0^T(z,\varphi')_\Gamma\,d\Gamma+\int_0^T a(z,\varphi)\,dt=\int_0^T(v,\varphi)_\Gamma\,dt+(y_0,\varphi(0))-(\chi,\varphi(T))_\Gamma. \tag{2.48}$$

Let φ be such that $\varphi(T)=0$. Then, from (2.48), we deduce that $z=y$ and further that

$$(\chi,\varphi(T))_\Gamma=(y(T),\varphi(T))_\Gamma,\quad \text{hence } \chi=y(T).$$

Consequently, (2.45), (2.46) give us

$$\left.\begin{aligned}y_\varepsilon&\to y &&\text{weakly in } L^2(0,T;H^1(\Omega)),\\ y_\varepsilon(T)&\to y(T) &&\text{weakly in } L^2(\Gamma).\end{aligned}\right\} \tag{2.49}$$

We now consider

$$X_\varepsilon=\int_0^T a(y_\varepsilon-y,y_\varepsilon-y)\,dt+\frac{\varepsilon}{2}\|y_\varepsilon(T)\|^2_{L^2(\Omega)}+\frac{1}{2}\|y_\varepsilon(T)-y(T)\|^2_{L^2(\Gamma)}.$$

Utilising the equations satisfied by y_ε and y, we obtain

$$X_\varepsilon = \int_0^T ((v(t), y_\varepsilon(t))_\Gamma + (v, y)_\Gamma)\, dt - \int_0^T [a(y_\varepsilon, y) + a(y, y_\varepsilon)]\, dt$$
$$- (y_\varepsilon(T), y(T))_\Gamma + \tfrac{1}{2} \|\hat{y}_0\|^2_{L^2(\Omega)} + \|y_0\|^2_{L^2(\Gamma)};$$

whence

$$X_\varepsilon \to 2 \int_0^T [(v, y)_\Gamma - a(y, y)]\, dt - \|y(T)\|^2_{L^2(\Gamma)} + \|y_0\|^2_{L^2(\Gamma)} = 0$$

whence (2.42). □

With the optimal control problem of Section 2.1, we associate the problem

$$\operatorname*{Inf}_{v \in \mathscr{U}_{ad}} J_\varepsilon(v) = J_\varepsilon(u_\varepsilon), \tag{2.50}$$

where

$$J_\varepsilon(v) = \int_\Sigma |y_\varepsilon(v) - z_d|^2\, d\Sigma + (Nv, v)_{L^2(\Sigma)}, \tag{2.51}$$

$y_\varepsilon(v)$ being given by (2.35), (2.36), (2.37).

From Theorem 2.3, using the same type of proof as in Theorem 1.3, we deduce

Theorem 2.4. *Let the hypotheses be the same as in Theorem 2.1. For each $\varepsilon > 0$, let $y_\varepsilon(v)$ be the state corresponding to the control v of the approximating Cauchy-Kowaleska system defined by (2.35), (2.36), (2.37) and let $J_\varepsilon(v)$ be given by (2.51). Then as $\varepsilon \to 0$, we have*

$$u_\varepsilon \to u \quad \text{in } \mathscr{U} = L^2(\Sigma), \tag{2.52}$$

$$y_\varepsilon(u_\varepsilon) \to y(u) \quad \text{in } L^2(0, T; H^1(\Omega)), \tag{2.53}$$

$$J_\varepsilon(u_\varepsilon) \to J(u). \tag{2.54}$$

We may apply the results of Chapter 3 to obtain the control u_ε. We introduce the adjoint state by

$$\left.\begin{aligned}
&-\varepsilon \frac{\partial}{\partial t} p_\varepsilon(u_\varepsilon) + A^* p_\varepsilon(u_\varepsilon) = 0 \quad \text{in } \Omega \times]0, T[, \\
&-\frac{\partial}{\partial t} p_\varepsilon(u_\varepsilon) + \frac{\partial}{\partial \nu_{A^*}} p_\varepsilon(u_\varepsilon) = y(u_\varepsilon) - z_d \quad \text{on } \Sigma, \\
&p_\varepsilon(u_\varepsilon) = 0 \quad \text{for } x \in \Omega, \quad t = T,
\end{aligned}\right\} \tag{2.55}$$

and we obtain

Theorem 2.5. *The optimal control u_ε for the approximate Cauchy-Kowaleska type of problem is given by* (2.35), (2.36), (2.37) *(with $v=u_\varepsilon$),* (2.55) *and*

$$\int_\Sigma (p_\varepsilon(u_\varepsilon)+Nu_\varepsilon)(v-u_\varepsilon)d\Sigma \geqslant 0 \quad \forall v\in\mathscr{U}_{\mathrm{ad}}. \tag{2.56}$$

Let us also note that by applying Theorem 2.3 (with time reversed) to the system (2.55) we obtain

$$p_\varepsilon(u_\varepsilon)\to p(u) \quad \text{in } L^2(0,T;H^1(\Omega)). \tag{2.57}$$

Example 2.3. Take $\mathscr{U}_{\mathrm{ad}}=\mathscr{U}$.

Then u_ε is given by the solution of

$$\left.\begin{aligned} \varepsilon\frac{\partial y_\varepsilon}{\partial t}+Ay_\varepsilon&=0,\\ -\varepsilon\frac{\partial p_\varepsilon}{\partial t}+A^*p_\varepsilon&=0 \quad \text{in } Q,\\ \frac{\partial y_\varepsilon}{\partial t}+\frac{\partial y_\varepsilon}{\partial \nu_A}+N^{-1}p_\varepsilon&=0 \quad \text{on } \Sigma,\\ -\frac{\partial p_\varepsilon}{\partial t}+\frac{\partial p_\varepsilon}{\partial \nu_{A^*}}&=y_\varepsilon-z_d \quad \text{on } \Sigma,\\ y_\varepsilon(x,0)=\hat{y}_0(x),\quad p_\varepsilon(x,T)&=0,\quad x\in\Omega, \end{aligned}\right\} \tag{2.58}$$

and then

$$u_\varepsilon=-N^{-1}p_\varepsilon. \tag{2.59}$$

As $\varepsilon\to 0$, $\{y_\varepsilon,p_\varepsilon\}\to\{y,p\}$ (solution of (2.44) in $L^2(0,T;H^1(\Omega))^2$.

We may decouple system (2.58) in the manner of Chapter 3, Sections 4 and 5. We obtain

$$p_\varepsilon(t)=P_\varepsilon(t)y_\varepsilon(t)+r_\varepsilon(t), \tag{2.60}$$

P_ε satisfying an integro-differential equation of Riccati type which in this case is

$$\left.\begin{aligned} -\varepsilon(P'_\varepsilon\varphi,\psi)-(P'_\varepsilon\varphi,\psi)_\Gamma+a(\varphi,P_\varepsilon\psi)+a^*(P_\varepsilon\varphi,\psi)&\\ +(N^{-1}P_\varepsilon\varphi,P_\varepsilon\psi)_\Gamma=(\varphi,\psi)_\Gamma \quad \forall\varphi,\psi\in H^1(\Omega),& \end{aligned}\right\} \tag{2.61}$$

with

$$P_\varepsilon(T)=0;\ ^{11} \tag{2.62}$$

---

[11] When establishing (2.61), note that P_ε satisfies
$$\varepsilon(P_\varepsilon\varphi,\psi)+(P_\varepsilon\varphi,\psi)_\Gamma=\varepsilon(\varphi,P_\varepsilon\psi)+(\varphi,P_\varepsilon\psi)_\Gamma.$$

r_ε is then given by

$$\left.\begin{aligned}&-\varepsilon(r'_\varepsilon,\psi)-(r'_\varepsilon,\psi)_\Gamma+a^*(r_\varepsilon,\psi)+(N^{-1}r_\varepsilon,P_\varepsilon\psi)_\Gamma=-(z_d,\psi)_\Gamma,\\&\forall\psi\in H^1(\Omega),\end{aligned}\right\}\quad(2.63)$$

and

$$r_\varepsilon(T)=0. \quad (2.64)$$

As $\varepsilon\to 0$, we have in particular

$$P_\varepsilon(t)\psi\to P(t)\psi \quad \text{in } L^2(\Omega)\ \forall\psi\in H^1(\Omega). \quad (2.65)$$

Remark 2.4. It would be interesting to obtain a direct proof of the existence and uniqueness of a solution of (2.61), (2.62) in $[0,T]$ as well as a proof of (2.65). It should be possible to improve the last result.

Example 2.4. Let $\mathscr{U}_{ad}$ be as in Example 2.2. u_ε is then obtained by solving the unilateral problem

$$\left.\begin{aligned}&\varepsilon\frac{\partial y_\varepsilon}{\partial t}+Ay_\varepsilon=0,\\&-\varepsilon\frac{\partial p_\varepsilon}{\partial t}+A^*p_\varepsilon=0 \quad \text{in } Q,\\&\frac{\partial y_\varepsilon}{\partial t}+\frac{\partial y_\varepsilon}{\partial \nu_A}\geqslant 0 \quad \text{on } \Sigma,\\&\left(p_\varepsilon+N\left[\frac{\partial y_\varepsilon}{\partial t}+\frac{\partial y_\varepsilon}{\partial \nu_A}\right]\right)\geqslant 0 \quad \text{on } \Sigma,\\&\left(p_\varepsilon+N\left[\frac{\partial y_\varepsilon}{\partial t}+\frac{\partial y_\varepsilon}{\partial \nu_A}\right]\right)\left(\frac{\partial y_\varepsilon}{\partial t}+\frac{\partial y_\varepsilon}{\partial \nu_A}\right)=0 \quad \text{on } \Sigma,\\&-\frac{\partial p_\varepsilon}{\partial t}+\frac{\partial p_\varepsilon}{\partial \nu_{A^*}}=y_\varepsilon-z_d \quad \text{on } \Sigma,\\&y_\varepsilon(x,0)=\hat{y}_0(x),\quad p_\varepsilon(x,T)=0,\quad x\in\Omega,\end{aligned}\right\}\quad(2.66)$$

and hence

$$u_\varepsilon=\frac{\partial y_\varepsilon}{\partial t}+\frac{\partial y_\varepsilon}{\partial \nu_A} \quad \text{on } \Sigma. \quad (2.67)$$

As $\varepsilon\to 0$, we can again prove,

$$\{y_\varepsilon,p_\varepsilon\}\to\{y,p\} \quad \text{(solution of 2.32) in } L^2(0,T;H^1(\Omega))^2.$$

Remark 2.5. In the case where the state is given by (2.33), we may consider the following approximation (of Cauchy-Kowaleska type in Ω):

$$\left.\begin{aligned} &\varepsilon \frac{\partial^2}{\partial t^2} y_\varepsilon(v) + A y_\varepsilon(v) = 0 \quad \text{in} \quad Q, \\ &\frac{\partial^2}{\partial t^2} y_\varepsilon(v) + \frac{\partial}{\partial v_A} y_\varepsilon(v) = v \quad \text{on} \quad \Sigma, \\ &y_\varepsilon(x,0;v) = \tilde{y}_0(x), \quad \frac{\partial}{\partial t} y_\varepsilon(x,0;v) = \tilde{y}_1(x), \quad x \in \Omega, \end{aligned}\right\} \tag{2.68}$$

where $A\tilde{y}_0 = 0$, $\tilde{y}_0 = y_0$ on Γ (and analogous definition for $\tilde{y}_1$).

2.3. Linearized Navier-Stokes Equation

Let us consider the system whose state is given by

$$\left.\begin{aligned} &\frac{\partial}{\partial t} \boldsymbol{y}_0(v) - \Delta \boldsymbol{y}_0(v) + \operatorname{grad} y_{n+1}(v) = \boldsymbol{f} + \boldsymbol{v} \quad \text{in} \quad \Omega \times]0,T[, \\ &\operatorname{div} \boldsymbol{y}_0(v) = 0, \\ &\boldsymbol{y}_0(v) = \{y_{01}(v), \ldots, y_{0n}(v)\} = 0 \quad \text{on} \quad \Sigma, \\ &\boldsymbol{y}_0(x,0;v) = \boldsymbol{y}_0(x), \end{aligned}\right\} \tag{2.69}$$

where

$$\boldsymbol{y}(v) = \{\boldsymbol{y}_0(v), y_{n+1}(v)\}.$$

This is the linearised Navier-Stokes equation.

We may formulate problem (2.69) in a precise manner as follows: we consider the spaces

$$\left.\begin{aligned} V &= \{\boldsymbol{\varphi} \mid \boldsymbol{\varphi} \in (H_0^1(\Omega))^n, \operatorname{div} \boldsymbol{\varphi} = 0\}, \\ H &= \{\boldsymbol{\varphi} \mid \boldsymbol{\varphi} \in (L^2(\Omega))^n, \operatorname{div} \boldsymbol{\varphi} = 0\} \end{aligned}\right\} \tag{2.70}$$

which are closed sub-spaces of $(H_0^1(\Omega))^n$ and of $(L^2(\Omega))^n$.

If we set

$$(\boldsymbol{\varphi}, \boldsymbol{\psi}) = \sum_{i=1}^{n} \int_\Omega \varphi_i \psi_i \, dx,$$

$$a(\boldsymbol{\varphi}, \boldsymbol{\psi}) = \sum_{i,j=1}^{n} \int_\Omega \frac{\partial \varphi_i}{\partial x_j} \frac{\partial \psi_i}{\partial x_j} \, dx,$$

problem (2.69) may be reformulated as follows: we wish to find $y_0(v)$ as the solution of

$$\frac{d}{dt}(y_0(v),\psi)+a(y_0(v),\psi)=(f+v,\psi) \quad \forall\psi\in V, \tag{2.71}$$

$$y_0(0;v)=y_0 \quad \text{in } H, \tag{2.72}$$

where f, v are given functions in $(L^2(Q))^n$.

The passage to a "variational" formulation "eliminates" $y_{n+1}(v)$.

According to Theorem 1.2, Chapter 3, problem (2.71), (2.72) admits a unique solution satisfying

$$y_0(t;v)\in L^2(0,T;V). \tag{2.73}$$

If the cost function is given by

$$\left.\begin{aligned} J(v) &= \int_Q |y_0(x,t;v)-z_d|^2\,dx\,dt+(Nv,v)_{\mathscr{U}}, \\ \mathscr{U} &= (L^2(Q))^n, \end{aligned}\right\} \tag{2.74}$$

the results of Chapter 3, Section 2 are applicable.

The optimal control u is determined by the solution of the system of equations (2.71), (2.72) together with that of

$$\left.\begin{aligned} &-\frac{d}{dt}(p_0(u),\psi)+a(p_0(u),\psi)=(y(u)-z_d,\psi) \quad \forall\psi\in V, \\ &p_0(T;u)=0, \\ &p_0(u)\in L^2(0,T;V), \end{aligned}\right\} \tag{2.75}$$

and of

$$(p_0(u)+Nu,v-u)\geqslant 0 \quad \forall v\in\mathscr{U}_{\mathrm{ad}}. \tag{2.76}$$

System (2.69) may now be approximated by a Cauchy-Kowaleska system as follows: for $\varepsilon>0$, we define, $\{y_0^\varepsilon, y_{n-1}^\varepsilon\}$ as the solution of

$$\left.\begin{aligned} &\frac{\partial}{\partial t}y_0^\varepsilon(v)-\Delta y_0^\varepsilon(v)+\operatorname{grad} y_{n+1}^\varepsilon=f+v \quad \text{in } Q, \\ &\varepsilon\frac{\partial}{\partial t}y_{n+1}^\varepsilon(v)+\operatorname{div} y_0^\varepsilon(v)=0 \quad \text{in } Q, \\ &y_0^\varepsilon(v)\in L^2(0,T;(H_0^1(\Omega))^n), \\ &y_{n+1}^\varepsilon(v)\in L^\infty(0,T;L^2(\Omega)), \\ &y_0^\varepsilon(x,0;v)=y_0(x), \quad y_{n+1}^\varepsilon(x,0;v)=g(x), \quad x\in\Omega, \end{aligned}\right\} \tag{2.77}$$

where g is any function in $L^2(\Omega)$.

We may prove (by using methods similar to that of Theorem 1.2, Chapter 3) that problem (2.77) admits a unique solution and we may also show that as $\varepsilon \to 0$,

$$y_0^\varepsilon(v) \to y_0(v) \quad \text{in } L^2(0,T;(H_0^1(\Omega))^n). \tag{2.78}$$

If we then define

$$J_\varepsilon(v) = \int_Q |y_0^\varepsilon(x,t;v) - z_d|^2\,dx\,dt + (Nv,v)_{\mathscr{U}}, \tag{2.79}$$

it may be verified that as $\varepsilon \to 0$,

$$\operatorname*{Inf}_{v\in\mathscr{U}_{\text{ad}}} J_\varepsilon(v) = J_\varepsilon(u_\varepsilon) \to J(u) = \operatorname*{Inf}_{v\in\mathscr{U}_{\text{ad}}} J(v), \tag{2.80}$$

$$u_\varepsilon \to u \quad \text{in } \mathscr{U} = (L^2(Q))^n, \tag{2.81}$$

$$y_0^\varepsilon(u_\varepsilon) \to y_0(u) \quad \text{in } L^2(0,T;(H_0^1(\Omega))^n). \tag{2.82}$$

Thus we have obtained results of the same type as in Section 2.3.

Remark 2.6. Similar results may be obtained for non-linear Navier-Stokes equations (cf. Lions [1], for the case where there are no controls).

3. Penalization

Let us consider the following example (Chapter 3, Section 3.2.2): the state $y(v)$ is given by

$$\frac{\partial}{\partial t} y(v) + A\,y(v) = f \quad \text{in } Q = \Omega \times]0,T[, \quad f \in L^2(Q), \tag{3.1}$$

$$\frac{\partial y}{\partial \nu_A}(v) = v \quad \text{on } \Sigma, \quad v \in L^2(\Sigma), \tag{3.2}$$

$$y(x,0;v) = y_0(x), \quad y_0 \in L^2(\Omega), \tag{3.3}$$

where $A = A(x,t,\partial/\partial x)$ is a second order elliptic operator in x (cf. Chapter 3, Section 3.2.2). We assume that

$$\left.\begin{aligned} &\mathscr{U} = L^2(\Sigma), \quad v \in \mathscr{U}, \\ &\mathscr{U}_{\text{ad}} = \text{closed, convex subset of } \mathscr{U}. \end{aligned}\right\} \tag{3.4}$$

We assume that (observation of the final state)

$$\left.\begin{aligned} &J(v) = \int_Q (y(x,T;v) - z_d(x))^2\,dx + (N v, v)_{L^2(\Sigma)}, \\ &N \in \mathscr{L}(\mathscr{U};\mathscr{U}), \quad N \geqslant \nu \ \text{(identity)}, \quad \nu > 0. \end{aligned}\right\} \tag{3.5}$$

We seek

$$\operatorname*{Inf}_{v\in\mathscr{U}_{ad}} J(v). \tag{3.6}$$

We shall approximate this problem by a family of problems in which y and v become the independent variables.

We introduce the following notation:

$$Y=\left\{y \mid y\in L^2(0,T;H^1(\Omega)),\ \frac{\partial y}{\partial t}+Ay\in L^2(Q),\ \frac{\partial y}{\partial \nu_A}\in L^2(\Sigma)\right\}. \tag{3.7}$$

Note that the definition makes sense since if $y\in L^2(0,T;H^1(\Omega))$ and $\frac{\partial y}{\partial t}+Ay\in L^2(Q)$, then we may define $\partial y/\partial\nu_A$ (with the aid of Green's formula—cf. Lions-Magenes [1], Chapter 4, Vol. 2) and hence the condition "$\partial y/\partial\nu_A\in L^2(\Sigma)$" makes sense. It may be verified that y endowed with the norm

$$\|y\|_Y=\left(\|y\|^2_{L^2(0,T;H^1(\Omega))}+\left\|\frac{\partial y}{\partial t}+Ay\right\|^2_{L^2(Q)}+\left\|\frac{\partial y}{\partial \nu_A}\right\|^2_{L^2(\Sigma)}\right)^{\frac{1}{2}} \tag{3.8}$$

is a Hilbert space.

Further, set

$$\varepsilon=\{\varepsilon_1,\varepsilon_2,\varepsilon_3\},\quad \varepsilon_i>0, \tag{3.9}$$

and define the functional

$$\left.\begin{aligned}\mathscr{J}_\varepsilon(y,v)={}&\int_\Omega (y(x,T)-z_d(x))^2\,dx+(Nv,v)_{\mathscr{U}}\\&+\frac{1}{\varepsilon_1}\left\|\frac{\partial y}{\partial t}+Ay-f\right\|^2_{L^2(Q)}+\frac{1}{\varepsilon_2}\int_\Sigma\left(\frac{\partial y}{\partial\nu_A}-v\right)^2 d\Sigma\\&+\frac{1}{\varepsilon_3}\int_\Omega (y(x,0)-y_0(x))^2\,dx\end{aligned}\right\} \tag{3.10}$$

on $Y\times\mathscr{U}$.

We now consider the new problem (which is said to have been obtained from problem (3.6) by *penalization*[12])

$$\text{find}\quad \operatorname*{Inf}_{\substack{y\in Y\\ v\in\mathscr{U}_{ad}}} \mathscr{J}_\varepsilon(y,v)=j_\varepsilon. \tag{3.11}$$

---

[12] The factors $\frac{1}{\varepsilon_i}$ introduce a *penalty* if the constraints (3.1), (3.2), (3.3) are not satisfied.

Theorem 3.1. *The (penalized) problem* (3.11) *admits a unique solution*

$$\{y_\varepsilon, u_\varepsilon\}. \tag{3.12}$$

Proof. Consider the second order homogeneous part of the quadratic form $J_\varepsilon(y,v)$ denoted by $\beta(y,v)$ which is given by

$$\beta(y,v) = \int_\Omega y(x,T)^2\,dx + (Nv,v)_{\mathcal{U}} + \frac{1}{\varepsilon_1}\left\|\frac{\partial y}{\partial t} + Ay\right\|^2_{L^2(Q)} + \frac{1}{\varepsilon_2}\int_\Sigma\left(\frac{\partial y}{\partial \nu_A} - v\right)^2 d\Sigma + \frac{1}{\varepsilon_3}\int_\Omega y(x,0)^2\,dx.$$

It suffices to prove

$$\beta(y,v) \geqslant c(\|y\|_Y^2 + \|v\|_{\mathcal{U}}^2). \tag{3.13}$$

Since $(Nv,v)_{\mathcal{U}} \geqslant \nu\|v\|^2_{\mathcal{U}}$, we have

$$\beta(y,v) \geqslant \nu\|v\|^2_{\mathcal{U}} + \frac{1}{\varepsilon_2}\left\|\frac{\partial y}{\partial \nu_A} - v\right\|^2_{\mathcal{U}} + \frac{1}{\varepsilon_1}\left\|\frac{\partial y}{\partial t} + Ay\right\|^2_{L^2(Q)} + \frac{1}{\varepsilon_3}\int_\Omega y(x,0)^2\,dx + \int_\Omega y(x,T)^2\,dx,$$

and hence

$$\beta(y,v) \geqslant \left(\nu + \frac{1}{\varepsilon_2}\right)\|v\|^2_{\mathcal{U}} + \frac{1}{\varepsilon_2}\left\|\frac{\partial y}{\partial \nu_A}\right\|^2_{\mathcal{U}} - \frac{2}{\varepsilon_2}\|v\|_{\mathcal{U}}\left\|\frac{\partial y}{\partial \nu_A}\right\|_{\mathcal{U}} + \frac{1}{\varepsilon_1}\left\|\frac{\partial y}{\partial t} + Ay\right\|^2_{L^2(Q)} + \frac{1}{\varepsilon_3}\int_\Omega y(x,0)^2\,dx + \int_\Omega y(x,T)^2\,dx,$$

from which we deduce (3.13) (with c independent of ε). □

We have the following theorem:

Theorem 3.2. *As* $\varepsilon = \{\varepsilon_1, \varepsilon_2, \varepsilon_3\} \to 0$, *we have*

$$j_\varepsilon \to \operatorname*{Inf}_{v \in \mathcal{U}_{\mathrm{ad}}} J(v) = j, \tag{3.14}$$

$$u_\varepsilon \to u \quad \text{in } \mathcal{U} \ (u = \text{optimal control for problem (3.6)}). \tag{3.15}$$

$$y_\varepsilon \to y(u) \quad \text{in } Y. \tag{3.16}$$

Proof. 1. We have

$$\mathcal{J}_\varepsilon(y_\varepsilon,u_\varepsilon)\leqslant \mathcal{J}_\varepsilon(y(u),u)=J(u)=j, \tag{3.17}$$

since the "penalty terms" in (3.10) are zero if $y=y(u)$, $v=u$.

2. From (3.17), $\mathcal{J}_\varepsilon(y_\varepsilon,u_\varepsilon)$ is bounded and from (3.10) we have

$$\mathcal{J}_\varepsilon(y_\varepsilon,u_\varepsilon)\geqslant \nu\,\|u_\varepsilon\|^2_{\mathcal{U}}$$

and

$$\mathcal{J}_\varepsilon(y_\varepsilon,u_\varepsilon)\geqslant \frac{1}{\varepsilon_1}\left\|\frac{\partial y_\varepsilon}{dt}+Ay_\varepsilon-f\right\|^2_{L^2(Q)}+\frac{1}{\varepsilon_2}\left\|\frac{\partial y_\varepsilon}{\partial\nu_A}-u_\varepsilon\right\|^2_{\mathcal{U}}$$

$$+\frac{1}{\varepsilon_3}\,\|y_\varepsilon(x,0)-y_0(x)\|^2_{L^2(Q)},$$

from which we obtain

$$\|u_\varepsilon\|\leqslant c, \tag{3.18}$$

$$\left\|\frac{\partial y_\varepsilon}{\partial t}+Ay_\varepsilon-f\right\|_{L^2(Q)}\leqslant c\sqrt{\varepsilon_1}, \tag{3.19}$$

$$\left\|\frac{\partial y_\varepsilon}{\partial\nu_A}-u_\varepsilon\right\|_{\mathcal{U}}\leqslant c\sqrt{\varepsilon_2}, \tag{3.20}$$

$$\|y_\varepsilon(x,0)-y_0\|_{L^2(\Omega)}\leqslant c\sqrt{\varepsilon_3}. \tag{3.21}$$

Hence in particular:

$$\left.\begin{aligned}&\frac{\partial y_\varepsilon}{\partial t}+Ay_\varepsilon && \text{is bounded in } L^2(Q),\\ &y_\varepsilon(x,0) && \text{is bounded in } L^2(\Omega),\\ &\frac{\partial y_\varepsilon}{\partial\nu_A} && \text{is bounded in } L^2(\Sigma),\end{aligned}\right\} \tag{3.22}$$

which implies that

$$y_\varepsilon \quad \text{is bounded in } L^2(0,T;H^1(\Omega)).\ ^{13} \tag{3.23}$$

But (3.23) combined with (3.19), (3.20), (3.21) proves that y_ε is bounded in Y. Hence we may extract a subsequence, denoted again by $\{y_\varepsilon,u_\varepsilon\}$ such that

$$\left.\begin{aligned}&y_\varepsilon\to\tilde{y} && \text{weakly in } V,\\ &u_\varepsilon\to\tilde{u} && \text{weakly in } \mathcal{U} \text{ (and hence } u\in\mathcal{U}_{\mathrm{ad}}).\end{aligned}\right\} \tag{3.24}$$

---

[13] Apply Theorem 1.2, Chapter 3, as in Section 3.2, Chapter 3.

From (3.19), (3.20), (3.21) we have

$$\frac{\partial \tilde{y}}{\partial t} + A\tilde{y} = f,$$

$$\frac{\partial \tilde{y}}{\partial \nu_A} = \tilde{u},$$

$$\tilde{y}(x,0) = y_0$$

giving us

$$\tilde{y} = y(\tilde{u}). \tag{3.25}$$

But we clearly have

$$\mathscr{I}_\varepsilon(y_\varepsilon, u_\varepsilon) \geqslant \int_\Omega (y_\varepsilon(x,T) - z_d)^2\, dx + (Nu_\varepsilon, u_\varepsilon)_{\mathscr{U}} \tag{3.26}$$

and

$$y_\varepsilon(x,T) \to \tilde{y}(x,T) \quad \text{weakly in } L^2(\Omega),$$

and hence (3.26) implies

$$\liminf \mathscr{I}_\varepsilon(y_\varepsilon, u_\varepsilon) \geqslant \int_\Omega (\tilde{y}(x,T) - z_d)^2\, dx + (N\tilde{u}, \tilde{u})_{\mathscr{U}},$$

which from (3.25) gives us

$$\liminf \mathscr{I}_\varepsilon(y_\varepsilon, u_\varepsilon) \geqslant J(\tilde{u}). \tag{3.27}$$

Now (3.27) combined with (3.17) proves (3.14) and also $\tilde{u} = u$. Hence we have (3.15), (3.16) where the convergence is weak convergence.

3. It remains to prove strong convergence. But

$$\mathscr{I}_\varepsilon(y_\varepsilon, u_\varepsilon) = \alpha_\varepsilon + \beta_\varepsilon - 2\int_\Omega y_\varepsilon(x,T)\, z_d(x)\, dx + \int_\Omega z_d^2(x)\, dx,$$

where

$$\alpha_\varepsilon = \int_\Omega y_\varepsilon(x,T)^2\, dx + (Nu_\varepsilon, u_\varepsilon)_{\mathscr{U}},$$

$$\beta_\varepsilon = \frac{1}{\varepsilon_1}\left\|\frac{\partial y_\varepsilon}{\partial t} + Ay_\varepsilon - f\right\|^2_{L^2(Q)} + \frac{1}{\varepsilon_2}\left\|\frac{\partial y_\varepsilon}{\partial \nu_A} - u_\varepsilon\right\|^2_{L^2(\Sigma)} + \frac{1}{\varepsilon_3}\int_\Omega (y_\varepsilon(x,0) - y_0(x))^2\, dx,$$

and since

$$\mathscr{I}_\varepsilon(y_\varepsilon, u_\varepsilon) \to j = J(u) = \varepsilon - 2\int_\Omega y(x,T;u)\, z_d(x)\, dx + \int_\Omega z_d^2(x)\, dx,$$

where

$$\alpha = \int_\Omega (y(x,T;u))^2\, dx + (Nu, u)_{\mathscr{U}},$$

we deduce that

$$\alpha_\varepsilon + \beta_\varepsilon \to \alpha.$$

Since $\liminf \alpha_\varepsilon \geqslant \alpha$ (by virtue of weak convergence) we deduce that

$$\alpha_\varepsilon \to \alpha, \tag{3.28}$$

$$\beta_\varepsilon \to 0. \tag{3.29}$$

From (3.28) we deduce that $u_\varepsilon \to u$ strongly in $\mathscr{U}$ and hence (3.15) is proved. Then $\partial y_\varepsilon / \partial \nu_A \to u$ strongly in $\mathscr{U} = L^2(\Sigma)$ and since we already know that

$$\frac{\partial y_\varepsilon}{\partial t} + A y_\varepsilon \to f \quad \text{strongly in } L^2(Q),$$

$$y_\varepsilon(x,0) \to y_0 \quad \text{strongly in } L^2(\Omega),$$

we obtain (3.16). □

Remark 3.1. We have implicity proved the relations (3.19), (3.20), (3.21)—and amongst other things according to (3.29) the above relations are true with c arbitrarily small for $|\varepsilon|$ sufficiently small.

Remark 3.2. We have presented the method of penalization by considering example (3.1), (3.2), (3.3)—but the *method is general.* With minor technical variations it is valid for linear elliptic, parabolic and other systems which we have considered in previous chapters.

Remark 3.3. The pair $\{y_\varepsilon, u_\varepsilon\}$ may be easily characterised by the methods used in the previous chapters. These lead to new unilateral problems involving the pair $\{y_\varepsilon, u_\varepsilon\}$.

Remark 3.4. Methods of the same type are also applicable to non-linear partial differential equations. For example, if the state is given as in Chapter 3, Section 15.2, and the cost function by (15.6), Chapter 3, we may consider

$$\mathscr{J}_\varepsilon(y,v) = \int_Q (y(x,T) - z_d(x))^2\, dx + \nu \|v\|_{L^\infty(Q)}$$

$$+ \frac{1}{\varepsilon_1} \left\| \frac{\partial y}{\partial t} + Ay + vy - f \right\|^2_{L^2(Q)} + \frac{1}{\varepsilon_2} \|y(x,0) - y_0(x)\|^2_{L^2(\Omega)}$$

on the vector space[14] of functions $v \in L^\infty(Q)$ and y such that

$$y \in L^2(0,T; H_0^1(\Omega)), \quad \frac{\partial y}{\partial t} + Ay \in L^2(Q).$$ [15]

[14] Even though the differential operator is non-linear, due to its special structure y belongs to a vector space.

[15] Hence $Y = H^{2,1}(Q) \cap L^2(0,T; H_0^1(\Omega))$ provided the coefficients are sufficiently regular.

We can then prove that if $\{y_\varepsilon, u_\varepsilon\}$ is an optimal pair for $\mathscr{J}_\varepsilon(y, v)$, we may extract a weakly convergent subsequence in $Y \times \mathscr{U}$ such that

$$\{y_\varepsilon, u_\varepsilon\} \xrightarrow{\text{weakly}} \{y(u), u\}.$$

Notes

The parabolic regularization procedure has been used in Lions-Magenes [1], Chapter 3. By using this procedure the system of equations determining the optimal control for problems in Chapter 4 may be deduced by passing to the limit in the results of Chapter 3. This is precisely what has been done in Section 1.2 and by decoupling in Section 1.3.

An idea similar to that of Theorem 1.5 (in a different framework) may be found in A. N. Tichonoff [1].

Problems of the type considered in Section 2.1 but in the absence of control have been studied in Lions [3], Chapter 6, Section 6 using different methods. Cf. also A. Friedman-M. Shinbrot [1], Lions-Magenes [1]. Problems of this type appear in applications cf. Garipov [1]. In order to utilize finite difference methods more easily, approximation by Cauchy-Kowaleska problems was introduced in Lions [11].

We restricted ourselves to *linear* (Navier-)Stokes equations in Section 2.3. As indicated in Remark 2.6, existence and approximation results (the question of uniqueness of optimal control is not clear) of Section 2.3 may be extended to the non-linear case by using the work of Leray [1], [2], [3] and Ladyzenskaya [2]. For results on the control of systems governed by the Navier-Stokes equation (in the stationary case), cf. J. P. Gossez [1].

The penalization method introduced in Section 3 appears to be novel. Generally, a penalty term is introduced for the constraint (cf. Courant [1] and numerous other papers on non-linear programming) whereas here we introduce a penalty term involving the state equation. The method indicated in Section 3 was introduced (Los Angeles, August 1967) with a view to numerical applications, for which we refer the reader to J. P. Yvon [1]. Other applications have been given by A. V. Balakrishnan [4], [5]. Cf. also A. Bensoussan-P. Kenneth [1]. An extension of this procedure for certain differential games is due to L. Tartar [1]. This kind of method seems particularly useful in connection with *stochastic* optimal control problems; see Bensoussan [1], Bensoussan-Lemarechal [1].

Bibliography

Agmon, S., Nirenberg, L.: (1) Properties of Solutions of Ordinary Differential Equations in Banach Space. Comm. Pure Applied Maths. **XVI**, 129—239 (1963); — (2) Lower Bounds and Uniqueness Theorems for Solutions of Differential Equations in a Hilbert Space. Comm. Pure Applied Maths. **XX**, 207—229 (1967).

Agranovitch, Visik: (1) Elliptic Problems with Parameters and Parabolic Problems of General Type. Uspekhi Mat. Nauk. SSR **19**, 53—161 (1964).

Ambarzumian, V.A.: (1) Diffuse Reflection of Light, Doklady Akad. Nauk **38**, 229—232 (1943).

Annin, B.D.: (1) Existence and Uniqueness of the Solution of the Elastic Torsion Problem for a Cylindrical Bar of Oval Cross Section. P. M. M. **29**, n° 5, 879—887 (1965).

Antosiewicz, H.A.: (1) Linear Control Systems. Archive for Rat. Mech. and Anal. **12**, 313—324 (1963).

Arbib, M.: Cf. Kalman, Falb and Arbib.

Aronszajn, N.: (1) An Unique Continuation Theorem for Solution of Elliptic Partial Differential Equations or Inequalities of Second Order. J. Math. Pures Appl. **36**, 235—249 (1957). (C. R. Acad. Sc. Paris **242**, 723—725 (1956).

Artola, M.: (1) Equations paraboliques à retardement. C. R. Acad. Sc. Paris **264** (1965); — (2) Sur les perturbations des équations d'évolution. Applications à des problèmes de retard. Annales E. N. S. **2** (1969), 137—253.

Athans, M., Falb, P. L.: (1) Optimal Control. Mc Graw Hill, New York, N. Y. 1966.

Aubin, J. P.: (1) Book to appear.

Auslender: Methodes numeriques pour la resolution des problemes d'optimisation avec contraintes. Thesis, University of Grenoble, 1969.

Axelband, E. I.: (1) A Solution to the Optimal Pursuit Problem for Distributed Parameter Systems. Journal of Computer and System Sciences **1**, 261—286 (1967).

Baiocchi, C.: (1) Sulle equazioni differenziale astratte lineari del primo e del secondo ordine negli spazi di Hilbert. Annali Mat. Pure ed Appl. **76**, 233—304 (1967); — (2) Soluzioni ordinarie e generalizzate del problema di Cauchy per equazioni differenziali astratte lineari del secondo ordine di spazi di Hilbert. Ricerche di Mat. **XVI**, 27—95 (1967).

Balakrishnan, A. V.: (1) Optimal Control Problems in Banach Spaces. J. SIAM Control **3**, 152—180 (1965); — (2) Semi Group Theory and Control Theory. Proc. IFIP, Washington D. C.: Spartan Books 1965; — (3) Foundations of the State Space Theory of Continuous Systems. I. Journal of Computer and system Sciences **I** (1967); — (4) A New Computing Technique in System Identification, Journal of Computer and System Sciences **2**, 102—116 (1968); — (5) On a New Computing Technique in Optimal Control, SIAM J. on Control **6**, 149—173 (1968).

— Lions, J. L.: State Estimation for Infinite Dimensional Systems, Journal of Computer and System Sciences **1**, 391—403 (1967).

Behn, R., Ho, Y. C.: (1) On a Class of Linear Stochastic Differential Games, I. E. E. E. Transactions on Automatic Control, Vol. AC 13, 1968.
Bellman, R.: Dynamic Programming. Princeton University Press, 1957.
— Kalaba, R., Wing, G. M.: (1) Invariant Imbedding and Mathematical Physics. I: Particle Processes. J. Math. Phys. **1**, 280—308 (1960).
— Glicksberg, I., Gross, O.: (1) On the Bang-Bang Control Problem. Quart. Appl. Math. **14**, 11—18 (1956).
Bensoussan, A.: (1) Une méthode d'identification de valeur initiale. C. R. Acad. Sc. Paris **265**, 724—727 (1967); — (2) C. R. Acad. Sciences Paris, deux notes 1968; — (3) Sur l'identification et le filtrage de systèmes gouvernés far des équations aux dérivées partielles. Cahiers IRIA **1**, 1969; — (4) Etude de certains problèmes de jeux pour des systèmes décrits par des équations differentielles opérationnelles. To appear. Bensoussan-Le Marechal, to appear.
— Bossavit, Nédelec [Cahiers IRIA **2**, 1970]
— Kenneth, P.: (1) [Cahiers IRIA **2**, 1970].
— Berkowitz, L. D., Pollard, H.: (1) A Non-Classical Variational Problem Arising from an Optimal Filter Problem. Archive Rat. Mech. Anal. **26**, 281—304 (1967).
Boltyanskii: Cf. Pontryagin.
Bossavit: (1) Thèsis. Paris, 1970.
— Bensoussan, Nédelec: Cf. Bensoussan, Bossavit, Nedelec.
Bourbaki, N.: (1) Intégration, Chap. 1 à 4. Act. Sc. Ind., Paris: Hermann 1966.
Brézis, H.: (1) Equations et inéquations non linéaires dans les espaces vectoriels en dualité. Annales. Inst. Fourier **18**, 115—175 (1968); — (2) Thèse, Paris 1970; — (3) Fonctions duales relativement à une forme bilinéaire C. R. Acad. Sc. Paris **264**, 284—286 (1967).
— Lions, J. L.: Sur certains problèmes unilatéraux hyperboliques, C. R. Acad. Sc. Paris **264**, 928—931 (1967).
— Sibony, M.: (1) Méthodes d'approximation et d'itération pour les opérateurs monotones. Archive Rat. Mech. Anal **28**, 59—82 (1968).
— Stampacchia, G.: (1) Sur la régularité de la solution d'inéquations elliptiques. Bull. Soc. Math. F. **96**, 153—180 (1968).
Brogan, W. L.: (1) Optimal Control Theory Applied to Systems Described by Partial Differential Equations. Ph. D. Thesis, U. C. L. A., 1965.
Browder, F.: (1) Non Linear Monotone Operations and Convex Sets in Banach Spaces. Bull. Amer. Math. Soc. **71**, 780—785 (1965); — (2) On the Unification of the Calculus of Variations and the Theory of Monotone non Linear Operations in Banach Spaces. Proc. Nat. Acad. Sc. U.S.A. **56**, 419—425 (1966); — (3) Nonlinear Operators and Nonlinear Equations of Evolution in Banach Spaces. Proc. Symp. on Nonlinear Functional Analysis. Chicago: April 1968. In preparation.
Brunovsky, Pavol: (1) Controllability and Linear Closed-Loop Controls in Linear Periodic Systems, J. of Diff. Equs. **6**, 296—313 (1969).
Bucy, R. S., Joseph, P. D.: (1) Filtering for Stochastic Processes with Applications to Guidance. Interscience 1968.
— Bucy: Cf. Kalman-Bucy.
Butkovskii, A. G.: (1) Theory of optimal Control of Distributed Parameter Systems. Moscow 1965 [Russian]. English Translation. American Elsevier, 1969.
— Poltanvskii, L. N.: (1) Optimal Control of a Two Dimensional Distributed Oscillatory System. Automation and Remote Control **27**, 553—563 (1966).
Castaing, C.: (1) Les multi-applications mesurables. Thesis. Paris 1967.
Cea, J.: (1) Théorie de l'optimisation. Cours Fac. Sciences Rennes 1966—1967; — (2) Approximation de la solution d'un problème d'optimisation. In preparation.
Cecconi, J.: (1) Sulla teoria dei controlli. Colloque Analyse Fonctionnelle. Rome, March 1968.

Cesari, L.: (1) Multidimensional Lagrange Problems of Optimization in a Fixed Domain and an Application to a Problem of Magnetohydrodynamics. Archive Rat. Mech. and Analysis, 29 (2), (1968), 81—104; — (2) Existence Theorems for Multidimensional Problems of Optimal Control; in Differential Equations and Dynamical Systems. 115—132. Academic Press 1967.

Cirina, M.: (1) Boundary Controllability of Nonlinear Hyperbolic Systems. SIAM J on Control **7**, 198—212 (1969).

Chandrasekharan, S.: (1) Radiative transfer. Oxford: Clarendon Press 1950.

Conti, R.: (1) Contribution to Linear Control Theory. Journal of Diff. Equations **1**, 427—445 (1965); — (2) On Some Aspects of Linear Control Theory; in Mathematical Theory of Control, ed. by Balakrishnan and Neustadt. 285—300. Academic Press 1967.

Conway, E. D., Hopf, E.: (1) Hamilton's Theory and Generalized Solutions of the Hamilton-Jacobi Equation. J. Math. Mech. **13**, 939—986 (1964).

Cooper, J. F.: (1) In preparation.

Cordes, H. O.: (1) Über die Bestimmtheit der Lösungen elliptischer Differentialgleichungen durch Anfangsvorgaben. Nachr. Akad. Wiss. Göttingen **11**, 239—258 (1956).

Courant, R.: (1) Variational Methods for the Solution of Problems of Equilibrium and Vibrations. Bull. Amer. Math. Soc. **49**, 1—23 (1943).

Cristescu, N.: (1) Dynamic Plasticity. North Holland 1967.

Degtyarev, G. L., Sirazetdinov, T. K.: (1) On Optimal Control of One-Dimensional Distributed Parameter Systems. Automatika et Telemekanika 29—38 (1967).

Da Prato: (1) Equations d'évolution dans des algèbres d'opérateurs et applications. Journal de Math. Pures et Appliquées, **48** (1969), 59—107; — (2) Symposium Rome, May 1970.

Delfour, M. C., Mitter, S. K.: (1) Reachability of Perturbed Systems and Min-Sup Problems. SIAM J. on Control **7**, 521—533 (1969).

Demidov, Marchouk, G. I.: (1) An existence theorem in meteorological problems. Soviet Math. **7**, 1310—1312 (1966).

Dieudonné, J.: (1) Foundations of Modern Analysis. Academic Press 1960.

Dixmier, J.: (1) Les algèbres d'opérateurs dans les espaces hilbertiens. Paris: Gauthier-Villars 1957.

Donsker, M. D.: (1) On Function Space Integrals; in Analysis in Function Space, ed. by W. T. Martin and I. Segal, M. I. T. Press, 1964, 17—30.

— Lions, J. L.: (1) Frechet-Volterra Variational Equations, Boundary Value Problems and Function Space Integrals. Acta Math. **108**, 147—228 (1962).

Downing, A. C., Householder, A. S.: (1) Some inverse characteristic value problems. Journ. A. C. M. **3**, 203—207 (1956).

Dubovickii, A. Yu., Miljutin, A. A.: (1) Optimization problems with constraints. Journal of Numerical Math. and Math. Physics **5**, 395—453 (1965).

Dunford, N., Schwartz, J. T.: (1) Linear Operators. Part I. Interscience 1958.

Duvaut: Cf. Lanchon-Duvaut.

Egorov, A. I.: (1) Necessary Conditions of Optimality for Distributed Parameter Systems. Mat. Sbornik **69**, (III), 371—421 (1966).

Egorov, Yu. V.: (1) On Certain Optimal Control Problems. Journal of Numerical Mathematics and Mathematical Physics. **3**, 887—904 (1963); — (2) Sufficient Conditions for Optimality in Banach Spaces. Mat. Sbornik **64**, 79—101 (1964).

Ekeland, I.: (1) in preparation.

Erzberger, H., Kim, M.: (1) Optimum Boundary Control of Distributed Parameter Systems. Information and Control **9**, 265—278 (1966).

Faedo, S.: (1) Un nuovo metodo per l'analisi esistenziale e quantitativa dei problemi di propagazione. Ann. Scuola Norm. Sup. Pisa **1**, 1—40 (1949).

Falb, P. L.: (1) Infinite Dimensional Control Problems. I. On the Closure of the Set of Attainable States for Linear Systems. J. Math. Anal. Appl. **9**, 12—22 (1964).

— Kleinman, D. L.: (1) Remarks on the Infinite Dimensional Riccati Equation. I. E. E. E. Trans. Automatic Control, AC-11, 534—536 (1966).

Fattorini, H. O.: (1) Time Optimal Control of Solutions of Operational Differential Equations. J. SIAM. Control., Ser. A. 2, 54—59 (1964); — (2) Some Remarks on Complete Controllability. Journal Soc. Ind. Appl. Math. **4**, 686—693 (1966); — (3) On Complete Controllability of Linear Systems. Journal Diff. Equations **3**, 391—402 (1967); — (4) Controllability of Higher Order Linear Systems; in Mathematical Theory of Control. Ed. by Balakrishnan and Neustadt. 301—311. Academic Press (1967); — (5) Boundary Control Systems. SIAM J. Control, Vol. **6**, 399—385 (1968); — (6) Some Observations on a Paper of A. Friedmann. In preparation; — (7) A Remark on the "Bang-Bang" Principle for Linear Control Systems in Infinite-Dimensional Space. SIAM J. Control. Vol. **6**, 109—113 (1968); — (8) Control with Bounded Inputs. In Computing Methods in Optimization Problems. San Remo 1968; — (9) Ordinary Differential Equations in Linear Topological Spaces II, J. of Diff. Eqns. **6**, 50—70 (1969).

Faurre, P.: (1) in preparation; — (2) Sur les points conjugués en commande optimale. C. R. Acad. Sc. Paris. In preparation.

Fenchel, W.: (1) On Conjugate Convex Functions. Canadian J. of Math. **1**, 73—77 (1949).

Fichera, G.: (1) Problemi elastostatici con vincoli unilaterali: il problema di Signorini con ambigue condizioni al contorno. Memorie dell'Acad. Naz. dei Lincei, S VIII, **VII**, 91—140 (1964).

Fleming, W. H.: (1) Some Markovian Optimization Problems. J. Math. and Mech. **12**, 131—140 (1963); — (2) Duality and a Priori Estimates in Markovian Optimization Problems. J. Math. Anal. Appl. **16**, 254—279 (1966); — (3) Stochastic Lagrange Multipliers. Proc. Conf. Math. Control Theory, Los Angeles, January 1967; — (4) Optimal Control of Partially Observable Diffusions. SIAM J. Control, **6** (1968), 194—214; — (5) The Cauchy Problem for a Non Linear First Order Partial Differential Equation. J. Diff. Equations. **5** (1969), 515—530.

Fomin, S. V.: Cf. Gelfand-Fomin.

Friedman, A.: (1) Optimal Control in Banach Spaces. Journal of Math. Analysis and Appl. **18**, 35—55 (1967); — (2) Optimal Control for Parabolic Equations. Journal of Math. Analysis and Appl. **18**, 479—491 (1967); — (3) Partial Differential Equations of Parabolic Type: Prentice Hall 1964; — (4) Optimal Control in Banach Space with Fixed End Points. J. of Math. Anal. and Applns. **24**, 161—181 (1968); — (5) Differential Games of Pursuit in Banach Space. Journal of Math. Analysis and Applications **25**, 93—113 (1969).

— Shinbrot, M.: (1) The Initial Value Problem for the Linearized Equations of Water Waves. J. Math. Mech. **27**, 107—180 (1967).

Gamkrelidze: (1) On Some Extremal Problems in the Theory of Differential Equations with Applications to the Theory of Optimal Control. J. SIAM Control **3**, 106—128 (1965). See also: Pontryagin.

Garipov, R. M.: (1) On the Linear Theory of Gravity Waves: The Theorem of Existence and Uniqueness. Archive for Rat. Mech. Anal. **24**, 352—362 (1957).

Garnir, H. G.: (1) Les problèmes aux limites de la physique mathématique. Stuttgart: Birkhäuser 1958.

Gelfand, I. M., Silov, G. E.: (1) Fonctions généralisees **1**, **2**, **3**, Moscow 1958.

— Fomin, S. V.: (1) Calculus of variations. Prentice Hall 1963.

Glicksberg: Cf. Bellmann-Glicksberg-Gross.

Gossez, J.P.: (1) in preparation.

Green, J.W.: (1) An Expansion Method for Parabolic Partial Differential Operators. J. Res. Nat. Bur. Stand **51**, 127—132 (1953).

Grisvard, P.: (1) Problèmes aux limites résolus par le calcul opérationnel. C.R. Acad. Sc. Paris **262**, 1306—1308 (1966); — (2) Equations opérationnelles abstraites dans les espaces de Banach et problèmes aux limites dans des ouverts cylindriques. Ann. Sc. Norm. Pisa **21** (1967).

Gross: Cf. Bellmann-Glicksberg-Gross.

Hadamard, J.: (1) Le problème de Cauchy et les équations aux dérivées partielles linéaires hyperboliques. Paris: Hermann 1932.

Hadeler, K. P.: (1) Ein inverse eigenwertproblem. Linear algebra and its apcations **1**, 83—101 (1968).

Halanay, A.: (1) Differential Equations. Academic Press 1966.

Halkin, H.: (1) A Generalization of Lasalle's "Bang-Bang" Principle. J. SIAM Control. **2**, 199—202 (1965); — (2) Non Linear Non Convex Programming in an Infinite Dimensional Space; in Mathematical Theory of Control. Ed. by Balakrishnan-Neustadt. 10—25. Academic Press 1967.

— Neustadt, L.: (1) General Necessary Conditions for Optimization Problems. Proc. Nat. Acad. Sciences **56**, 1066—1071 (1966).

Hartman, P., Stampacchia, G.: (1) On Some Non Linear Elliptic Differential Functional Equations. Acta Math. **115**, 271—310 (1966).

Haugazeau, Y.: (1) Sur des inéquations variationnelles. C.R. Acad. Sc. Paris (1967); — (2) Thesis. Paris 1968.

Heinz, E.: (1) Über die Eindeutigkeit beim Cauchyschen Anfangswertproblem einer elliptischen Differentialgleichung zweiter Ordnung. Akad. Wiss. Göttingen **1**, 1—12 (1955).

Hestenes, M.: (1), Calculus of Variations and Optimal Control Theory. Wiley 1966.

Hille, E., Philipps, R.S.: (1) Functional Analysis and Semi-groups: Amer. Math. Soc. Coll. Pub. **XXXI**, 1957.

Ho, Y. C.: Cf. Behn and Ho.

Hopf, E.: (1) The Partial Differential Equation $u_t+uu_x=uu_{xx}$. Comm. Pure Appl. Math. **3**, 201—230 (1950).

— Cf. Conway.

Horwath, J.: (1) Topological Vector Spaces and Distributions **1**. Addison-Wesley 1966.

Householder, A. S.: Cf. Downing-Householder.

Huet, D.: (1) Phénomènes de perturbation singulière dans les problèmes aux limites. Ann. Inst. Fourier **10**, 1—96 (1960).

Hummer, D.G., Rybicki, G.: (1) Computational Methods for Non-LTE Line-Transfer Problems, in Methods in Computational Physics, vol. **7**, 53—127. Academic Press 1967.

Il'in, A.M: (1) On a class of Ultra-Parabolic Equations. Doklady Akad. Nauk **159**, 1214—1217 (1964).

— Cf. Oleinik-Il'in-Kalashnikov.

Ioffe, A.D., Tikomirov, V.M.: (1) Relaxed Variational Problems. Troudi Mosk. Mat. Obv. **18**, 187—266 (1968).

John, F.: (1) Extremum Problems with Inequalities as Subsidiary Conditions. Courant Anniversary Volume, 187—204 (1948).

Joseph, P.D.: Cf. Bucy-Joseph.

Kalashnikov: Cf. Oleinik-Il'in-Kalashnikov.

Kalaba, R.: Cf. Bellman-Kalaba-Wing.

Kalman, R.E.: (1) Contributions to the Theory of Optimal Control. Bol. Soc. Mat. Mexicano, 102—119 (1960); — (2) The Theory of Optimal Control and the Calculus of Variations; in Mathematical Optimization Techniques, ed. by R. Bellman, 209—331. Univ. of Calif. Press 1963; — (3) On the General Theory of Control Systems. Proc. I. F. A. C. Moscow, 1960. 491—192. Butterworths 1961.

— Bucy, R.S.: (1) New Results in Linear Filtering and Prediction Theory. Journal Basic Engineering, 95—107 (1961).

— Falb, P.L., Arbib, M.A.: (1) Topics in Mathematical System Theory. McGraw Hill, New York, N.Y. 1969.

Kato, T.: (1) Perturbation Theory for Linear Operators. Berlin-Heidelberg-New York: Springer 1966.

Kenneth, P.: Cf. Bensoussan-Kenneth.

Kim: Cf. Erzberger-Kim.

Kleinman: Cf. Falb-Kleinman.

Krein, S.G.: (1) On a Certain Class of Well-Posed Boundary Value Problems. Doklady Akad. Nauk. **114**, 1162—1165 (1957); — (2) Equations différentielles linéaires dans les espaces de Banach. In preparation.

Kuhn, H.W., Tucker, A.W.: (1) Non Linear Programming. Proc. Symp. Math., 481—492. Stat. Univ. Cal. Press 1951.

Kushner, H. J.: (1) On the optimal control of linear distributed parameter systems with white noise input. SIAM J. Control, **6** (1968), 596—614.

Laborde, F.: (1) Sur un problème inverse d'un problème de valeurs propres. C. R. Acad. Sc. Paris (1968).

Ladyzenskaya, O.A.: (1) Mixed Problems for Hyperbolic Equations. Moscow: 1953; — (2) Linear Differential Equations in Banach Spaces. Moscow: 1967.

Laurent, P.J.: (2) Cônes de déplacements admissibles, sons gradients et approximation convexe dans un espace normé. Summer School OTAN, Venezia (June 1968).

Ladyzenskaya, O.A., Solonnikov, V.A., Ouraltseva, N.N.: (1) Linear and Quasi-linear Parabolic Equations (in Russian). Moscow, 1967.

Lasalle, J.: (1) The time optimal control problem. Contrib. to Non linear Oscillations **5**, 1—24 (1960).

Lanchon, H., Duvaut, G.: (1) Sur la solution du problème de torsion élasto-plastique d'une barre cylindrique de section quelconque. C. R. Acad. Sc. Paris, t. **264**, 520—523 (1967).

Landis, E.M.: (1) On Certain Qualitative Properties in the Theory of Elliptic and Parabolic Equations [in Russian]. Uspehi Mat. Nauk. **14**, 21—85 (1959).

Lattès, R., Lions, J.L.: (1) Méthode de quasi réversibilité et applications. Paris: Dunod 1967; English translation: Elsevier 1969.

Laurent, P.J.: (1) Approximation convexe. Séminaire analyse numérique, Grenoble 1967.

Lawruk, B., Rolewicz, S.: (1) The minimum time control problem for a certain class of linear parabolic partial differential equations controlled by the boundary condition. Acad. Polonaise Sc., Ser. Math. **16**, nr. 6 (1968).

Lax, P.D., Milgram, N.: (1) Parabolic equations. Contributions to the theory of partial differential equations. Annals of Math. Studies, n° 33, 167—190, Princeton 1954.

Lee, E.B., Markus, L.: (1) Foundations of Optimal Control Theory, J. Wiley, 1967.

Leray, J.: (1) Etude de diverses équations intégrales non linéaires et de quelques problèmes que pose l'hydrodynamique. J. Math. Pures Appl. **12**, 1—82 (1933); — (2) Essai sur les mouvements plans d'un liquide visqueux que limitent des parois. J. Math. Pures Appl. **13**, 331—418 (1934); — (3) Sur les mouvements d'un liquide visqueux emplissant l'espace. Acta Math. **63**, 193—248 (1934).

Leray, J., Lions, J. L.: (1) Quelques résultats de Visik sur les problèmes elliptiques non linéaires par les méthodes de Minty-Browder. Bull. Soc. Math. France **93**, 97—107 (1965).

Lions, J. L.: (1) Sur le contrôle optimal de systèmes décrits par des équations aux dérivées partielles linéaires. C. R. Acad. Sc. Paris **263**, (I), 661—663 (1966); (II), 713—715; (III), 776—779; — (2) Optimisation pour certaines classes d'équations d'évolution non linéaire. Annali di Mat. **LXXII**, 275—294 (1966); — (3) Équations différentielles opérationnelles et problèmes aux limites. Berlin-Heidelberg-New York: Springer 1961, 2^d^ Edition, Revised, in English, in preparation; — (3 bis) Operational Differential Equations and Boundary Value Problems. 2nd edition, revised and augmented, of (3). Springer 1970; — (4) Problèmes aux limites en théorie des distributions. Acta Math. **94**, 13—153 (1955); — (5) A paraitre; — (6) Quelques résultats d'existence dans des équations aux dérivées partielles. Bull. S. M. F. **87**, 245—273 (1959); — (7) Sur certaines équations paraboliques non linéaires. Bull. S. M. F. **93**, 155—175 (1965); — (8) On some optimization problems for linear parabolic equations; in Functional Analysis and Optimization, éd. E. Caianiello, Acad. Press, 1966, 115—131; — (9) Sur certaines équations aux dérivées partielles à coefficients opérateurs non bornés. Journal Analyse Math. **VI**, 333—335 (1958); — (10) Equations différentielles opérationnelles dans les espaces de Hilbert. Centro Int. Mat. Estivo, Varenna, 1963. (Equazioni differenziali astratte, Cremonese, Roma, 1963.); — (11) Approximation par des systèmes du type Cauchy-Kowaleska. École d'été CEA-EDF, 1965; cours CIME, juillet 1967; — (12) Sur le feedback non linéaire. IRIA, 1968; — (13) Quelques méthodes de résolution des problèmes aux limites non linéaires. Paris: Dunod and Gauthier Villars 1969; — (14) Sur certains problèmes non linéaires liés au contrôle optimal de systèmes gouvernés par des équations aux dérivées partielles. Colloque Bruxelles, CBRM, 1970, 99—107.

— Magenes, E.: (1) Problemes aux limites non homogenes et application, **1**, **2**, **3**. Paris. Dunod 1968; — (2) Note LINCEI, 1967.

— Malgrange, B.: (1) Sur l'unicité rétrograde. Math. Scand. **8**, 277—286 (1960).

— Stampacchia, G.: (1) Variational Inequalities. Comm. Pure Applied Math. **XX** (1967), 493—519; —(2) Inéquations variationnelles non coercives. C. R. Acad. Sc. Paris **261**, 25—27 (1965).

— Strauss, W.: (1) Some Non Linear Evolution Equations. Bull. Soc. Math. F. **93**, 43—96 (1965).

Littmann, W.: (1) The Wave Operator and L^p Norms. J. Math. Mech. **12**, 55—68 (1963).

Lukes, D. L.: (1) Optimal Regulation of Nonlinear Dynamical Systems. SIAM J. on Control **7**, 75—100 (1969).

Lukes, D. L., Russell, D. L.: (1) The Quadratic Criterion for Distributed Systems. SIAM J. on Control **7**, 101—121 (1969).

Luré, K. A.: (1) Optimum Control of Conductivity of a Fluid Moving in a Channel in a Magnetic Field. P.M.M. **28**, 258—267 (1964).

Mac Camy, R. C., Mizel, V. J., Seidman, T. I.: (1) Approximate Boundary Controllability for the Heat Equation I. J. of Math. Analysis and Applns. **23**, 699—703 (1968); — (2) Approximate Boundary Controllability for the Heat Equation II. J. of Math. Analysis and Applns. **28**, 482—492 (1969).

Magenes: Cf. Lions-Magenes.

Malanowski, K.: (1) On Time Optimal Contral of Vibrating String. In preparation; — (2) On Optimal Control of Vibrating String. SIAM J. on Control **7**, 260—271 (1969).

Malgrange: Cf. Lions-Malgrange.

Mandelbrojt, S.: (1) Sur les fonctions convexes. C.R. Acad. Sc. **209**, 977—978 (1939).

Markus, L.: (1) Controllability and Observability; in Functional Analysis and Optimization, ed. by E. Caianiello, 133—143. Academic Press 1966; — (2) The Bang-Bang Principle. Lecture Series in Diff. Equations. 25—45. Georgetown Univ., 1965.

— Cf. Lee-Markus.

McShane, E.J.: (1) Optimal Controls, Relaxed and Ordinary; in Mathematical Theory of Control, ed by Balakrishnan and Neustadt, 1—9. Academic Press 1967.

Miasnikov: Cf. Mosolov-Miasnikov.

Milgram: Cf. Lax-Milgram.

Minty, G.J.: (1) On the Generalization of a Direct Method of the Calculus of Variations. Bull. Amer. Soc. **73**, 315—321 (1967); — (2) Monotone (non linear) Operations in Hilbert Space. Duke Math. Journal **29**, 341—346 (1962).

Miranker, W.: (1) Approximate Controllability for Distributed Linear Systems. J. Math. Anal. Appl. **10**, 378—387 (1965).

Mischenko: (1) Work on differential games. In preparation. See also Pontryagin.

Mitter, S.K.: (1) Theory of Inequalities and the Controllability of Linear Systems; in Mathematical Theory of Control, ed. by Balakrishnan and Neustadt, 203—212. Academic Press 1967.

— Cf. Delfour-Mitter.

— Cf. Simon-Mitter.

— Phillipson-Mitter.

Mizel, V.J., Seidman, T.I.: (1) Observation and Prediction for the Heast Equation. J. of Math. Analysis and Applications **28**, 303—312 (1969).

Mizohata, S.: (1) Unicité du prolongement des solutions pour quelques opérateurs différentiels paraboliques. Mém Coll. Sc. Univ. Kyoto, Série A, **31**, 219—239 (1958).

Moiseiev, N.N.: (1) Optimisation; in Mathematics applied to Physics, Roubine Ed., Unesco Publication.

Moreau, J.J.: (1) Proximité et dualité dans un espace hilbertien. Bull. Soc. Math. France **93**, 273—299 (1965).

Mortensen, R.E.: (1) A Priori open Loop Optimal Control of Continuous Time Stochastic Systems. Int. J. Control **3**, 113—127 (1966); — (2) Stochastic Optimal Control with Noisy Observations. Int. J. Control **4**, 455—464 (1966).

Mosolov, P.P., Miasnikov, V.P.: (1) Variational Methods in the Theory of the Fluidity of a Viscous-Plastic Medium. P.M.M. **29**, 468—492 (1965).

Müller, C.: (1) On the Behavior of the Solutions of the Differential Equation $\Delta u = F(x,u)$ in the Neighborhood of a Point. Comm. Pure Applied Math. **7**, 505—515 (1954).

Sz.-Nagy: Cf. F. Riesz, B. Sz.-Nagy.

Necas, J.: (1) Les méthodes directes dans la théorie des équations elliptiques. Ed. Acad. Tchécoslovaque des Sciences, Prague: 1967.

Nédelec, J. C: (1) Thesis. Paris 1970.

Nédelec, Bensoussan, Bossavit: Cf. Bensoussan-Bossavit-Nédelec.

Nelson, E.: (1) L'équation de Schroedinger et les intégrales de Feynman. Coll. Int. C.N.R.S., Paris, 1962, 151—158.

Neustadt, L.: (1) An Abstract Variational Theory With Applications to a Broad Class of Optimization Problems (I) (II). SIAM J. Control **4**, 505—527 (1966); **5**, 90—137 (1967).

Neustadt, L.: Cf. Halkin-Neustadt.

Nirenberg: Cf. Agmon-Nirenberg.

Oleinik, O.A.: (1) On second order linear equations with non-negative characteristic forms. Mat. Sbornik **69**, 111—140 (1966) [Russian].

— Il'in, A.M., Kalashnikov, A.S.: (1) Second order linear equations of parabolic type. Uspekhi Mat. Nauk. **17**, 3—146 (1962).

Ouralts'eva: Cf. Ladyzenskaya-Solonnikov-Ouralts'eva.

Pearson, J.D.: (1) Duality and a Decomposition Technique. J. SIAM Control **4**, 164—172 (1966).

Pederson, R.N.: (1) On the Unique Continuation Theorem for Certain Second and Fourth Order Elliptic Equations. Comm. Pure Applied Math. **11**, 67—80 (1958).

Phillips: Cf. Hille-Phillips.

Phillipson, G.A., Mitter, S.K.: (1) Numerical Solution of a Distributed Identifikation Problem via a Direct Method in Computing Methods in Optimization Problems **2**, eds. L. A. Zadeh, L. W. Neustadt and A. V. Balakrishnan, Academic Press, New York, N. Y., 1969; — (2) State Identification of Distributed Parameter Systems. Proceeding 4th IFAC Congress, Warsaw 1969.

Pollard, H.: Cf. Berkowitz and Pollard.

Poltavskii: Cf. Butkovskii-Poltavskii.

Pontryagin, L.S., Boltyanskii, V.G., Gamkrelidze, R.V., Mischenko, E.F.: (1) The Mathematical Theory of Optimal Processes. Interscience 1962.

Psenichny, B.N.: (1) Linear Differential Games. Automatike Telemekhanika, 65—78 (1968); — (2) Teknika Cibernetika, 1968, 13—22; — (3) in preparation.

Pucci, C.: (1) Un problema variazionale per i coefficienti di equazioni differenziali di tipo ellittico. Ann. Scuola Norm. Pisa **XVI**, 159—172 (1962); — (2) Operatori ellittici estremanti. Annali di Mat. Pura ed Appl. **72**, 141—170 (1966).

Riesz, F., Sz-Nagy, B.: (1) Leçons d'Analyse Fonctionnelle. Acad. Sc. Hongrie, 1952.

Rockafellar, R.T.: (1) Minimax Theorem and Conjugate Saddle Functions. Math. Scand. **14**, 151—173 (1964); — (2) Extension of Fenchel's Duality Theorem for Convex Functions. Duke Math. Journal **33**, 81—90 (1966); — (3) Conjugates and Legendre Transforms of Convex Functions. Can. J. Math. **19**, 200—205 (1967).

Rolewicz: Cf. Lawruk-Rolewicz.

Rosenbrock, H.H.: (1) Distinctive Problems of Process Control. Chem. Eng. Prog. **58** (1962).

Russell, D.L.: (1) Optimal Regulation of Linear Symmetric Hyperbolic Systems with Finite Dimensional Controls, J. Soc. Ind. Appl. Math. **4**, 276—294 (1966); — (2) The Kuhn Tucker Conditions in Banach Space with an Application to Control Theory. J. of Math. Anal. **15**, 200—212 (1966); — (3) On Boundary Value Control of Linear Symmetric Hyperbolic Systems; in Mathematical theory of Control, ed. by Balakrishnan and Neustadt. 312—321. Academic Press 1967; — (4) Continuity in the Strong Topology of Operator-Valued Solutions of Nonlinear Differential Equations with an Application to Optimal Control. SIAM J. on Control **7**, 132—141 (1969); — (5) Linear Stabilization of the Linear Oscillator in Hilbert Space. J. of Math. Annal. and Applns. **3**, 663—675 (1969).

Cf. Lukes-Russell.

Rybicki, G.: Cf. Hummer-Rybicki.

Sakawa, Y.: (1) Solution of an Optimal Control Problem in a Distributed-Parameter System. I.E.E.E. Trans. on Automatic Control, AC-9, 420—426 (1964); — (2) Optimal Control of a Certain Type of Linear Distributed Parameter Systems. I.E.E.E. Trans. on Automatic Control, AC-11, 35—41 (1966).

Schmaedeke, W.: (1) Mathematical Theory of Optimal Control for Semi-Linear Hyperbolic Systems in Two Independent Variables. J. SIAM Control **5**, n° 1, 138—152 (1967).

Schwartz, J.: Cf. Dunford-Schwartz.

Schwartz, L.: (1) Théorie des distributions, t. 1. Paris: Hermann 1950; — (2) Théorie des distributions, t. 2. Paris: Hermann 1951; — (3) Théorie des distributions à valeurs vectorielles, I, II. Annales Institut Fourier **7**, 1—141 (1957); **8**, 1—209 (1959); — (4) Théorie des noyaux. Proceedings of the International Congress of Mathematicians **1**, 220—230 (1950).

Shinbrot: Cf. Friedman-Shinbrot.

Sibony: Cf. Brézis-Sibony.

Silov: Cf. Gelfand-Silov.

Simon, J.D., Mitter, S.K.: (1) A Theory of Modal Control. Information and Control **13**, 316—353 (1968).

Sirazetdinov, T.K.: (1) Optimum Control of Elastic Aircraft. Automation and Remote Control **27**, n° 7, 1139—1152 (1966).

— Cf. Degetyarev, Sirazetdinov.

Sobolev, S.L.: (1) Applications of Functional Analysis to Equations of Mathematical Physics. Leningrad 1950.

Solonnikov, V.A.: (1) Parabolic Equations in L^p, Troudi Steklov (1966).

— Cf. Ladyzenskaya-Solonnikov-Ouralts'eva.

Stampacchia: Cf. Brézis-Stampacchia; Lions-Stampacchia.

Stampacchia, G.: (1) Formes bilinéaires coercitives sur les ensembles convexes. C.R. Acad. Sc. Paris **258**, 4413—4416 (1964); — (2) Le problème de Dirichlet pour les équations elliptiques du second ordre à coefficients discontinus. Ann. Inst. Fourier **15**, 189—258 (1965).

Strauss, W.A.: (1) The Initial Value Problem for Certain Non Linear Evolution Equations. Amer. J. Math., **89** (1967), 249—259.

— Cf. Lions-Strauss.

Tanabe, H.: (1) On Differentiability and Analyticity of Solutions of Weighted Elliptic Boundary Value Problems, Osaka Math. Journal **2**, 163—190 (1965).

Tartar, L.: (1) Cahiers IRIA **2** (1969).

Temam, R.: (1) Etude directe d'une équation d'évolution du type de Riccati, associée à des opérateurs non bornés. C. R. Acad. Sc. Paris (1969); — (2) Sur l'équation de Riccati associée à des opérateurs non bornés. To appear in J. Funct. Analysis; — (3) Sur la stabilité et la convergence de la méthode des pas fractionnaires. Annali di Mat. Pure ed Appl. **LXXIX**, 191—380 (1968); — (4) Sur les équations de Carleman. Archive for Rat. Mech. Analysis. 1969.

Tichonoff, A.N.: (1) Methods for the Regularization of Optimal Control Problems, Doklady Akad. Nauk. **162** (1965), Soviet Maths., 761—763.

Treves, F.: (1) Linear Partial Differential Operators with Constant Coefficients. Gordon and Breach 1966.

Tsujioka, K.: (1) Remarks on Controllability of Second Order Evolution Equations in Hilbert Spaces. SIAM J. on Control. Vol. **8**, 90—99 (1970).

Tucker, A.W.: Cf. Kuhn-Tucker.

Tzafestas, S.G., Nightingale, J.M.: (1) Optimal Filtering, Smoothing and Prediction in Linear Distributed Systems. Proc. I.E.E.E. **115**, 1207—1212 (1968). Optimal Control of a Class of Linear Stochastic Distributed Parameter Systems, Ibid., 1213—1220.

Vallee, A.: (1) Un problème de controle optimum dans certaines équations différentielles d'évolution. Ann. Sc. Norm. Sup. Pisa **20**, 25—30 (1966).

Varaiya, P.P.: (1) On the Existence of Solutions to a Differential Game. J. SIAM Control **5**, 153—162 (1967); — (2) An Extremal Problem in Banach Space with Application to Optimal Control. In preparation.

Visik, I.M.: (1) On Strongly Elliptic Differential Equations. Mat. Sbornik **29**, 615—676 (1951); — (2) Resolution of Boundary Value Problems for Quasi-

Linear Parabolic Equations of Arbitrary Order. Mat. Sbornik **59**, 289—335 (1962).

— Cf. Agranovich-Visik.

Wang, P.K.C.: (1) Control of Distributed Parameter Systems; in Advances in Control Systems, ed. by C.T. Leondes, Academic Press **1**, 75—172 (1964); — (2) Optimal Control of a Class of Linear Symmetric Hyperbolic Systems with Applications to Plasma Confinement. J. of Math. Analysis and Applns. **28**, 594—608 (1969).

Warga, J.: (1) Relaxed Variational Problems. J. Math. Anal. Appl. **4**, 111—128 (1962).

Wiberg, D.M.: (1) Feedback Control of Linear Distributed Systems. Journal of Basic Engineering **89** (D), 379—384 (1967).

Wing, G.M.: Cf. Bellman-Kalaba-Wing.

Wonham, W.M.: (1) On Pole-Assignment in Multi-Input Controllable Linear Systems. I.E.E.E. Trans. on Automatic Control, AC-12, 660—665 (1967); — (2) On a Matrix Riccati Equation of Stochastic Control. SIAM J. on Control **6**, 681—697 (1968).

Yanenko, N.N.: (1) Fractional Steps Methods. Novosibirsk 1966 [Russian]; translation in French by A. Colin. Paris 1968.

Yosida, K.: (1) Functional Analysis. Berlin-Heidelberg-New York: Springer 1965.

Young, L.C.: (1) Generalized Curves and the Existence of an Attained Absolute Minimum in the Calculus of Variations. C.R. Soc. Sc. et Lettres Varsovie, Cl. III, **30**, 212—234 (1937).

Yvon, J.P.: (1) Cahiers IRIA **2** (1969).

Zaidman, S.: (1) Convexity Properties for Weak Solutions of Some Differential Equations in Hilbert Spaces. Can. Journal of Math. **XVII**, 802—807 (1965).

Subject Index

Die Grundlehren der mathematischen Wissenschaften in Einzeldarstellungen mit besonderer Berücksichtigung der Anwendungsgebiete

2. Knopp: Theorie und Anwendung der unendlichen Reihen. DM 48,—; US $ 13.20
3. Hurwitz: Vorlesungen über allgemeine Funktionentheorie und elliptische Funktionen. DM 49,—; US $ 13.50
4. Madelung: Die mathematischen Hilfsmittel des Physikers. DM 49,70; US $ 13.70
10. Schouten: Ricci-Calculus. DM 58,60; US $ 16.20
14. Klein: Elementarmathematik vom höheren Standpunkte aus. 1. Band: Arithmetik, Algebra, Analysis. DM 24,—; US $ 6.60
15. Klein: Elementarmathematik vom höheren Standpunkte aus. 2. Band: Geometrie. DM 24,—; US $ 6.60
16. Klein: Elementarmathematik vom höheren Standpunkte aus. 3. Band: Präzisions- und Approximationsmathematik. DM 19,80; US $ 5.50
20. Pólya/Szegö: Aufgaben und Lehrsätze aus der Analysis II: Funktionentheorie, Nullstellen, Polynome, Determinanten, Zahlentheorie. DM 38,—; US $ 10.50
22. Klein: Vorlesungen über höhere Geometrie. DM 28,—; US $ 7.70
26. Klein: Vorlesungen über nicht-euklidische Geometrie. DM 24,—; US $ 6.60
27. Hilbert/Ackermann: Grundzüge der theoretischen Logik. DM 38,—; US $ 10.50
30. Lichtenstein: Grundlagen der Hydromechanik. DM 38,—; US $ 10.50
31. Kellogg: Foundations of Potential Theory. DM 32,—; US $ 8.80
32. Reidemeister: Vorlesungen über Grundlagen der Geometrie. DM 18,—; US $ 5.00
38. Neumann: Mathematische Grundlagen der Quantenmechanik. DM 28,—; US $ 7.70
40. Hilbert/Bernays: Grundlagen der Mathematik I. DM 68,—; US $ 18.70
43. Neugebauer: Vorlesungen über Geschichte der antiken mathematischen Wissenschaften. 1. Band: Vorgriechische Mathematik. DM 48,—; US $ 13.20
50. Hilbert/Bernays: Grundlagen der Mathematik II. DM 84,—; US $ 23.10
52. Magnus/Oberhettinger/Soni: Formulas and Theorems for the Special Functions of Mathematical Physics. DM 66,—; US $ 16.50
57. Hamel: Theoretische Mechanik. DM 84,—; US $ 23.10
58. Blaschke/Reichardt: Einführung in die Differentialgeometrie. DM 24,—; US $ 6.60
59. Hasse: Vorlesungen über Zahlentheorie. DM 69,—; US $ 19.00
60. Collatz: The Numerical Treatment of Differential Equations. DM 78,—; US $ 19.50
61. Maak: Fastperiodische Funktionen. DM 38,—; US $ 10.50
62. Sauer: Anfangswertprobleme bei partiellen Differentialgleichungen. DM 41,—; US $ 11.30
64. Nevanlinna: Uniformisierung. DM 49,50; US $ 13.70
66. Bieberbach: Theorie der gewöhnlichen Differentialgleichungen. DM 58,50; US $ 16.20
67. Byrd: Handbook of Elliptic Integrals for Engineers and Scientists. In preparation
68. Aumann: Reelle Funktionen. DM 68,—; US $ 18.70
69. Schmidt: Mathematische Gesetze der Logik I. DM 79,—; US $ 21.80
71. Meixner/Schäfke: Mathieusche Funktionen und Sphäroidfunktionen mit Anwendungen auf physikalische und technische Probleme. DM 52,60; US $ 14.50
73. Hermes: Einführung in die Verbandstheorie. DM 46,—; US $ 12.70
74. Boerner: Darstellungen von Gruppen. DM 58,—; US $ 16.00
75. Rado/Reichelderfer: Continuous Transformations in Analysis, with an Introduction to Algebraic Topology. DM 59,60; US $ 16.40
76. Tricomi: Vorlesungen über Orthogonalreihen. DM 68,—; US $ 18.70

77. Behnke/Sommer: Theorie der analytischen Funktionen einer komplexen Veränderlichen. DM 79,—; US $ 21.80
78. Lorenzen: Einführung in die operative Logik und Mathematik. DM 54,—; US $ 14.90
80. Pickert: Projektive Ebenen. DM 48,60; US $ 13.40
81. Schneider: Einführung in die transzendenten Zahlen. DM 24,80; US $ 6.90
82. Specht: Gruppentheorie. DM 69,60; US $ 19.20
84. Conforto: Abelsche Funktionen und algebraische Geometrie. DM 41,80; US $ 11.50
86. Richter: Wahrscheinlichkeitstheorie. DM 68,—; US $ 18,70
87. Waerden, van der: Mathematische Statistik. In Vorbereitung
88. Müller: Grundprobleme der mathematischen Theorie elektromagnetischer Schwingungen. DM 52,80; US $ 14.60
89. Pfluger: Theorie der Riemannschen Flächen. DM 39,20; US $ 10.80
90. Oberhettinger: Tabellen zur Fourier-Transformation. DM 39,50; US $ 10.90
91. Prachar: Primzahlverteilung. DM 58,—; US $ 16.00
93. Hadwiger: Vorlesungen über Inhalt, Oberfläche und Isoperimetrie. DM 49,80; US $ 13.70
94. Funk: Variationsrechnung und ihre Anwendung in Physik und Technik. DM 120,—; US $ 33.00
95. Maeda: Kontinuierliche Geometrien. DM 39,—; US $ 10.80
97. Greub: Linear Algebra. DM 39,20; US $ 9.80
98. Saxer: Versicherungsmathematik. 2. Teil. DM 48,60; US $ 13.40
99. Cassels: An Introduction to the Geometry of Numbers. DM 69,—; US $ 19.00
100. Koppenfels/Stallmann: Praxis der konformen Abbildung. DM 69,—; US $ 19.00
101. Rund: The Differential Geometry of Finsler Spaces. DM 59,60; US $ 16.40
103. Schütte: Beweistheorie. DM 48,—; US $ 13.20
104. Chung: Markov Chains with Stationary Transition Probabilities. DM 56,—; US $ 14.00
105. Rinow: Die innere Geometrie der metrischen Räume. DM 83,—; US $ 22.90
106. Scholz/Hasenjaeger: Grundzüge der mathematischen Logik. DM 98,—; US $ 27.00
107. Köthe: Topologische lineare Räume I. DM 78,—; US $ 21.50
108. Dynkin: Die Grundlagen der Theorie der Markoffschen Prozesse. DM 33,80; US $ 9.30
110. Dinghas: Vorlesungen über Funktionentheorie. DM 69,—; US $ 19.00
112. Morgenstern/Szabó: Vorlesungen über theoretische Mechanik. DM 69,—; US $ 19.00
113. Meschkowski: Hilbertsche Räume mit Kernfunktion. DM 58,—; US $ 16.00
114. MacLane: Homology. DM 62,—; US $ 15.50
115. Hewitt/Ross: Abstract Harmonic Analysis. Vol. 1: Structure of Topological Groups, Integration Theory, Group Representations. DM 76,—; US $ 20.90
116. Hörmander: Linear Partial Differential Operators. DM 42,—; US $ 10.50
117. O'Meara: Introduction to Quadratic Forms. DM 68,—; US $ 18.70
118. Schäfke: Einführung in die Theorie der speziellen Funktionen der mathematischen Physik. DM 49,40; US $ 13.60
119. Harris: The Theory of Branching Processes. DM 36,—; US $ 9.90
120. Collatz: Funktionalanalysis und numerische Mathematik. DM 58,—; US $ 16.00
121.
122. Dynkin: Markov Processes. DM 96,—; US $ 26.40
123. Yosida: Functional Analysis. DM 66,—; US $ 16.50
124. Morgenstern: Einführung in die Wahrscheinlichkeitsrechnung und mathematische Statistik. DM 38,—; US $ 10.50
125. Itô/McKean: Diffusion Processes and Their Sample Paths. DM 58,—; US $ 16.00
126. Lehto/Virtanen: Quasikonforme Abbildungen. DM 38,—; US $ 10.50

127. Hermes: Enumerability, Decidability, Computability. DM 39,—; US $ 10.80
128. Braun/Koecher: Jordan-Algebren. DM 48,—; US $ 13.20
129. Nikodým: The Mathematical Apparatus for Quantum-Theories. DM 144,—; US $ 36.00
130. Morrey: Multiple Integrals in the Calculus of Variations. DM 78,—; US $ 19.50
131. Hirzebruch: Topological Methods in Algebraic Geometry. DM 38,—; US $ 9.50
132. Kato: Perturbation Theory for Linear Operators. DM 79,20; US $ 19.80
133. Haupt/Künneth: Geometrische Ordnungen. DM 68,—; US $ 18.70
134. Huppert: Endliche Gruppen I. DM 156,—; US $ 42.90
135. Handbook for Automatic Computation. Vol 1/Part a: Rutishauser: Description of ALOGL 60. DM 58,—; US $ 14.50
136. Greub: Multilinear Algebra. DM 32,—; US $ 8.00
137. Handbook for Automatic Computation. Vol. 1/Part b: Grau/Hill/Langmaack: Translation of ALOGL 60. DM 64,—; US $ 16.00
138. Hahn: Stability of Motion. DM 72,—; US $ 19.80
139. Mathematische Hilfsmittel des Ingenieurs. Herausgeber: Sauer/Szabó. 1. Teil. DM 88,—; US $ 24.20
140. Mathematische Hilfsmittel des Ingenieurs. Herausgeber: Sauer/Szabó. 2. Teil. DM 136,—; US $ 37.40
141. Mathematische Hilfsmittel des Ingenieurs. Herausgeber: Sauer/Szabó. 3. Teil. DM 98,—; US $ 27.00
142. Mathematische Hilfsmittel des Ingenieurs. Herausgeber: Sauer/Szabó. 4. Teil. DM 124,—; US $ 34.10
143. Schur/Grunsky: Vorlesungen über Invariantentheorie. DM 32,—; US $ 8.80
144. Weil: Basic Number Theory. DM 48,—; US $ 12.00
145. Butzer/Berens: Semi-Groups of Operators and Approximation. DM 56,—; US $ 14.00
146. Treves: Locally Convex Spaces and Linear Differential Equations. DM 36,—; US $ 9.90
147. Lamotke: Semisimpliziale algebraische Topologie. DM 48,—; US $ 13.20
148. Chandrasekharan: Introduction to Analytic Number Theory. DM 28,—; US $ 7.00
149. Sario/Oikawa: Capacity Functions. DM 96,—; US $ 24.00
150. Iosifescu/Theodorescu: Random Processes and Learning. DM 68,—; US $ 18.70
151. Mandl: Analytical Treatment of One-dimensional Markov Processes. DM 36,—; US $ 9.80
152. Hewitt/Ross: Abstract Harmonic Analysis. Voll. II. Structure and Analysis for Compact Groups. Analysis on Locally Compact Abelian Groups. DM 140,—; US $ 38.50
153. Federer: Geometric Measure Theory. DM 118,—; US $ 29.50
154. Singer: Bases in Banach Spaces I. DM 112,—; US $ 30.80
155. Müller: Foundations of the Mathematical Theory of Electromagnetic Waves. DM 58,—; US $ 16.00
156. van der Waerden: Mathematical Statistics. DM 68,—; US $ 18.70
157. Prohorov/Rozanov: Probability Theory. DM 68,—; US $ 18.70
158. Constantinescu/Cornea: Potential Theory on Harmonic Spaces. In preparation
159. Köthe: Topological Vector Spaces I. DM 78,—; US $ 21.50
160. Agrest/Maksimov: Theory of Incomplete Cylindrical Functions and their Applications. In preparation
161. Bhatia/Szegö: Stability Theory of Dynamical Systems. DM 58,—; US $ 16.00
162. Nevanlinna: Analytic Functions. DM 76,—; US $ 21.00
163. Stoer/Witzgall: Convexity and Optimization in Finite Dimensions I. DM 54,—; US $ 14.90
164. Sario/Nakai: Classification Theory of Riemann Surfaces. DM 98,—; US $ 27.00

165. Mitrinović: Analytic Inequalities. DM 88,—; US $ 26.00
166. Grothendieck/Dieudonné: Eléments de Géométrie Algébrique I. DM 84,—; US $ 23.10
167. Chandrasekharan: Arithmetical Functions. DM 58,—; US $ 16.00
168. Palamodov: Linear Differential Operators with Constant Coefficients. DM 98,—; US $ 27.00
169. Rademacher: Topics in Analytic Number Theory. In preparation
170. Lions: Optimal Control of Systems Governed by Partial Differential Equations. DM 78,—; US $ 21.50
171. Singer: Best Approximation in Normed Linear Spaces by Elements of Linear Subspaces. DM 60,—; US $ 16.50
172. Bühlmann: Mathematical Methods in Risk Theory. DM 52,—; US $ 14.30
173. F. Maeda/S. Maeda: Theory of Symmetric Lattices. DM 48,—; US $ 13.20
174. Stiefel/Scheifele: Linear and Regular Celestial Mechanics. DM 68,—; US $ 18.70
175. Larsen: An Introduction to the Theory of Multipliers. DM 84,—; US $ 23.10
176. Grauert/Remmert: Analytische Stellenalgebren. DM 64,—; US $ 17.60
177. Flügge: Practical Quantum Mechanics I. In preparation
178. Flügge: Practical Quantum Mechanics II. In preparation
179. Giraud: Cohomologie non abélienne. En préparation

MIX
Papier aus verantwortungsvollen Quellen
Paper from responsible sources
FSC® C105338

If you have any concerns about our products,
you can contact us on
ProductSafety@springernature.com

In case Publisher is established outside the EU,
the EU authorized representative is:
Springer Nature Customer Service Center GmbH
Europaplatz 3, 69115 Heidelberg, Germany

Printed by Libri Plureos GmbH
in Hamburg, Germany